Content and Strategy Guide for the

SAT®

MATH

- 2 Full-Length SAT® Math Practice Tests
- 400+ SAT® Math Practice Questions
- Preliminary Diagnostic Exam
- 12 Topic-Specific Problem Sets
- 4 Domain-Based Exams
- Complete Content Review
- Key Insights and Explanations

2022 | SCHWANN'S REVIEW

CREATED BY

Bo Nalsen

ISBN: 978-0-578-81769-9

Schwann's Review LLC

SchwannsReview.com

Letter to the Reader

Dear Reader,

First and foremost, thank you for your interest in the Schwann's Review Content and Strategy Guide for the SAT® Math. Our team understands that your SAT® scores play a crucial role in both the college admissions and scholarship approval processes. In this book, we strive to give you key insights into the SAT® Math exam from two different angles: content and strategy. To earn your highest score, you will need to know how to tackle each question in the most efficient way possible. As you will see, we approach math questions using methods that combine theory with application to give you all of the most up-to-date and effective problem-solving tools. Our review focuses on sensitizing test-takers and teachers to the nuances of the SAT® Math test by recreating questions that reflect the scope of the exam and providing solutions that deliver on the promise of facilitating perfect scores.

While the math section spans several content areas that are designed to challenge students on their college readiness, we know that the SAT® Math Test presents obstacles that may not assess the overall ability of every test-taker. We, the Schwann's Review Team, want all students to demonstrate their full potential on this important standardized test, which is why we have broken down the exam into its fundamental parts, offering some of the most reliable organizational tools, solving strategies, and test-taking tips that are available. By condensing the subject matter to reflect high-yield material, we aim to reveal exactly what you need to know to answer every question on the SAT® Math test quickly and correctly. The content and strategies you will see in this book are designed to give you a leg up on the SAT® Math, and we firmly believe that our review will put you in position to achieve your dream score in less time than any other testing resource.

So – without further ado – welcome aboard! We are now on the same team, together on this important journey. We will soon introduce you to the structure of the test, organization of this book, and then all of the content and strategies you need to raise your score. Again, thank you for joining us during this critical time of your academic career. We are fully committed to the task at hand and are cheering you on every step of the way!

Sincerely,

Bo Nalsen
Primary Contributor
Schwann's Review LLC

Table of Contents

INTRODUCTION

Test Information

General Test Outline

The SAT® is a standardized test owned by the College Board and used by many colleges and universities in the college admissions process. In its latest form, the SAT® contains four separate testing sections, an optional essay, and the possibility of an experimental section. Test-takers will receive a full breakdown of their performance in their score report, including topic-specific subscores and cross-test scores that reveal student proficiencies in greater detail. Importantly, students will be given section scores ranging from 200 to 800 in 1) Evidenced-Based Reading and Writing and 2) Math. The total score is the sum of these two section scores and can range from 400 to 1600 as a result. Overall, the test will take three hours, plus additional time for the essay or experimental section.

Math Test Details

Since this review guide is designed only for the math portion of the exam, we will spend time now to discuss the specifics of the math test only. As it is outlined, the SAT® Math Test contains content relevant to four domains: Heart of Algebra, Passport to Advanced Math, Problem Solving and Data Analysis, and Additional Topics. You can expect to see these questions appear in two distinct math sections on exam day, which occur in sections three and four of the test. The first of these is the No Calculator portion, which gives students 25 minutes to complete 20 questions, of which 5 are grid-in. The second is the Calculator portion that gives students 55 minutes to complete 38 questions, of which 8 are grid-in. There is no guessing penalty. Based on their performance in these two sections, test-takers will receive subscores ranging from 1 to 15 in the first three testing domains. Additionally, a test score ranging from 10 to 40 and a section score ranging from 200 to 800 will be given.

Understanding Your Performance

To better gauge your performance, you will need to know a few things, the first being any recent scores you have on relevant exams like the PSAT® and Official SAT® Practice Tests. Having these scores will allow you to set reasonable goals and monitor your progress. Another action that may help you set target scores is to analyze the score ranges of the middle 50% of accepted students at the college and universities that interest you most. Please visit their admission websites to see their latest incoming class profiles so that you have a better idea of the score ranges you might need to gain acceptance into one of your top-choice schools. Additionally, some scholarships set minimum score requirements in order to qualify. Visit the websites of each scholarship that you are interested in to set target scores that will qualify you for their aid package. On your report, each of your scores will be given a percentile rank that reflects your performance relative to the difficulty of the test. Use the chart below to see the general score and percentile ranges that are typical for correctly answering any number of the 58 math questions presented on the test. These are rough estimates based on prior exams, and the SAT® may issue scoring guidelines that differ for any particular test.

Section Score	< 510	510 - 600	610 - 700	710 - 800
Percentile	< 45	45 - 75	77 - 92	93 - 99+
Number Correct	< 26	26 - 39	40 - 49	50 - 58

Using this Book

Starting Point

Take the Schwann's Review Diagnostic Exam to determine your current ability on the math portion of the test, then use the roadmap that follows to target your prep. Every question on the diagnostic exam will be discussed in the content portion of our review. Past performances on the SAT®, Official SAT® Practice Tests, and PSAT® should also be used to help you decide which sections of our review to focus your energy. However, many students will find it useful to start at the beginning of the book and follow it through in order all the way to the end. Doing so will give us the chance to introduce new strategies that will decrease the workload and increase your chances of answering each question correctly. Still, this is your book, and you may use it in whichever manner you see fit.

Content and Strategy

The first phase of our review is the content section that introduces every topic you will need to know to raise your score. We follow the four testing domains outlined by the College Board® similarly in our own breakdown of the test to ensure that every student who uses the Schwann's Review Content and Strategy Guide for the SAT® Math can attain their highest score. The most important information test-takers will need is provided in tables designed for quick access and ease-of-use. Additionally, formulas are appropriately designated in the text so that they are apparent even for those of us giving a page just a casual glance. This book is designed for you in mind, and we have aimed to make all essential material fully transparent so that you can get the most out of your prep. Along the way, you will also see example problems designed to better illustrate main testing points. Some of these come from our initial Diagnostic Exam, while others are new problems to showcase other dimensions of the exam and prepare students for future iterations of the test. Targeted problem sets and topic exams will give you the opportunity to practice content-specific skills at every stage of the book. Our strategy section will then go over key shortcuts and relevant testing methods that seek to give you the upper hand come test day.

Practice Tests

Following the content and strategy portion of the review, you will have a chance to tackle two full-length math tests. Compare your results to your earlier performance on the Diagnostic Exam and any prior scores you may have had coming into starting our review. After seeing your progress, adjust your target scores so that you have a clear goal when you go in to take the real test. We understand that you will want as much practice as possible so long as it helps you improve. Visit SchwannsReview.Com for more questions, information, and resources so that you can prepare for the test to the best of your ability. Good luck as you move forward in our review and your prep.

DIAGNOSTIC EXAM

DIRECTIONS

For questions 1-15, solve every problem and either circle or record the best answer from the answer choices provided. **For questions 16-20**, solve every problem and record your answer in the appropriate format. On the real test, you must grid in your answers according to the directions. You may use any available space in your workbook for scratch work.

NOTES

1. Calculator usage is NOT allowed in Part 1.
2. Assume real numbers unless otherwise indicated.
3. Figures may not be drawn to scale.
4. All figures are planar, unless otherwise indicated.
5. Domains represent all real numbers of for which f(x) is also a real number.

REFERENCE

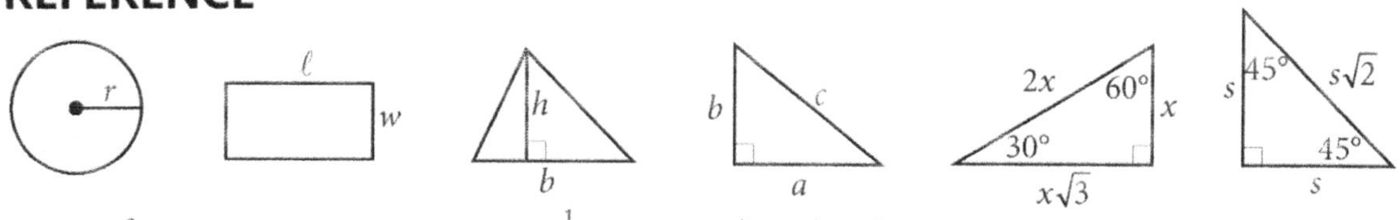

$A = \pi r^2$ $C = 2\pi r$ | $A = \ell w$ | $A = \frac{1}{2}bh$ | $c^2 = a^2 + b^2$ | Special Right Triangles

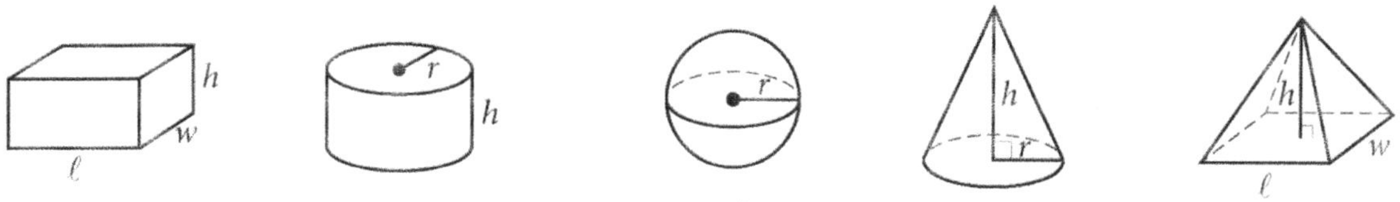

$V = \ell wh$ | $V = \pi r^2 h$ | $V = \frac{4}{3}\pi r^3$ | $V = \frac{1}{3}\pi r^2 h$ | $V = \frac{1}{3}\ell wh$

The arc of a circle contains 360 degrees.

The arc of a circle contains 2π radians.

The interior angles of a triangle add up to 180 degrees.

Part 1: No Calculator

25 Minutes, 20 Questions

1

Rob works at a marketing company where each week he receives a list of potential clients to call. The number of clients he has left to call at the end of each day can be found using the equation $N = 182 - 36d$, where N is the number of clients left and d is the number of days Rob worked that week. What is the meaning of 182 in this equation?

A. Rob will complete his calls within 182 days

B. Rob starts each week with 182 clients to call

C. Rob completes 182 calls per day

D. Rob spends a total of 182 hours on call

2

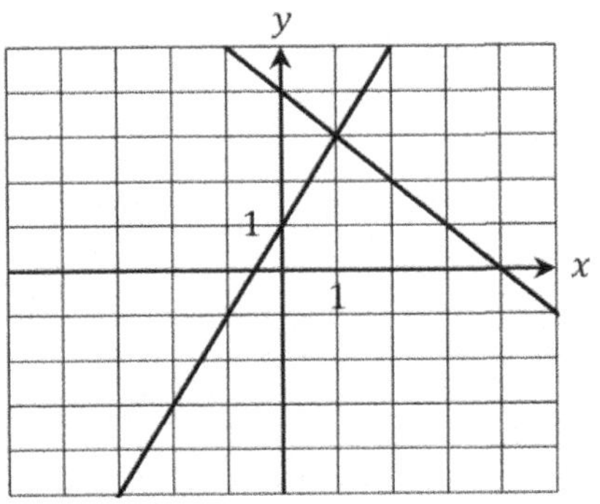

Two linear equations are graphed on xy-plane above. What is the (x, y) solution to the system formed by these equations?

A. (0,1)

B. (1,3)

C. (3,1)

D. (4,0)

3

If the expression $5x^2 - 10$ is rewritten in the form $5(x - a)(x + a)$, what is the value of the positive constant a?

A. $\sqrt{2}/2$

B. $\sqrt{2}$

C. 2

D. 5

4

A chemist works with solutions of hydrochloric acid (HCl) with acid concentrations expressed as a percentage of total volume. Which expression can be used to determine how many milliliters (mL) of 5% HCl solution must be added to 45mL of 8% HCl solution to obtain p mL of 6% HCl solution?

A. $\frac{6p - 360}{5}$

B. $\frac{0.6p - 360}{0.5}$

C. $\frac{6p - 3.60}{5}$

D. $\frac{6p + 360}{5}$

5

$$2x^2 - qx = -4$$

In the equation above, x is a variable and q is a positive constant. What value of q ensures that the equation has exactly one solution?

A. 0

B. $2\sqrt{2}$

C. $4\sqrt{2}$

D. 32

6

$$T = 2\pi\sqrt{\frac{L}{g}}$$

The period of a pendulum can be measured using the equation above, where T is the period in seconds, L is the length of the pendulum in meters, and g is the acceleration due to gravity in meters per second squared. If the length increases by a factor of 4, by what factor does the period increase?

A. 2

B. 4

C. 8

D. 16

7

Which of the following is equivalent to the difference between $b^2 + 2b$ and $3b + 2$?

A. $b^2 - b + 2$

B. $b^2 - b - 2$

C. $b^2 + 5b + 2$

D. $b^2 + b - 2$

8

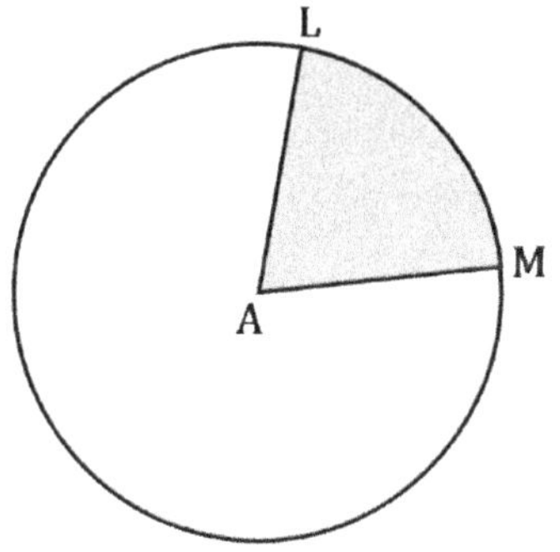

Circle A has a radius of 2 meters and arcLM has a length of 3 meters. What is the area in square meters of the shaded sector?

A. 1

B. 2

C. 3

D. 6

9

A distance runner training for a 5 kilometer race wants to run at least 20 laps around a track in at most 35 minutes as a workout. The runner decides to run r laps at race pace j laps at a jog. If a lap at race pace takes the runner 80 seconds and a lap at a jog takes the runner 100 seconds, which system of inequalities represents the runner's workout?

A. $\frac{5}{3}j + \frac{4}{3}r < 35$

$r + j \geq 20$

B. $\frac{5}{3}j + \frac{4}{3}r \leq 35$

$r + j \geq 20$

C. $100j + 80r \leq 35$

$r + j > 20$

D. $80j + 100r \geq 20$

$r + j \leq 35$

10

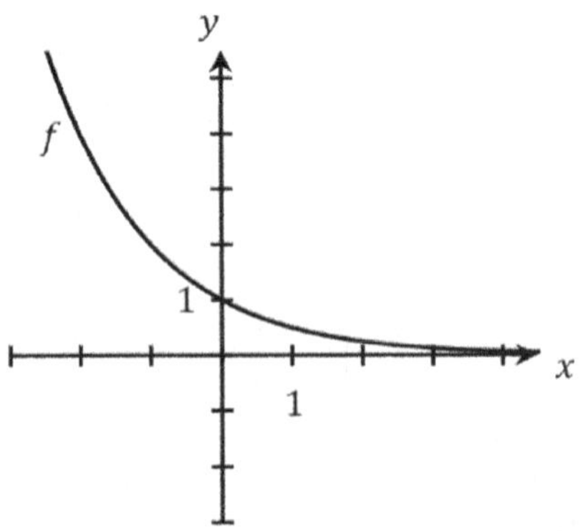

The function f is graphed on the xy-plane above. If the function g is the reflection of f across the y-axis, which of the following can represent the equation of $g(x)$?

A. $g(x) = 2x$

B. $g(x) = 2x + 1$

C. $g(x) = 2^x$

D. $g(x) = 2^x + 1$

11

An environmental committee wants to restore the greatest number of trees possible in a given acreage of forest over the long run. Which plan fits the needs of this committee?

A. Every month, 20 trees are added

B. Every month, 10% of the original number of trees is added

C. Every month, 15% of the original number of trees is added

D. Every month, 5% of the current number of trees is added

12

The volume of a right square pyramid is 42 cubic meters. If the side lengths to the base double, what is the volume, in cubic meters, of the pyramid?

A. 21

B. 42

C. 84

D. 168

13

$$f(x) = -2x^2 - 4x + 8$$

The function f provided above is graphed on the xy-plane, where $y = f(x)$. Which of the following is an equivalent form of the function f that contains constants locating its maximum?

A. $f(x) = -2x(x + 2) + 8$

B. $f(x) = -2x(x + 2) + 10$

C. $f(x) = -2(x + 1)^2 + 8$

D. $f(x) = -2(x + 1)^2 + 10$

14

$$\frac{2x^3 + 5x^2 + 3x + 4}{x + 1}$$

Which of the following is equivalent to the expression above?

A. $2x^2 + 7x + 10 + \frac{14}{x + 1}$

B. $2x^2 + 7x + 10 - \frac{6}{x + 1}$

C. $2x^2 + 3x + \frac{4}{x + 1}$

D. $2x^2 + 3x - \frac{4}{x + 1}$

15

$$(2 - 3i) - 4(5 - 2i)$$

Which of the following is equivalent to the expression above given that $i = \sqrt{-1}$?

A. $-18 + 5i$

B. $-18 - 5i$

C. $-18 - 11i$

D. $-22 + 11i$

16

When graphed on the xy-plane, line r contains point A $(1,3)$ and line s contains point B $(0, a)$. Lines r and s intersect at $(4,9)$. If a is a constant, then what must its value be to make lines r and s perpendicular to each other?

17

If an angle measure of 330° is written in radians as $n\pi$, what is the value of n?

18

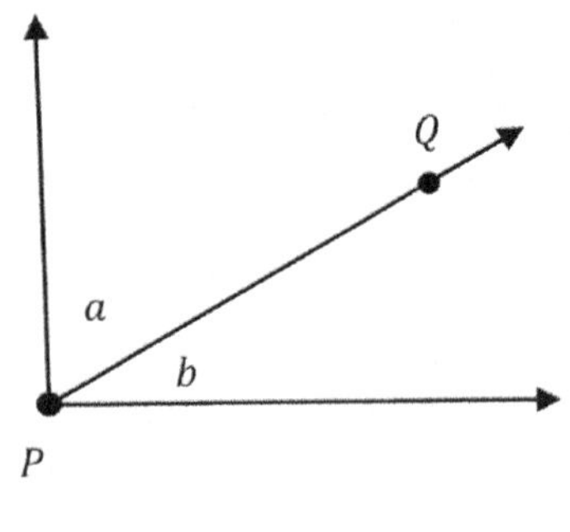

In the diagram above, the ray $\overrightarrow{PQ}$ divides the right angle at P into two different angles, a and b. What is the value of $\sin a - \cos b$

19

Line segment $\overline{AB}$ contains endpoints $A(7, -3)$ and $B(-5,9)$. If (a, b) is the midpoint of $\overline{AB}$, what is the value of $a + b$?

20

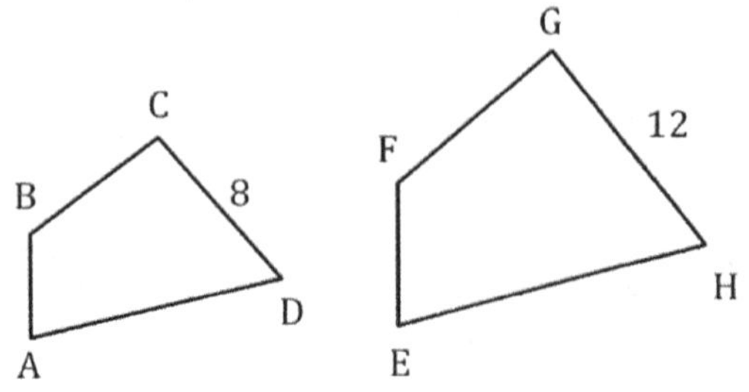

In the diagram to the above, quadrilateral ABCD is similar to quadrilateral EFGH. If quadrilateral ABCD has an area equal to 40 square meters, then what is the area, also in square meters, of the quadrilateral EFGH?

PART 2: CALCULATOR

DIRECTIONS

For questions 1-30, solve every problem and either circle or record the best answer from the answer choices provided. **For questions 31-38**, solve every problem and record your answer in the appropriate format. On the real test, you must grid in your answers according to the directions. You may use any available space in your workbook for scratch work.

NOTES

1. Calculator usage is allowed in Part 2.
2. Assume real numbers unless otherwise indicated.
3. Figures may not be drawn to scale.
4. All figures are planar, unless otherwise indicated.
5. Domains represent all real numbers of for which f(x) is also a real number.

REFERENCE

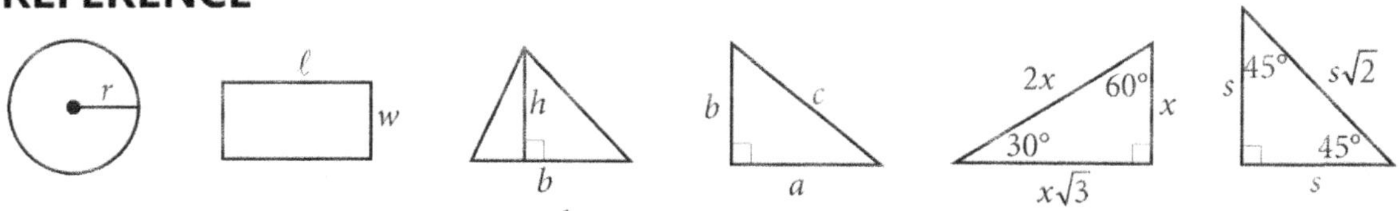

$A = \pi r^2$ $C = 2\pi r$ | $A = \ell w$ | $A = \frac{1}{2}bh$ | $c^2 = a^2 + b^2$ | Special Right Triangles

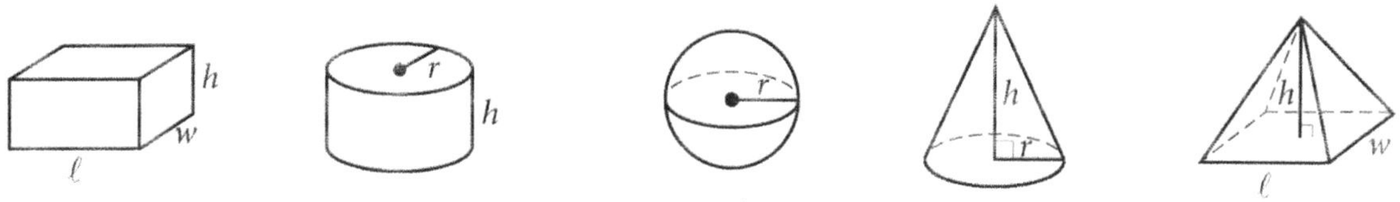

$V = \ell wh$ | $V = \pi r^2 h$ | $V = \frac{4}{3}\pi r^3$ | $V = \frac{1}{3}\pi r^2 h$ | $V = \frac{1}{3}\ell wh$

The arc of a circle contains 360 degrees.

The arc of a circle contains 2π radians.

The interior angles of a triangle add up to 180 degrees.

Part 2: Calculator

55 Minutes, 38 Questions

1

When line k is graphed on the xy-plane, it crosses through the points $(-a, 0)$ and $(0, 2a)$, where a is a positive constant. Which equation could represent line k?

A. $2x + y = -10$

B. $2x - y = -10$

C. $2x + y = 10$

D. $2x - y = 10$

2

Marcelo has to buy a textbook for his fall semester chemistry class. He finds out he can get a discount that takes 18% off the original price. He will still have to pay a 7% sales tax on the discounted price. If he pays h dollars at the register, what expression represents the original price of the textbook?

A. $0.89h$

B. $\dfrac{h}{0.89}$

C. $(0.82)(1.07)h$

D. $\dfrac{h}{(0.82)(1.07)}$

3

Which of the following is equivalent to the expression $(5x^2 + 1) - (2x^2 - 8x - 3)$?

A. $(3x + 2)(x + 2)$

B. $(3x + 2)(x - 2)$

C. $(3x - 1)(x + 2)$

D. $(3x + 2)(x - 2)$

4

$$\frac{2m}{\frac{2}{m-2} + \frac{1}{m+1}}$$

Which of the following is equivalent to the expression above for all values of $m > 2$?

A. $\dfrac{m^3 - m^2 - 2m}{3}$

B. $\dfrac{3}{2m^2 - 2m - 4}$

C. $\dfrac{m^2 - m - 2}{2m}$

D. $\dfrac{2m^2 - 2m - 4}{3}$

5

$$y = x + 4$$
$$(x-2)^2 + (y-4)^2 = 9$$

How many solutions are there to the system of equations provided above?

A. 0

B. 1

C. 2

D. More than 2

6

A car traveled at an average speed of 24 miles per hour for 1.9 hours and consumed fuel at a rate of 17 miles per gallon. Approximately how many gallons of fuel did the car use for the entire 1.9-hour trip?

A. 1.96

B. 2.54

C. 2.68

D. 3.19

7

$$5x - a(x-b) = 2x + 5$$

In the equation above, a and b are constants. If the equation has infinitely many solutions, what is the value of b?

A. 0

B. 3/5

C. 5/3

D. 5

8

$$\frac{27^{2x}}{9^y}$$

If $3x - y = 2$, what is the value of the expression above?

A. 9

B. 27

C. 81

D. 243

9

The area of a certain equilateral triangle with side lengths of 8 centimeters (cm) is $s^2\sqrt{3}\text{ cm}^2$. What is the perimeter of this triangle in terms of the constant s?

A. $3s$

B. $6s$

C. $9s$

D. $12s$

10

$$f(x) = x + 2$$
$$g(x) = x^2 - 3$$

The functions $f(x)$ and $g(x)$ provided above are graphed on the xy-plane and intersect at the points (m, n) and (p, q). What is the value of $m + p$?

A. -1

B. 0

C. 1

D. 2

Questions 11-12 use the following information

A survey of 280 randomly selected 12th grade students attending high school in County A was conducted to gather data on their college plans. The results of the survey are categorized by student GPA and preferred college location as shown in the two-way table below.

Number of 12th Grade Students in County A

	In-State	Out-of-State	Total
≥ 3.77 GPA	81	45	126
< 3.77 GPA	x	y	154

11

If the ratio of x to y is 5 to 2, how many students in the survey plan to attend an In-State college?

A. 110

B. 154

C. 176

D. 191

12

What is the probability that a student with a GPA of at least 3.77 plans to attend an Out-of-State college?

A. 5/14

B. 5/9

C. 9/14

D. 14/9

13

$$my - 2x = 8$$

$$ay + 6x = 2n$$

In the system of equations above, $a, m,$ and n are constants. If the system has infinitely many solutions, what is the value of the expression $\frac{3m}{n}$ in terms of a?

A. $a/12$

B. $-12a$

C. $4a$

D. $a/4$

14

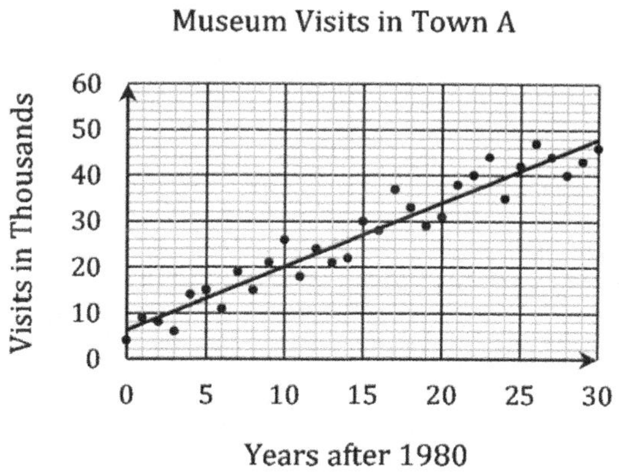

The number of annual museum visits in Town A in the years following 1980 was recorded and graphed in scatterplot above. A line of best fit was drawn to help estimate the number of museum visits that the city would receive for any particular year. In 2004, what was the percent error to the nearest hundredth between the estimated and actual number of museum visits Town A received?

A. 0.14

B. 2.52

C. 5.00

D. 14.29

15

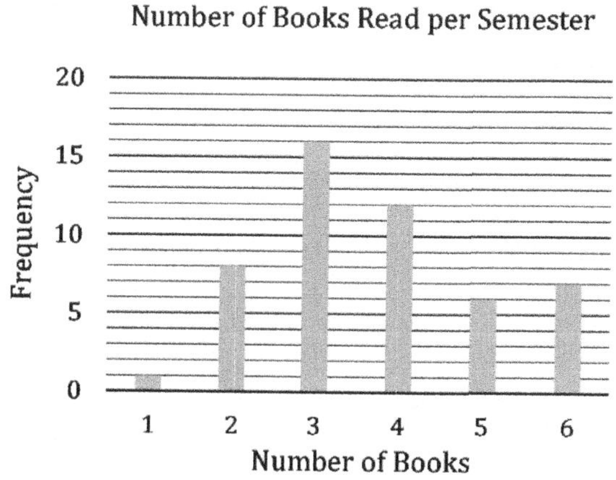

The frequency chart above summarizes the number of books read per semester by a random sample of fifty students attending a local college. What is the difference between the mean and median number of books read in this sample?

A. 0.2

B. 0.7

C. 1.2

D. 1.5

16

$$f(x) = \frac{1}{\sqrt{x-3} - 2}$$

The function above is graphed on the xy-plane, where $y = f(x)$. The curve of function f contains the coordinate pair (m, n). If m and n are constants, which of the following can be a value of m?

A. 0

B. 2

C. 6

D. 7

17

A survey was conducted in a district to determine the percentage of registered voters who plan to participate in the next local election. The survey only sampled registered voters in the district, recording each respondent's level of education and whether or not the respondent would vote in the next local election. The results of the survey concluded that 63 ± 2.9 percent of those holding at most a Bachelor's degree and that 78 ± 6.1 percent of those holding at least a Master's degree planned to vote in the next local election. If the survey results were calculated using the same level of confidence for both groups, which of the following is true?

A. Between 60.1 and 65.9 percent of registered voters holding at most a Bachelor's Degree in the district will vote in the in the next local election.

B. No more than 84.1 percent of registered voters holding at least a Master's Degree in the district will vote in the next local election.

C. The sample size of registered voters holding at most a Bachelor's degree was likely larger than the sample size of registered voters holding at least a Master's degree.

D. The sample size of registered voters holding at most a Bachelor's degree was likely smaller than the sample size of registered voters holding at least a Master's degree.

18

The dot plots above represent the salaries in US dollars of the 25 entry-level employees currently working at either Company A or Company B. Which of the following correctly compares the standard deviation of the salaries among the entry-level employees in each of the two companies?

A. The standard deviation of the entry-level salaries at Company A is greater.

B. The standard deviation of the entry-level salaries at Company B is greater.

C. The standard deviations of the entry-level salaries at Company A and Company B are equal.

D. No relationship can be determined from the provided information.

Questions 19-21 refer to the table below

x	$f(x)$	$q(x)$	$z(x)$
-2	8	-2	-9
-1	1	0	7
0	2	4	-1
1	-5	5	2
2	4	1	6

The functions $f, q,$ and z are each continuous functions of the input x over the domain of all real numbers. The chart details their value at particular quantities of x in the interval $[-2,2]$. Each function is plotted on the xy-coordinate axis (not shown).

19

For which of the following is $x + 1$ a factor?

I. $f(x)$

II. $q(x)$

III. $z(x)$

A. I only

B. II only

C. I and II

D. I and III

20

What is the value of $z\left(\frac{1}{4}f(2)\right)$?

A. -1

B. 2

C. 6

D. 7

21

Given that $f(x)$ is a continuous function, in what interval would it have an x-intercept?

A. $(-0.5,1.5)$

B. $(-2,0)$

C. $(1.5,2.5)$

D. None of the above

22

$$f(x) = 2ax^3 + 8bx^2 + 3x - 10$$

If $(x + 1)$ is a factor of $f(x)$ given above, what is the value of $4b - a$?

A. -1

B. 0

C. 1/2

D. 13/2

23

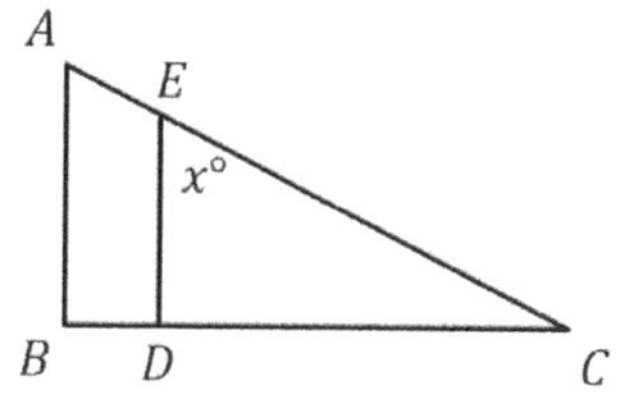

In triangle ABC above, $\angle B$ is 90, $AB = 10$, and $BC = 24$. If $\overline{AB} \parallel \overline{DE}$, $CD = 18$, and $\angle DEC = x°$, what is the value of $\cos x°$?

A. 5/12

B. 5/13

C. 12/5

D. 12/13

24

Given that $\sin\theta = 2T$ and $\cos\theta = T$, what could be the value of T?

A. $1/2$

B. $\sqrt{2}/2$

C. $\sqrt{3}/2$

D. $\sqrt{5}/5$

25

$$2a - 7b = 5$$
$$a + b = 1$$

Given the system of equations above, what is the value of $a - 2b$?

A. 0

B. 1

C. 2

D. 3

26

$$x + y < -2$$
$$-x + 2y < 6$$

Given the inequalities above, which could be a possible value for y?

A. 1

B. 2

C. 3

D. 4

27

$$(x + 2)^2 + (y - 6)^2 = 20$$

In the xy-plane, a circle with the equation above contains the point $A(0,10)$. If $\overline{AB}$ is the diameter of the circle, what coordinate pair represents B?

A. $(-4, 2)$

B. $(4, -2)$

C. $(0, 2)$

D. $(2, 0)$

28

$$2^{3x-1} = 5w$$

What is the value of 8^x in terms of w using the equation above?

A. $4w$

B. $6w$

C. $8w$

D. $10w$

29

An animal feed company sells a brand of dog food that contains a ratio of 2 grams of fat to 9 grams of carbohydrate and 8 grams of carbohydrate to 5 grams of protein. What is the ratio of fat to protein in grams?

A. $2 : 5$

B. $5 : 2$

C. $16 : 45$

D. $45 : 16$

30

The ratio of boys to girls in a high school physics class is $4 : 5$. If there are 27 students total in the class, how many are girls?

A. 12

B. 15

C. 18

D. 21

31

A local café sells two different brews of coffee: Wake-Up and Chill-Out. The Wake-Up coffee contains 5.83mg of caffeine in every 10mL whereas the Chill-Out coffee contains 3.96mg of caffeine in 12mL. How many more milligrams of caffeine to the nearest integer are in a 12oz cup of Wake-Up than there are in a 12oz cup of Chill-Out?

$$8\text{oz} = 240\text{mL}$$

32

The radius of a right cylinder increases by 15%. If the height remains constant, by what factor rounded to the nearest tenth does the volume change?

33

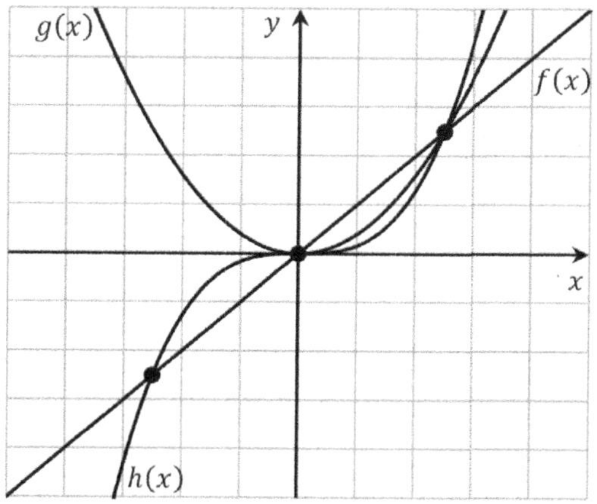

The functions $f, g,$ and h are graphed on the xy-plane above. Given that all points of intersection are marked, how many possible values of the constant a ensure that $f(a) = g(a) = h(a)$?

34

$$f(x) = |x - 2|$$

$$g(x) = \frac{1}{2}x$$

Functions f and g are provided above, and each contain the domain of all real numbers. When the functions are graphed on the xy-plane, they both contain the coordinate points (a, b) and (c, d) where $a, b, c,$ and d are all rational constants. If b is less than d, what is the value of b?

35

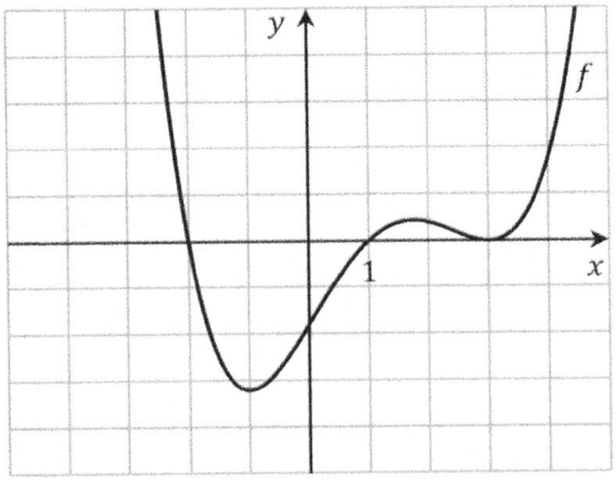

The function f is graphed on the xy-plane above where $y = f(x)$. It can be modeled in its factored form using the equation:

$$f(x) = (x + a)(x - b)^2(x - c)$$

Provided that $a, b,$ and c are positive integer constants, what is the value of b?

36

Track and Field Events by Category

	Sprints	Distance	Field	Total
Male	12	b	8	37
Female	a	23	11	d
Total	26	40	c	85

A high school track team has 85 students. The coach has organized the students using the chart above, where letters a-d represent constants. How many times more likely is it that a female track team member focuses on distance events than it is a male track team member focuses on distance events? Round final answer to the nearest hundredth.

37

$$(mx + 2)(nx - 1) = 28x^2 + cx - 2$$

In the equation above $c, m,$ and n are constants. If the equation is true for all values of x and $m - n = 3$, then what could be a positive integer value for c?

38

$$f(x) = (x + 2)^2 - 6$$

$$g(x) = (x - 4)^2 + 2$$

The functions f and g are defined above and graphed on the xy-plane. Function g is the image formed after each point on function f has been translated several units along the vertical and horizontal axes. What is the overall displacement that each point on f must undergo to form g?

Answer Key and Roadmap

Now that you have completed the Schwann's Review Math Diagnostic Exam, we can move onto the next part of your journey to a perfect score. Check your answers using the following charts to assess your overall performance. Be sure to highlight the questions causing the most trouble and identify the sections to which they belong. That way you have a better understanding as to which sections you should dedicate your time and energy so that you can earn the score you want come test day.

While you can certainly go straight to the pages listed to see the full solution outline for any particular question, we recommend reviewing the preliminary material of that section, retrying the question in the workspace, and then following the solution outline to see an efficient solving method for that question, taking notes as needed. If you felt like the Diagnostic Exam was challenging – it was! As we move forward together, we will review all relevant topics – complete with problem sets and section exams – so that you can ace not only Practice Tests 1 and 2 that close out our review book, but also the Official SAT® brought to you by the College Board®.

Without further ado, the solutions to the No Calculator and Calculator Diagnostic Exam are presented in the two tables that follow.

No Calculator

Question	Answer	Section	Page
1	B	Linear Equations	41
2	B	Linear Systems	56
3	B	Expressions and Equations	114
4	A	Linear Equations	45
5	C	Function Properties	141
6	A	Expressions and Equations	120
7	B	Expressions and Equations	113
8	C	Circle Properties	279
9	B	Linear Inequalities	73
10	C	Exponential Properties and Models	103
11	D	Exponential Properties and Models	105
12	D	Geometry and Trigonometry	256
13	D	Function Properties	136
14	C	Function Properties	153
15	A	Complex Numbers	243
16	11	Linear Equations	35
17	11/6	Geometry and Trigonometry	263
18	0	Geometry and Trigonometry	262
19	4	Linear Equations	38
20	90	Geometry and Trigonometry	266

Calculator

Question	Answer	Section	Page
1	B	Linear Equations	39
2	D	Rates and Probability	196
3	A	Expressions and Equations	115
4	D	Expressions and Equations	117
5	C	Circle Properties	283
6	C	Rates and Probability	195
7	C	Function Properties	151
8	C	Exponential Properties and Models	100
9	B	Geometry and Trigonometry	260
10	C	Function Properties	140
11	D	Rates and Probability	204
12	A	Rates and Probability	205
13	A	Linear Systems	60
14	D	Rates and Probability	202
15	A	Data Trends and Statistics	213
16	C	Advanced Functions and Equations	165
17	C	Data Trends and Statistics	218
18	A	Data Trends and Statistics	217
19	B	Function Properties	129
20	B	Function Properties	130
21	A	Function Properties	130
22	D	Function Properties	155
23	B	Geometry and Trigonometry	261
24	D	Geometry and Trigonometry	262
25	C	Linear Systems	62
26	A	Linear Inequalities	72
27	A	Circle Properties	281
28	D	Exponential Properties and Models	99
29	C	Rates and Probability	194
30	B	Rates and Probability	194
31	91	Rates and Probability	195
32	1.3	Geometry and Trigonometry	257
33	2	Advanced Functions and Equations	166
34	2/3	Advanced Functions and Equations	167
35	3	Function Properties	145
36	1.04	Rates and Probability	206
37	1	Function Properties	152
38	10	Function Properties	149

LINEAR ALGEBRA

Linear Equations

Properties of Lines

Forms of Linear Equations

Interpreting Linear Models

Expressions and Equations

Linear Systems

Solving Methods

Number of Solutions

Advanced Strategies

Linear Inequalities

Solution Sets

Constructing Inequalities

Introduction

The focus of this section is in using linear equations and inequalities algebraically, graphically, and contextually to relate information and find unknown values. Consequently, test-takers will often have to create their own linear equations and systems based on the description of a unique scenario. Heart of Algebra topics, as they are defined by College Board, appear in both the calculator and no calculator sections of the test, and often challenge students in their critical reading skills just as much as in their math fluency. That is to say that these questions are often the wordiest ones on the test, costing students vital time that could be put to better use elsewhere. To get around this, we have outlined exactly what you need to know to answer these questions correctly in as little time as possible.

1.1 Linear Equations

a. Properties of Lines

Lines are used in algebra to represent linear relationships between an input and output. The input and output values form points on the line when they are graphed on a coordinate plane and can be represented in equations using variables or function notation. Linear equations are often displayed in slope-intercept form, or $y = mx + b$. A key point in writing linear equations is that variables can only occur as independent terms with a degree – or exponent – of one. Lines are different from other functions because they always have a constant slope, or rate of change. Without modification, lines will always have an unrestricted domain and range, crossing each axis at one point, called an **intercept**. These attributes of lines have earned them a distinguished place on the test. Of most importance, students must be able to interpret, create, and solve linear equations.

Variables and Function Notation

As noted, linear equations are often expressed in the form $y = mx + b$. This is called slope-intercept form because it contains the slope, m, and the y-intercept, b; however, there are other common linear representations as well that are summarized in the tables that follow. But first, it is crucial to understand how to work with linear equations expressed using both variables and function notation, as well as how to represent coordinate pairs correctly and substitute them into equations of all types. Table 1.11 summarizes how variables and function notation can be used to represents linear equations in slope-intercept form.

Table 1.11 | Variables and Function Notation

	Variables	Function Notation
Equation	$y = mx + b$	$f(x) = mx + b$
Input	x	x
Output	y	$f(x)$
Coordinate Pair	(x, y)	$(x, f(x))$

The main takeaway in Table 1.11 is the idea that $f(x)$ gives us an alternative way to write the y-variable, with the added benefit that it is more descriptive because it also shows the input. Simply knowing the correct usage of function notation may be enough to answer some questions. There will be a more in-depth discussion on function notation in 2.3 Function Properties of the Advanced Mathematics section.

Q+A

Linear Equations and Function Notation **No Calculator**

$$f(x) = -0.5x + 3$$

Which of the following coordinate pairs satisfies the function f provided above?

A. $(-0.5, f(3))$ B. $(0, f(3))$ C. $(4, f(1))$ D. $(4, f(4))$

Workspace

Solution Outline

Step 1: Identify the point $(x, f(x))$. For functions given in function notation, (x, y) coordinates can instead be represented as $(x, f(x))$. Only choice D contains a consistent x-value, so it must be correct. Note, that the value of $f(4)$ is 1 because when 4 is plugged into the equation for x, the result of $-0.5(4) + 3$ is 1. Therefore, the point can also read as $(4,1)$.

Answer: D

Notes

Students are also tested on their ability to create linear equations and identify their unique features. The most important aspects of linear equations on the coordinate plane are slope and intercepts. Working with these features is essential to answering questions about linear relationships on the coordinate plane and linear models in context.

Discussion on Slope

Slope indicates the steepness of a line. It can be measured by mapping the "rise over run" required to move from one point on the line to the next. For any given line, the slope is constant. The slope of a line can be identified from its linear equation or solved for by using two known points on the line. There are four types of slope: positive, negative, zero, and undefined. Note that lines with a positive slope move up to the right, lines with a negative slope move down to the right, lines with zero slope are horizontal, and lines with undefined slope are vertical These ideas are summarized in Table 1.12 found below.

Table 1.12 | Four Types of Slope

Positive Slope		Negative Slope	
Form	$y = mx + b$	Form	$y = mx + b$
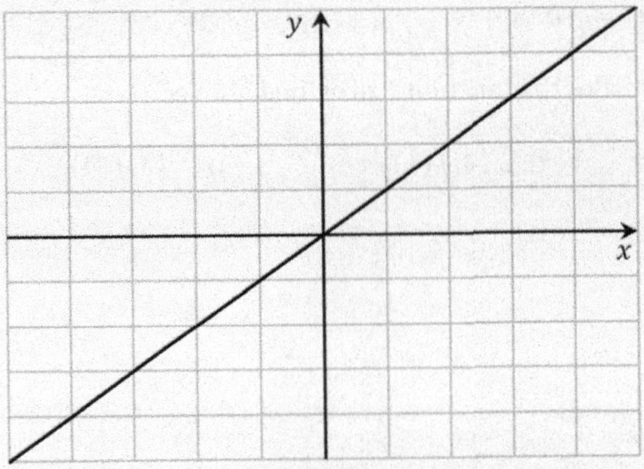		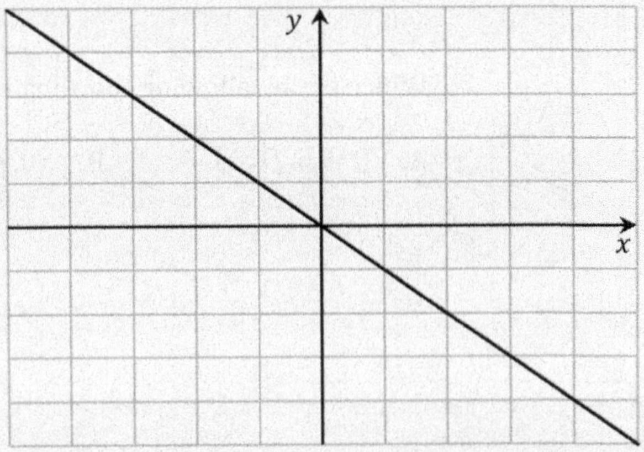	
Zero Slope		Undefined Slope	
Form	$y = c$	Form	$x = c$
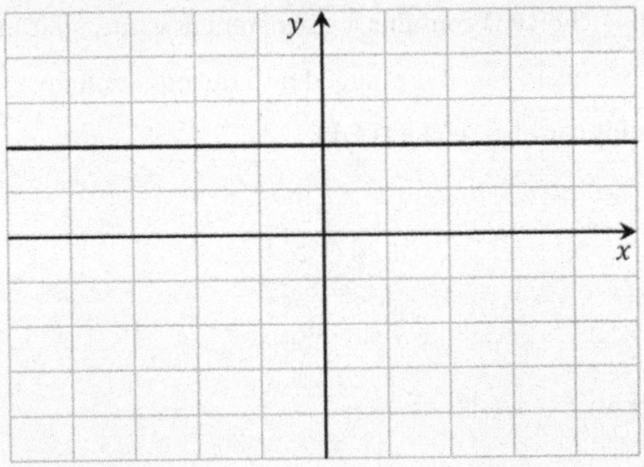		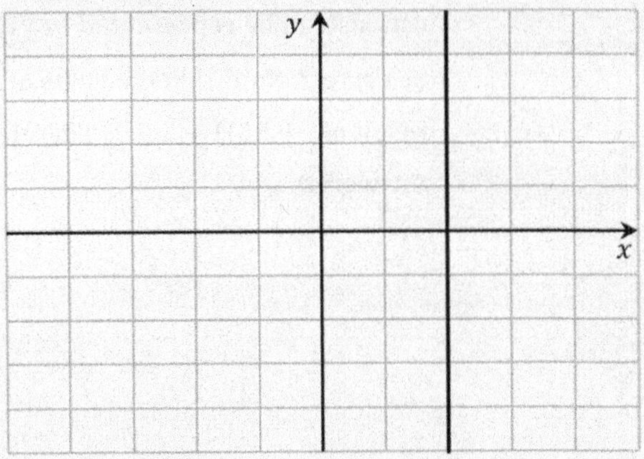	

Many test questions require the equation and general principles of slope. Recall that the sayings often used to describe slope are "rise over run" and "change in y over change in x". Both phrases indicate that slope relates to a line's steepness and provides a basic framework to calculate any function's average rate of change. Still, the slope equation will be used in many problems where there are known points on the coordinate plane. The slope between any two points (x_1, y_1) and (x_2, y_2) can be found by dividing the difference between the y-values by the difference between the x-values.

Slope Equation	$m = \dfrac{y_2 - y_1}{x_2 - x_1}$

Understand that in many cases, especially graphical models, the slope can often be found using a rough approximation. Simply estimating the rise and run may be sufficient in answering most questions.

Q+A

Application of Slope

No Calculator

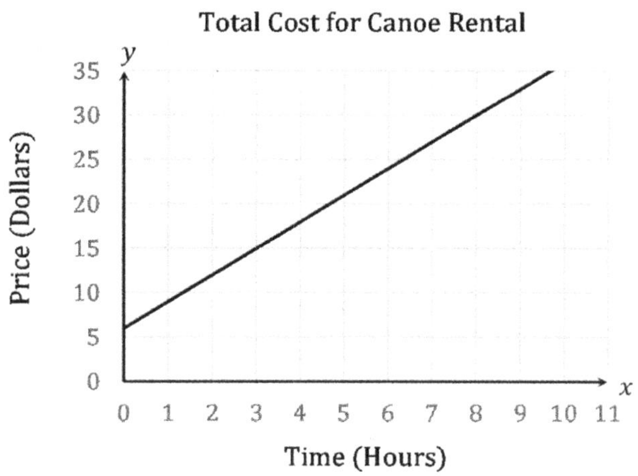

The xy-plane above contains a line that models the total cost for a canoe rental that is based off an initial deposit and fixed hourly rate. According to the model, how much does it cost to rent a canoe every hour?

A. $3.00

B. $6.00

C. $9.00

D. $12.00

Workspace

Solution Outline

Step 1: The cost per hour is equivalent to the slope of the line graphed on the xy-plane because it represents the rate of change between the price and number of hours. Identify two known points on the line or estimate the rise and run. The graph starts at (0,6) and crosses through the point (1,9). This is a rise of three and run of one. In context, the line goes up $3.00 every hour.

Answer: A

Notes

Using the slope equation is a prerequisite skill that will allow us to create a linear equation when only points on the line are known. Understanding the features of and calculations for slope is also essential for our success in being able to relate lines based on their slopes. The next section continues the discussion of slope by explaining two types of lines that are related to each other by their slopes: parallel and perpendicular. In addition, the examples that follow will help us recall how to use the slope equation using points when one coordinate has unknown constants. Recall that constants are numerical values that do not change; instead, they stay the same – or constant – within a particular context.

Parallel and Perpendicular Lines

In addition to finding and using slope, there are two important types of lines that are related to each other by their slopes: parallel and perpendicular. Parallel lines have the same slope and never cross when graphed on the same coordinate plane. Perpendicular lines, on the other hand, cross at right angles and have slopes that are opposite reciprocals. Keep in mind that the word "opposite" in this context means having different signs (e.g. positive and negative) and "reciprocal" means the flipped fraction of. This relationship is required for the lines to intersect at perfect 90-degree angles. Knowing how to relate their slopes will help us answer questions about parallel and perpendicular lines. Use Table 1.13 to see how these lines relate to each when graphed on the coordinate plane.

Table 1.13 | Graphs of Parallel and Perpendicular Lines

Parallel Lines

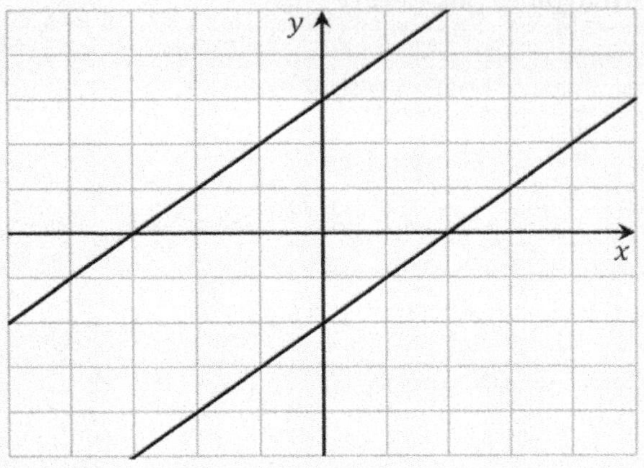

Perpendicular Lines

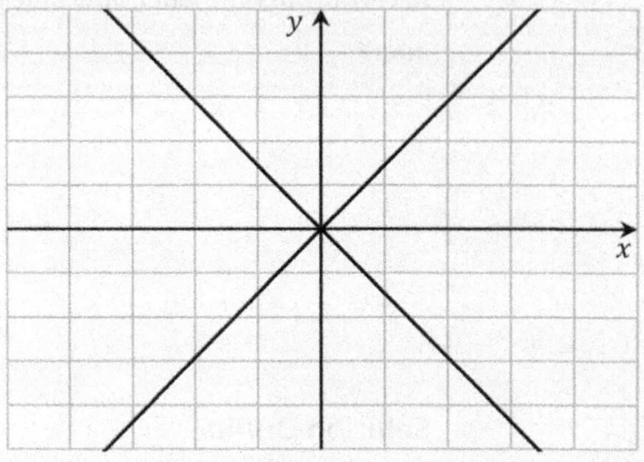

Q+A

Parallel Lines **No Calculator**

Line p is parallel to line q and contains the points $(-4,0)$ and $(0,5)$. Line q contains the points $(8,0)$ and $(0, b)$. What must be the value of the constant b?

A. -10 B. -5 C. 5 D. 10

Workspace

Solution Outline

Step 1: Understand that the lines must have the same slope because they are parallel.

Step 2: Use a simple sketch or the slope formula to find the value of b that makes the slopes equal. The slope of line p is $5/4$. If $b = -10$, the slope of line q becomes up ten and over eight, which is equivalent to up five and over four.

Answer: A

Notes

In the next example, we will look at the relationship between the slopes of perpendicular lines, as well as use the slope equation in greater detail. While using the answer choices as plug in values is an efficient process, it can be just as quick to solve completely by hand, which is necessary when in the grid-in portion of the exam.

Q+A

Perpendicular Lines — Diagnostic Exam | No Calculator Q16

When graphed on the xy-plane, line r contains point A $(1,3)$ and line s contains point B $(0, a)$. Lines r and s intersect at $(4,9)$. If a is a constant, then what must its value be to make lines r and s perpendicular to each other?

Workspace

Solution Outline

Step 1: Find the slope of line r because it contains two known points: $(1,3)$ and $(4,9)$.

$$m = \frac{9-3}{4-1} \quad \text{or} \quad 2$$

Step 2: Use the opposite reciprocal of the slope found in Step 1 to ensure line s is perpendicular to line r. Then use this slope to find the constant a. The opposite reciprocal of 2 is $-1/2$.

$$\frac{a-9}{0-4} = -\frac{1}{2}$$

$$a - 9 = 2$$

$$a = 11$$

Answer: 11

Notes

Discussion on Intercepts

Intercepts occur when a graphed curve crosses the coordinate axes. Therefore, lines contain two intercepts: the x and y. Be sure to recognize the x- and y-intercepts both given a graph on the xy-plane and listed out as coordinate pairs.

- The x-intercept always occurs when $y = 0$. The coordinate will take the shape $(a, 0)$ where a is the x-intercept.
- The y-intercept always occurs when $x = 0$. The coordinate will take the shape $(0, b)$ where b is the y-intercept.

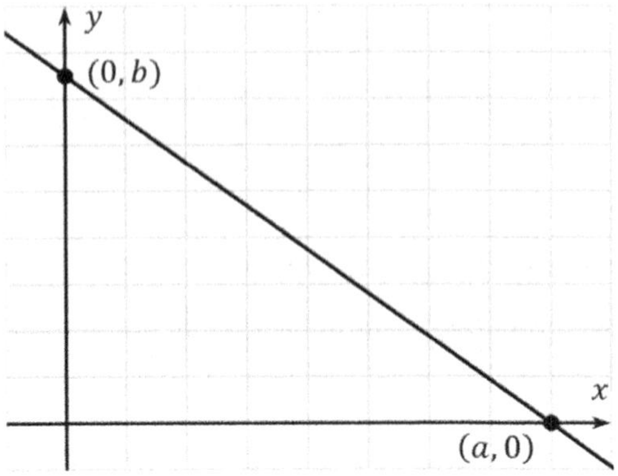

Q+A

Intercepts and Function Notation **No Calculator**

$$f(x) = 3x - 15$$

Which of the following coordinate pairs represents the x-intercept of the function f above when it is graphed on the xy-plane where $y = f(x)$?

A. $(5, f(0))$ B. $(5, f(5))$ C. $(0, f(5))$ D. $(0, f(0))$

Workspace

Solution Outline

Step 1: Identify the points $(x, f(x))$ that maintain consistent x-values. These are choices B and D.

Step 2: Ensure that the x-value used brings the function value to zero. Remember, x-intercepts are the x-values that make y or $f(x)$ equal zero. An x-value of 5 does this, since $3(5) - 15 = 0$.

Answer: B

Notes

Midpoint and Distance Formulas

A few other formulas are important to know: midpoint and distance. One helpful idea to remember is that we can take a simple average to find the middle of any two numbers. Apply this concept below to see how this relates to the midpoint formula using (x_1, y_1) and (x_2, y_2) as endpoints.

Midpoint Formula	$(x_m, y_m) = \left(\frac{x_1 + x_2}{2}, \frac{y_1 + y_2}{2}\right)$

In addition, the distance formula that describes the length between two points, (x_1, y_1) and (x_2, y_2), comes from the Pythagorean Theorem, $a^2 + b^2 = c^2$, which relates the three sides of a right triangle, where a and b are the lengths of the legs to the right triangle and c is the hypotenuse. In the distance formula, a is the Δx, b is the Δy, and c is the distance between the two points. Solving for c in the Pythagorean Theorem in this context yields the distance formula.

Distance Formula	$\text{Distance} = \sqrt{(x_2 - x_1)^2 + (y_2 - y_1)^2}$

Using the graph to the right, we can see that the change in x, or Δx, is equivalent to the gap, or difference, between the x-coordinates and that the change in y, or Δy, is equivalent to the gap, or difference, between the y-coordinates. By dashing in these lines, we can create a right triangle that makes it much easier to identify the distance formula as a unique application of the Pythagorean Theorem.

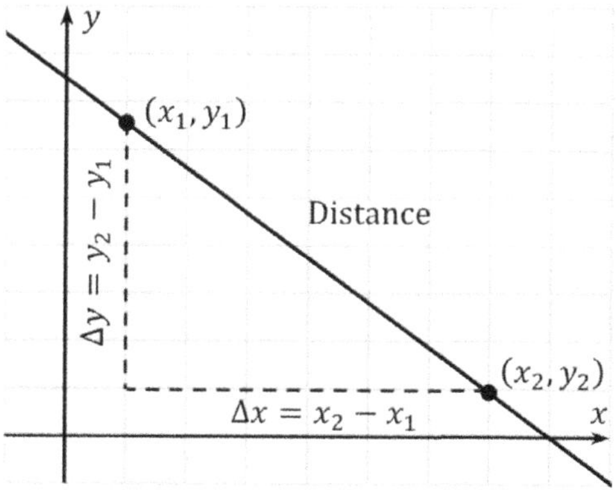

Q+A

Midpoint Formula **Diagnostic Exam | No Calculator Q19**

Line segment $\overline{AB}$ contains endpoints $A(7, -3)$ and $B(-5, 9)$. If (a, b) is the midpoint of $\overline{AB}$, what is the value of $a + b$?

Workspace

Solution Outline

Step 1: Find halfway between the two x-coordinates and the two y-coordinates by taking a simple average.

$$a = \frac{7 + (-5)}{2} \qquad b = \frac{-3 + 9}{2}$$

$$a = 1 \qquad b = 3$$

Step 2: Find the value of $a + b$.

$$a + b = 1 + 3$$

$$a + b = 4$$

Answer: 4

Notes

b. Forms of Linear Equations

Knowing the three most common forms of linear equations will help us in constructing linear equations, finding their unique characteristics, relating them to other linear functions, and interpreting their significance in context. These are the three main forms of linear equations: Slope-intercept, Point-slope, and Standard form. Table 1.14 provides a summary for each by describing their most important attributes.

Table 1.14 | Three Forms of Linear Equations

Slope-intercept Form	$y = mx + b$	m	Slope
		b	y-intercept
Point-slope Form	$y - y_1 = m(x - x_1)$	m	Slope
		(x_1, y_1)	Coordinate pair
Standard Form	$Ax + By = C$	$-A/B$	Slope
		A, B, C	Constants

Slope-intercept form is the most important form of linear equations both in the classroom and on standardized testing since the two main features – slope and y-intercept – of linear equations can be easily identified and applied. Almost all linear models that students must analyze in context on the test are provided in this form. Moreover, slope-intercept form can be a good starting place for generating equations in standard form, which does not have the same ease of use, as its constants do not often represent any particular characteristic of a line; rather it's the relationship between the constants that determine the line's slope and y-intercept. Still, students may elect to use point-slope or standard form to construct equations. Be sure to recognize variables and constants, even if the variables are not simply x and y, which is often the case. Variables in an equation may take on a range of different values, whereas constants just have one.

Q+A

Standard Form Equation **Diagnostic Exam | Calculator Q1**

When line k is graphed on the xy-plane, it crosses through the points $(-a, 0)$ and $(0, 2a)$, where a is a positive constant. Which equation could represent line k?

A. $2x + y = -10$

B. $2x - y = -10$

C. $2x + y = 10$

D. $2x - y = 10$

Workspace

Solution Outline

Step 1: Find the slope m of line k.

$$m = \frac{2a - 0}{0 + a}$$

$$m = \frac{2a}{a} \text{ or } 2$$

Step 2: Write the equation in slope-intercept form, then bring the equation into standard form.

$$y = 2x + b$$

$$2x - y = -b$$

Step 3: Recognize that $(0, 2a)$ is the y-intercept of line k and must be positive. Therefore, b is positive and $-b$ is in fact negative. The constant on the right-hand side of the equation then is also negative. Choice B satisfies all criteria.

Answer: B

Notes

c. Interpreting Linear Models

Linear models are used to represent specific situations that take on linear relationships. The questions about linear models on the test will often describe a scenario by laying out the context, providing a linear equation in slope-intercept form, and defining the variables used to create the equation. Test-takers will then be asked to determine the significance of either the slope or the y-intercept. These questions get very wordy, so we should look for three items in particular to save time: **Value, Units, and Key-terms**. Generally, the question and answer choice together will define all three parts.

The value is the number (often given as part of the question) that can be located in the provided equation and represents either the slope or y-intercept. Some questions may instead ask for the value of the slope or y-intercept indirectly by referencing its units and key terms. Use the given equation to first determine the "x" and "y" variables, then apply their units to interpret their meaning in context.

The value of the slope will have units that reflect "rise over run" and the y-intercept will have the same units as the y-variable, as summarized in Table 1.15 that follows. The most frequently used key terms are also summarized. These are great words to look for to narrow down answer choices or understand what the question is asking for if it is not already clear. One way to answer problems on Interpreting Linear Models effectively is by circling the variables and the units that describe them, then go to the question to see whether the slope or y-intercept is being referenced. We can then look at all the answer choices to see if any of the appropriate key terms appear. By only searching for concrete phrases, we can answer questions correctly and in less time.

Table 1.15 | Linear Models Summary

Value	Units	Key Terms
m	$\frac{\text{Units of "}y\text{" Variable}}{\text{Units of "}x\text{" Variable}}$	Each, Every, Per
b	Units of "y" Variable	Initial, Beginning, Starting, One-Time, Original, Once, Fixed Amount

In some cases, words like hourly, weekly, and annually can be stated as per hour, each week, and every year to describe the slope. Also, be wary of instances when the y-intercept may not actually represent a beginning, such as in countdown situations where the y-intercept may instead represent a final value.

When given a linear equation in context, as we'll see in the following example, the units of each variable are clearly defined. More importantly, these variables are always "y"-acting and "x"-acting, meaning that we will see a relationship in the form $y = mx + b$, but the variables may be represented by letters other than x and y.

Q+A

Linear Models **Diagnostic Exam | No Calculator Q1**

Rob works at a marketing company where each week he receives a list of potential clients to call. The number of clients he has left to call at the end of each day can be found using the equation $N = 182 - 36d$, where N is the number of clients left and d is the number of days Rob worked that week. What is the meaning of 182 in this equation?

A. Rob will complete his calls within 182 days

B. Rob starts each week with 182 clients to call

C. Rob completes 182 calls per day

D. Rob spends a total of 182 hours on call

Workspace

Solution Outline

Step 1: Identify the value presented in the question (182) as the y-intercept.

Step 2: Identify the units of the y-variable (N) as the number of clients.

Step 3: Look for key terms of y-intercepts, like "starts".

Answer: B

Notes

d. Expressions and Equations

Linear expressions and equations have a prominent place on the current test. Students must be able to quickly write and evaluate linear expressions as well as create and solve linear equations. Questions that challenge students on these grounds may already provide an algebraic equation and framework or, instead, outline a lengthy word problem that test-takers must then interpret and solve. While these questions may range in difficulty and focus, there are many points of commonality that bring them all together. Among these places of overlap are solving strategies that will make getting the right answer a more efficient and reliable process.

Among the most important techniques students can perform is reaching expression or equation multiples. Manipulating expressions or equations through multiplication or division can make them simpler to work with, leaving less room for potential errors. For example, the following equations are equivalent; however, Equation 1 contains fractions that may be difficult to work with, whereas Equation 2 contains integers. To move from Equation 1 to Equation 2, simply multiply by 12 since all the denominators of Equation 1 go into 12 evenly.

Equation 1	Equation 2
$\frac{1}{2}x + \frac{1}{3}y = \frac{1}{4}$	$6x + 4y = 3$

We can also manipulate equations to find "hidden" expressions and their multiples. This idea is important because it helps you learn to recognize whole expressions instead of first having to determine independent variables. This ties into one of the main testing points: Analysis of Structure. Another key idea that will shine in the following example is that we can find the value of expressions without knowing the value of the variables that comprise it.

Q+A

Value of an Expression **No Calculator**

If $6k - 8 = 10$, what is the value of $3k - 4$?

A. 5 B. 6 C. 7 D. 8

Workspace

Solution Outline

Step 1: Identify the multiple between the left side of the given equation and the expression.

Step 2: Multiply the equation by one-half to reach the desired expression and its value.

$$6k - 8 = 10$$

$$\frac{1}{2}(6k - 8 = 10)$$

$$3k - 4 = 5$$

Answer: A

Notes

Some questions may offer students information about the line and the points that the line contains to then ask about the value of an expression. In the next example, we will see how creating a linear equation and making substitutions will help us find these unknown quantities.

Q+A

Equation of a Line

No Calculator

Line k is graphed on the xy-plane and crosses through the points $(1,4)$ and $(3,-2)$. If the point (a,b) is also contained by line k, then what is the value of $3a + b$?

A. -7

B. -3

C. 3

D. 7

Workspace

Solution Outline

Step 1: Write the equation of line k by first using its points to solve for the slope and then plugging that slope and one of the points into point-slope form.

Slope Equation

$$m = \frac{-2-4}{3-1}$$

$$m = -3$$

Point-Slope Form

$$y - 4 = -3(x - 1)$$

$$y = -3x + 7$$

Step 2: Substitute the point (a, b) into the equation and generate the desired expression. Recall that (a, b) is an (x, y) point; therefore, $x = a$ and $y = b$.

$$y = -3x + 7$$

$$b = -3a + 7$$

$$3a + b = 7$$

Answer: D

Notes

Word problems may also be described to have students find the values of unknown variables and expressions. The most powerful step we can take to set up a general structure for an expression or equation is define our variables and pair them with any pertinent quantitative information. Many questions centered around linear expressions and equations rely on unit rates. **Unit rates** describe how the quantity of a given measurement corresponds to just *one* of another measurement. For example, x miles per hour and y dollars per month are both unit rates because the former is relative to *one* hour and the latter to *one* month. Unit rates can help us determine the total amount that is generated after a number of units are applied.

Unit Rate Application	Total Amount = Number of Units × Unit Rate

The following examples will help us understand how the test expects students to use unit rates in writing linear expressions and equations.

Q+A

Linear Expressions **No Calculator**

A shipping container is loaded with x small boxes and y large boxes. Every small box weighs 65 kilograms and every large box weighs 105 kilograms. If the shipping container is itself 1,200 kilograms, what expression represents the total weight of the shipping container and the boxes in it?

A. $65x + 105y$

B. $170xy$

C. $65x + 105y + 1{,}200$

D. $170xy + 1{,}200$

Workspace

Solution Outline

Step 1: Find the total weight applied by the small boxes and large boxes independently. Since each small box is 65 kilograms and there are x small boxes, the weight of just the small boxes is $65x$. Similarly, since each large box is 105 kilograms and there are y large boxes, the weight of just the large boxes is $105y$.

Step 2: Understand that the total weight in question is the sum of all boxes and the container. Therefore, the total weight is $65x + 105y + 1{,}200$.

Answer: C

Notes

In other problems, it may be helpful to first write an equation to then find the expression that represents the value of an unknow variable. Still, having good organization will aid us in our evaluation.

Q+A

Linear Expressions and Equations **Diagnostic Exam | No Calculator Q4**

A chemist works with solutions of hydrochloric acid (HCl) with acid concentrations expressed as a percentage of total volume. Which expression can be used to determine how many milliliters (mL) of 5% HCl solution must be added to 45mL of 8% HCl solution to obtain p mL of 6% HCl solution?

A. $\frac{6p-360}{5}$ B. $\frac{0.6p-360}{0.5}$ C. $\frac{6p-3.60}{5}$ D. $\frac{6p+360}{5}$

Workspace

Solution Outline

Step 1: Define and pair information. Each solution is expressed as a percentage of the total volume. Recall that "of" often indicates multiplication. Multiply the percent and total volume to reach the volume of acid present.

Step 2: Create and solve the equation for the desired variable. See to it that the 5% and 8% solutions are added to form the 6% solution.

$$\frac{5x}{100}+\frac{8(45)}{100}=\frac{6p}{100}$$

$$5x+360=6p$$

$$x=\frac{6p-360}{5}$$

Answer: A

Solution	Amount	Percent	Acid Total
5% HCl	$x\ mL$	$\frac{5}{100}$	$\frac{5x}{100}$
8% HCl	$45\ mL$	$\frac{8}{100}$	$\frac{8(45)}{100}$
6% HCl	$p\ mL$	$\frac{6}{100}$	$\frac{6p}{100}$

Notes

Other helpful words and phrases are summarized in Tables 1.16a and 1.16b, which organize the most common ways to describe general arithmetic and more advanced operations.

Table 1.16a | General Arithmetic

Operation	Symbol	Notes
Addition	$+$	Plus, added to, more than, sum, greater than, increased by
Subtraction	$-$	Minus, subtracted from, less than, difference, taken away from
Multiplication	$\times$	Times, of, multiplied by, product of, groups of
Division	$\div$	Divided by, quotient of, per, out of
Equal	$=$	Is, equals, makes, equivalent to

Table 1.16b | Advanced Operations

Operation	Symbol	Notes
Percent	%	A percent is always out of 100
Absolute Value	$\vert\ \vert$	Always positive, magnitude
Power	x^n	Squared ($n=2$), Cubed ($n=3$), Raised to the nth
Consecutive Integers	$x, x+1, x+2$ ...	Integers include whole numbers and their negatives
Consecutive Odd/Even	$x, x+2, x+4$ ...	Odd/even integers skip by twos
Always Even Integer	$2k$	Doubling an integer has an even product
Always Odd Integer	$2k+1$ ...	Odd integers occur one after even integers

Problem Set: Linear Equations

14 Questions

1

A line that passes through the points $(2,3)$ and (a, b) has a slope of -4. If $b = 7$, what is the value of a?

A. -1

B. 0

C. 1

D. 3

2

$$y = x + 7$$

If the line modeled by the equation above contains the point (m, n), which of the following is the value of $m - n$?

A. -7

B. 1

C. 5

D. 7

3

A pilot uses the equation $E = 250 - 2m$ to determine the elevation E in hundreds of feet of his aircraft m minutes after he begins the descent. What is the meaning of 250 in this equation?

A. The pilot descends at 250 feet every minute.

B. It will take 250 minutes for the aircraft to reach the ground.

C. The pilot begins his descent at 25,000 feet in elevation.

D. The aircraft covers 25,000 feet for every 2 minutes of travel.

4

$$2x - 3y = 10$$

$$-7x - ay = 15$$

What value of the constant a would make the lines formed by the equations above perpendicular?

A. $-21/2$

B. $-14/3$

C. $14/3$

D. $21/2$

5

A real estate agent earns a weekly income of \$1,500 based on her commission c and hourly wage w. If she works 30 hours a week for \$22 per hour, how many dollars does she earn every week in commission?

A. 660

B. 750

C. 820

D. 840

6

A line segment has endpoints P and Q. If $(1, -4)$ is the midpoint and Q is $(3,5)$, what is the y-coordinate of point P?

A. -13

B. -10

C. -5

D. -4

7

A line that passes through the points $(-4, -5)$ and $(4,4)$ has ordered pairs in all quadrants except for which of the following?

A. I

B. II

C. III

D. IV

8

Dom wants to create an equation to model the total monthly cost C (in dollars) of running his restaurant. He calculates the fixed costs total \$863.97 every month. If Dom also needs to pay \$14 per hour to z workers for every h hours that they work, which equation represents Dom's total monthly cost.

A. $C = 14h + z + 863.97$

B. $C = 14hz - 863.97$

C. $C = 863.97(14hz)$

D. $C = 14hz + 863.97$

9

A landscaping company is tasked with planting trees that are equally spaced along the perimeter of a golf course. The company uses the equation $m = 2.4 - 0.15t$ to model the remaining length m in kilometers of land around the golf course to cover after planting t number of trees. What is the meaning of 0.15 in this equation?

A. There are 0.15 kilometers between every tree.

B. There will be an average of 0.15 trees per kilometer.

C. After 0.15 kilometers, an average of 2.4 trees will have been planted.

D. The company begins the project with enough trees for a 0.15 kilometer stretch of land.

10

A grocery store sells almonds for \$4.99 per pound and cashews for \$6.50 per pound. If a customer spends \$5.52 per pound on a mixture containing 0.68 pounds of almonds and c pounds of cashews, what expression can be used to determine the combined number of pounds purchased?

A. $\dfrac{0.68(4.99) + 6.50c}{0.68 + c}$

B. $\dfrac{0.68(4.99) + 6.50c}{5.52(0.68)c}$

C. $\dfrac{0.68(4.99) + 6.50c}{5.52}$

D. $\dfrac{0.68(4.99) + 6.50c}{5.52(0.68 + c)}$

11

An urban planner estimates that a business will need to provide 55 parking spaces for x vistors per hour and another 15 parking spaces for the business's employees. If the business only expects three-fifths of x visitors per hour, how many parking spaces should it have in total?

A. 33

B. 48

C. 55

D. 70

12

$$\frac{3x - k}{8} = \frac{5}{2}$$

What is value of x in the equation described above if $k = 7$?

A. 3

B. 9

C. 18

D. 27

13

$$2(y + 5y) - (2y - 10)$$

Which of the following is equivalent to the expression above?

A. $10(y - 1)$

B. $10(y + 1)$

C. $2(7y + 5)$

D. $2(7y - 5)$

14

The point $(c, c - 2)$ is on the line $y = -3x + 1$. What is the value of c?

A. $-1/2$

B. $-1/3$

C. $3/4$

D. $4/3$

Explanations

Problem Set: Linear Equations

1. C

Use the slope equation to determine the unknown constant, a. Recall that the first ordered pair acts as (x_1, y_1) and the second as (x_2, y_2).

$$m = \frac{y_2 - y_1}{x_2 - x_1}$$

$$-4 = \frac{7 - 3}{a - 2}$$

$$-4(a - 2) = 4$$

$$a - 2 = -1$$

$$a = 1$$

2. A

Plug in the point (m, n) into the equation to generate the desired expression $m - n$. Since the point operates as an (x, y) pair, $x = m$ and $y = n$

$$y = x + 7$$

$$n = m + 7$$

$$m - n = -7$$

3. C

The value of 250 is the y-intercept, which maintains the units of the y-variable. The meaning of the y-variable described in the problem is that of elevation measured in hundreds of feet. Therefore 250 in the equation actually means 25,000 feet in elevation. Recall that a key word for the y-intercept is "begins".

4. C

In standard form, slope is found using the ratio, $-A/B$. The slope of the first equation is $2/3$. Perpendicular lines have slopes that are opposite reciprocals. So, the second line should have a slope of $-3/2$.

$$-\frac{3}{2} = -\frac{A}{B}$$

$$-\frac{3}{2} = -\frac{-7}{-a}$$

$$-\frac{3}{2} = -\frac{7}{a}$$

$$a = 14/3$$

5. D

Generate an equation for the weekly income as described in the question. Then solve for the commission, c.

$$\text{Weekly Income} = c + w \times \text{hours}$$

$$1{,}500 = c + 22(30)$$

$$c = 840$$

6. D

The midpoint (x_m, y_m) represents the middle between two endpoints (x_1, y_1) and (x_2, y_2). Therefore, y_m is between y_1 and y_2. The middle of any two values can be found by taking a simple average.

$$y_m = \frac{(y_1 + y_2)}{2}$$

$$-4 = \frac{(5 + y_2)}{2}$$

$$-8 = 5 + y_2$$

$$y_2 = -13$$

Another way to think about this is to understand that the midpoint is equidistant to the two endpoints. This means, the gap between the midpoint and either endpoint is the same. The second endpoint can be found by applying the same change that is required to go from the first endpoint to the midpoint.

Endpoint 1	5
Change	*minus* 9
Midpoint	−4
Change	*minus* 9
Endpoint 2	−13

7. B

A quick sketch of a line connecting the two points shows it going through all quadrants except II.

8. D

The equation for the total cost should include the sum of the variable and fixed costs. Cost is measured in dollars, so use dimensional analysis to prove each term added has final units of just dollars.

Choice A is incorrect because z represents the number of workers. Workers cannot be added to dollars. Choice B is incorrect because the costs are subtracted from each other, rather than added together. Choice C is incorrect because the final units are not in dollars.

Choice D is correct because it adds the costs together in dollars.

The $14hz$ term:

$$\frac{\$14}{\cancel{\text{hour}} * \cancel{\text{worker}}} * h\ \cancel{\text{hours}} * z\ \cancel{\text{workers}}$$

9. A

The value of -0.15 represents the slope of the line. Slope has units that follow rise over run.

$$\text{Units of slope} = \frac{\text{units of } y \text{ variable}}{\text{units of } x \text{ variable}}$$

$$\frac{-0.15}{1} = \frac{-0.15 \text{ kilometers}}{1 \text{ tree}}$$

Find the answer choices with 0.15 kilometers and 1 tree (or every tree). Remember, key terms for slope are each, every, and per. These words replace the fraction bar in the slope ratio.

10. C

Generate an equation for the total cost of the mixture using variables to represent any unknowns, then solve.

$$\text{Total cost} = \text{pounds} \times \text{cost per pound}$$

$$0.68(4.99) + 6.50c = 5.52x$$

$$x = \frac{0.68(4.99) + 6.50c}{5.52}$$

11. B

Find the new number of visitor spaces, then tally up the total number of spaces for both visitors and employees.

$$\frac{3}{5}(55 \text{ spaces} = x \text{ visitors})$$

$$33 \text{ spaces} = \frac{3}{5}x \text{ visitors}$$

There are 33 spaces for the visitors and an additional 15 spaces for the employees leave 48 spaces in total.

12. B

Plug in 7 for k, then manipulate the equation to solve for x.

$$\frac{3x - k}{8} = \frac{5}{2}$$

$$\frac{3x - 7}{8} = \frac{5}{2}$$

$$3x - 7 = 20$$

$$3x = 27$$

$$x = 9$$

13. B

Expand out the equation, then factor to reach final form.

$$2(y + 5y) - (2y - 10)$$

$$2(6y) - 2y + 10$$

$$12y - 2y + 10$$

$$10y + 10$$

$$10(y + 1).$$

14. C

The point $(c, c - 2)$ is the point (x, y). Therefore, $x = c$ and $y = c - 2$. Substitute these values into the given equation.

$$y = -3x + 1$$

$$c - 2 = -3c + 1$$

$$4c = 3$$

$$c = 3/4$$

1.2 Linear Systems

a. Solving Methods

Linear systems are collections of two or more linear equations. The test often requires students to create, solve, and relate linear systems of just two equations. Such linear systems can be solved algebraically through elimination and substitution, as well as graphically through mapping the equations on the coordinate plane and finding their point of intersection. Although test-takers should stick to strategies that they feel most comfortable with, it is worth noting when one method may be easier to employ than another. For example, when the equations of a system have all variables to one side, it is often easier to use elimination; however, when one equation is already solved for a variable, then it may be quicker to use substitution. And finally, when a question asks about a general point of intersection of two lines graphed on the xy-plane, such as the quadrant in which the intersection occurs, then we can rely on a rough sketch to get an approximate location. To summarize, linear systems can be solved using **elimination, substitution, and graphing**.

Elimination Method

Solving a system of equations through elimination entails adding or subtracting equations together so as to zero out, or eliminate one, or more, variable term. Doing so will leave us with an equation of just one variable, which can then be solved for. As noted in the opening statement, elimination will likely be the most efficient solving strategy when all equations of the system have the same structure, with all variables to one side. Ideally, the equations have matching variable terms, such as $\pm 4x$, but if they do not then at least one equation will have to be multiplied or divided so that the equations have a term of the same magnitude. The following example demonstrates a case where we must manipulate both equations of a linear system to create terms that can be eliminated.

Q+A

Linear Systems: Elimination **No Calculator**

$$3x + 2y = 5$$
$$-2x + 5y = 22$$

What is the solution (x, y) to the system of equations above?

A. $(3, -4)$

B. $(-1, 2)$

C. $(-1, 4)$

D. $(-4, 4)$

Workspace

Solution Outline

Step 1: Identify a common multiple for one the variables between the two equations. The x- terms both go into $\pm 6x$.

Step 2: Multiply the equations to reach the common multiple.

Equation 1:	Equation 2:
$3x + 2y = 5$	$-2x + 5y = 22$
$2[3x + 2y = 5]$	$3[-2x + 5y = 22]$
$6x + 4y = 10$	$-6x + 15y = 66$

Step 3: Add or subtract the equations to eliminate the "matched" variable term.

$$\begin{array}{r} 6x + 4y = 10 \\ + \quad -6x + 15y = 66 \\ \hline 19y = 76 \\ y = 4 \end{array}$$

Step 4: Plug $y = 4$ into one of the equations and solve for x.

$3x + 2(4) = 5$

$3x + 8 = 5$

$3x = -3$

$x = -1$

$(-1,4)$

Answer: C

Notes

Substitution Method

Solving a system through substitution relies on solving for a variable or expression in one equation that can then be used, or substituted into, another equation. The goal here is to create an equation of just one variable, whose quantity can then be evaluated. What should be emphasized here is the fact that expressions can also be substituted into equations, not just simply variables. The reason why this idea has so much value is that it often keeps us from having to grit our way into and out of the most granular details that often lead to unnecessary hurdles, like creating fractions or undertaking repetitive steps. In the following example, we will see how to use substitution effectively to quantify an unknown variable. Note that this example will demonstrate how to substitute in an expression, but students may also proceed in the more traditional route of solving all the way down to the singular variable.

Q+A

Linear Systems: Substitution

No Calculator

$$x + 8 = 2y$$
$$4y + 2x = 8$$

What value of x satisfies the system of equations above?

A. -4

B. -2

C. 1

D. 5

Workspace

Solution Outline

Step 1: Identify a variable or expression to substitute. Since the question asks specifically for the x-variable, time can be saved by replacing a y-term with an expression of x. Since the first equation is already solved for $2y$, it will be easy to multiply that equation by 2 to determine what $4y$ is equal to. Then we can substitute in the value of $4y$.

Step 2: Solve for the variable or expression to prepare for substitution.

$$x + 8 = 2y$$
$$2(x + 8 = 2y)$$
$$2x + 16 = 4y$$

Step 3: Make the substitution and solve.

$$(4y) + 2x = 8$$
$$(2x + 16) + 2x = 8$$
$$4x = -8$$
$$x = -2$$

Answer: B

Notes

Graphing Method

It is important to note that solutions to systems occur at their points of intersection. Some questions will provide a graph where these points of intersection are clear. However, even when graphs are not provided, a quick sketch can go a long way when the quadrant where the intersection takes place is sufficient to determine the correct answer. At the very least, the process of graphing will eliminate some very wrong answers choices. In the following example we will see how straightforward some questions on the test can be if they already provide a graph of the system on the coordinate plane. All we need to do is locate the intersection of the two curves. In cases where there are more curves than just two, the solutions to the system only occur where all lines intersect.

Q+A

Linear Systems: Graphing **Diagnostic Exam | No Calculator Q2**

Two linear equations are graphed on xy-plane below. What is the (x, y) solution to the system formed by these equations?

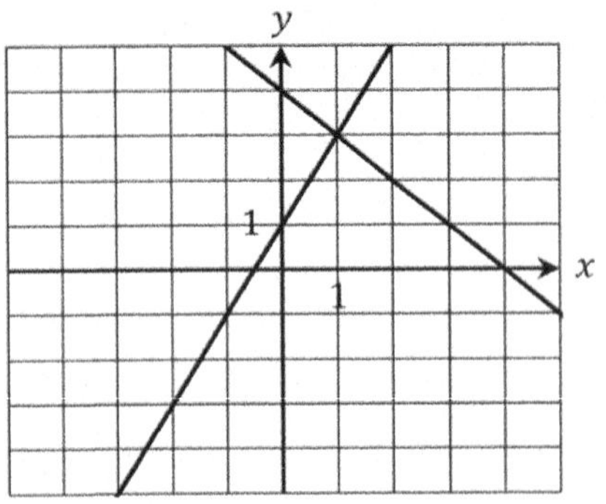

A. (0,1)

B. (1,3)

C. (3,1)

D. (4,0)

Workspace

Solution Outline

Step 1: Find the point of intersection. The x-coordinate is one and the y-coordinate is three.

Answer: B

Notes

Understand also that creating a rough sketch of the lines on the xy-plane when one is not already provided can also be an effective tool when answering questions that ask for the quadrant in which the intersection occurs or when all answer choices fall in different quadrants. Recall that the location of the four quadrants – I, II, III, IV – work their way counterclockwise beginning from top right to bottom right. We will see all quadrants labeled in the solution outline of the following example.

Q+A

Linear Systems: Graphing

No Calculator

$$y = x + 4$$
$$y = -2x$$

What quadrant on the xy-plane contains the solution to the system above?

A. I

B. II

C. III

D. IV

Workspace

Solution Outline

Step 1: Graph the equations. Since both of the provided equations are in slope-intercept form, create each graph by first plotting its y-intercept and then using its slope.

Answer: B

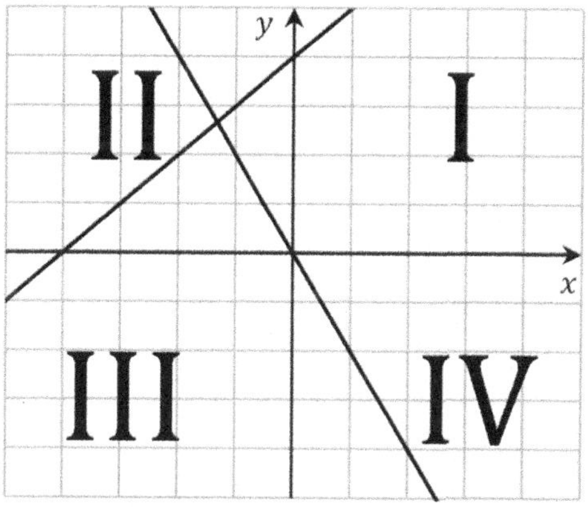

Notes

b. Number of Solutions

A system of two linear equations can have no solution, one solution, or infinitely many solutions. Lines that have different slopes will always have one point of intersection. Lines that have the same slope may be either parallel lines or identical lines. Parallel lines have no solution and identical lines have infinitely many. These cases are summarized in Table 1.21 that follows. The most difficult questions on the test about the number of solutions to a linear system involve lines given in standard form when the slope cannot be identified as a constant. Table 1.21 emphasizes solving strategies that we will discuss in the examples that follow.

Table 1.21 | Number of Solutions Summary

One Solution	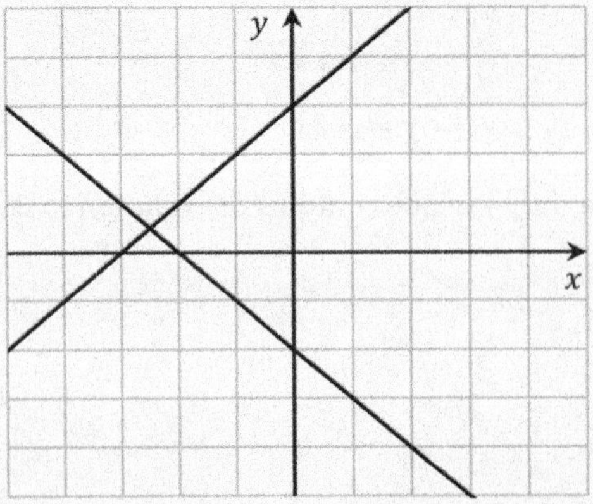	$m_1 \neq m_2$ $\frac{A_1}{A_2} \neq \frac{B_1}{B_2}$ $A_1x + B_1y = C_1$ $A_2x + B_2y = C_2$
No Solution	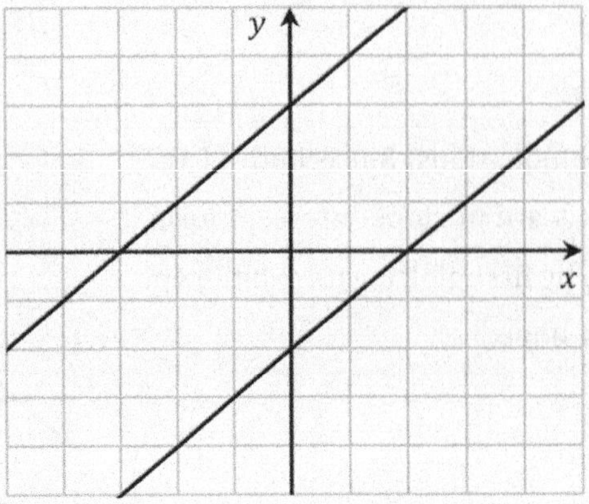	$m_1 = m_2$ $\frac{A_1}{A_2} = \frac{B_1}{B_2}$ $A_1x + B_1y = C_1$ $A_2x + B_2y = C_2$
Infinitely Many	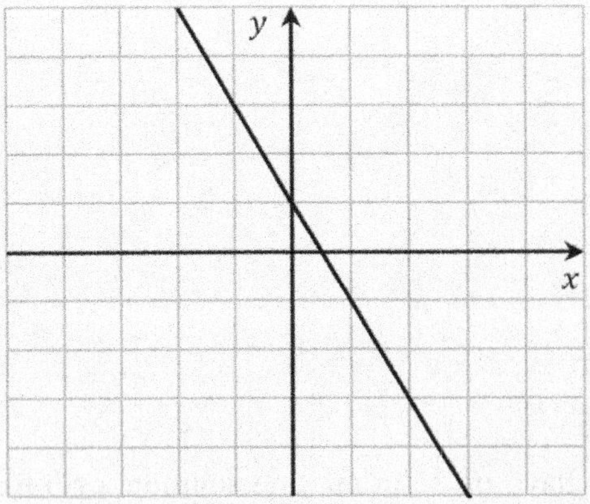	$m_1 = m_2$ $\frac{A_1}{A_2} = \frac{B_1}{B_2} = \frac{C_1}{C_2}$ $A_1x + B_1y = C_1$ $A_2x + B_2y = C_2$

Since many of the systems that students are asked to compare in finding the number of solutions are given in standard form, it is useful to note that we can see right away whether or not the lines will have the same slope – and, thus, if the system has one solution or not – by identifying the relationship of A and B – coming from $Ax + By = C$ form – between the equations. The slopes will be the same if the ratios are the same. We can see this through two strategies: proportions and matching. In the following example, we will see how we can set up a proportion to ensure two lines in standard form have the same slope and, therefore, have no solution. This is a more efficient solving strategy than the traditional approach of changing each equation into slope-intercept form to ensure $m_1 = m_2$. To reiterate, for any two **linear equations in standard form** to have the **same slope**, the **ratio of A to B** must be **equivalent.**

Q+A

Number of Solutions

No Calculator

In the system of equations below, c is a constant whereas x and y are variables. For what value of c will the system of equations have no solution?

$$cx - 7y = 5$$

$$5x - 4y = 3$$

A. 4/5

B. 5/4

C. 4/35

D. 35/4

Workspace

Solution Outline

Step 1: Use a proportion to ensure the slopes are equal because parallel lines have no solution.

$$\frac{A_1}{A_2} = \frac{B_1}{B_2}$$

$$\frac{c}{5} = \frac{-7}{-4}$$

$$c = 35/4$$

Answer: D

Notes

Another strategy we can use to solve for the number of linear equations is matching. This process requires using the structure of the equation to identify, create, and match like-terms. Since parallel lines have the same slope, but are different lines, they will ultimately have different equations; however, the Ax and By between the two equations can be matched by reaching a common multiple. Lines that have infinitely many solutions represent the same line and would ultimately have the same equation after matching one of the terms; thus, every corresponding term between the equations can be matched. The following example will go over the matching strategy as it relates to linear systems.

Q+A

Number of Solutions **Diagnostic Exam | Calculator Q13**

$$my - 2x = 8$$
$$ay + 6x = 2n$$

In the system of equations above, a, m, and n are constants. If the system has infinitely many solutions, what is the value of the $\frac{3m}{n}$ in terms of a?

A. $a/12$

B. $-12a$

C. $4a$

D. $a/4$

Workspace

Solution Outline

Step 1: Find a point of reference between the two equations. This means finding a term that can be matched between them. Since only the x-term in each equation has a known coefficient, we will make this our reference point. Multiply the top equation by negative three so that the x-terms of both equations are equal. In doing so, we get the following equations:

$$-3my + 6x = -24$$
$$ay + 6x = 2n$$

Step 2: Compare the equations. Since the system has infinitely many solutions and we have created a point of reference, every corresponding term between the equations must be equal.

$$2n = -24 \quad \text{and} \quad a = -3m$$

Step 3: Put it all together. The question asks for the value of $3m/n$ in terms of a. From Step 2, we can see that $3m = -a$ and $n = -12$. Therefore:

$$\frac{3m}{n} = \frac{a}{12}$$

Answer: A

Notes

c. Advanced Strategies

The test often asks for the value of expressions that include unknown variables. An important idea is that we do not always need to know the value of every – or any– variable to know the value of an expression. When confronted with a system of equations where the questions ask to identify an expression, look within or between the equations first to determine if you even need to find the "unknowns" to answer the question.

Q+A

Value of an Expression **No Calculator**

$$3y + 4x = 185$$
$$4y = 7x$$

In the xy-plane, the lines that form the system of equations above intersect at the point (b, c). What is the value of $\frac{b}{c}$?

A. 4/7

B. 7/4

C. 5/2

D. 35/8

Workspace

Solution Outline

Step 1: Use the second equation and the point of intersection to find the desired expression.

$$4y = 7x$$
$$4c = 7b$$
$$\frac{b}{c} = \frac{4}{7}$$

Answer: A

Notes

In the following example, we will have to use both equations of the system to reach the desired expression. By adding or subtracting two equations, we can sometimes reach either the expression or one of its multiples. Though this process can save time, understand that not every system that asks for the value of an expression will pan out using this technique. Consequently, we should take the first ten to fifteen seconds in our problem solving to see whether or not this process will be the best route forward. Since in many cases it can be, it shows that taking time in the beginning can save us time in the end.

Q+A

Value of an Expression

Diagnostic Exam | Calculator Q25

$$2a - 7b = 5$$
$$a + b = 1$$

Given the equations above, what is the value of $a - 2b$?

A. 0

B. 1

C. 2

D. 3

Workspace

Solution Outline

Step 1: Add the two equations together.

$$\begin{array}{rr} & 2a - 7b = 5 \\ + & a + b = 1 \\ \hline & 3a - 6b = 6 \end{array}$$

Step 2: Reach the value of the desired expression.

$$\frac{1}{3}(3a - 6b = 6)$$
$$a - 2b = 2$$

Answer: C

Notes

Problem Set: Linear Systems

12 Questions

1

A total of 153 people set out on a day-long hiking excursion, separating into a total of 40 groups. Some groups had three people each, and the rest had six people each. If every person on the excursion belonged to a group, how many groups had just three people?

A. 11

B. 14

C. 26

D. 29

2

Calculator Allowed

$$\text{Density (g/cm}^3\text{)} = \frac{\text{Total Mass (g)}}{\text{Total Volume (cm}^3\text{)}}$$

Brass is an alloy made from a mixture of copper and zinc metals. Copper and zinc have densities of 8.96 g/cm^3 and 7.14 g/cm^3, respectively, and no mass is lost or gained when they are mixed to form brass alloy. Which equation can be used to determine the grams of copper x and grams of zinc y that are in a 100-cubic centimeter sample of brass with a density of 8.31 g/cm^3?

A. $x + y = \frac{100}{8.31}$

B. $\frac{x}{8.96} + \frac{y}{7.14} = \frac{100}{8.31}$

C. $x + y = 8.31(100)$

D. $8.96x + 7.14y = 8.31(100)$

3

Lines l and k share the same y-intercept when graphed on the xy-plane. Line l has a slope of 2 and contains the point (m, n). Line k has a slope of 3 and contains the point $(2m, 3n)$. What is the value of $\frac{m}{n}$?

A. 1/2

B. 2/3

C. 3/2

D. 2

4

$$6x - 9y = 12$$
$$-4x + 6y = -8$$

How many solutions does the system of equations above contain?

A. Zero

B. Only One

C. Infinitely Many

D. Cannot be Determined

5

Lines a and b are perpendicular to each other and share the point $(1, -4)$ when graphed on the xy-plane. If line a has the equation $x + 2y = -7$, then which of the following is the equation of line b?

A. $2x + y = -7$

B. $2x + y = -2$

C. $2x - y = -7$

D. $2x - y = 6$

6

$$y = \frac{1}{2}x - 5$$

$$3y + 15 = x - 8$$

What value of x satisfies the system of equations above?

A. -16

B. -12

C. 8

D. 12

7

$$x - 3y = -8$$

$$2x - 6y = 4a$$

The system of equations above has no solution. Which CANNOT be the value of the constant, a?

A. -4

B. -2

C. 2

D. 4

8 Calculator Allowed

A summer camp program charges \$16.24 per day for children under the age of nine and \$14.29 per day for children from nine to fifteen years old. If the camp has a total of forty-six kids fifteen and under and earns \$5,106.43 in a seven-day period, how many are under the age of nine in the program?

A. 9

B. 18

C. 37

D. 41

9

A game requires players to roll a six-sided die to gain points. Thirty points are earned for rolling an even number and 6 points are lost for rolling an odd number. If a player rolled the die seven times and earned 102 points, how many odd numbers did player roll?

A. 1

B. 2

C. 3

D. 4

10

Line j passes through the points $(-1,0)$ and $(2,3)$ while line k has a slope of -3 and an x-intercept of 4. If the two lines intersect at (g, h) what is the value of $h - g$?

A. 1

B. 4

C. 6

D. 7

11

$$5a + 2b = 10$$
$$3a = 2$$

Which of the following can represent the value of $a + b$ in the system of equations above?

A. 3

B. 4

C. 5

D. 6

12 Calculator Allowed

$$7y = 9x + 42$$
$$1.75y = 2.25x + 10.5$$

How many solutions are there to the system of equations above?

A. Zero

B. One

C. Two

D. More than Two

Explanations

Problem Set: Linear Systems

1. D

Set up a system of equations, where one equation represents the total number of groups and the other represents the total number of hikers. Let x equal the number of three-person groups and y equal the number of six-person groups. Then use elimination or substitution to solve for x.

Equation 1: Number of Groups

$$x + y = 40$$

Equation 2: Number of People

$$3x + 6y = 153$$

The system works only when $x = 29$ and $y = 11$.

2. C

Since the volume and density of the brass sample is known, we can use its mass to help solve for the unknown variables, x and y. Rearranging the given density formula shows:

$$\text{Mass} = \text{Density} \times \text{Volume}$$

The masses of the copper and zinc samples must add to the mass of the brass sample. Putting that all together leaves us with:

$$x + y = 8.31(100)$$

3. A

Write the equations of line l and k in slope-intercept form with the understanding that the y-intercept is the same, unknown constant, b. Then plug in the given points into their respective equations in replace of x and y. Subtract the two equations to eliminate b, then create the desired expression.

Line l: Slope of 2 and point (m, n)

$$y = 2x + b$$

$$n = 2m + b$$

Line k: Slope of 3 and point $(2m, 3n)$

$$y = 3x + b$$

$$3n = 6m + b$$

Elimination: Line k – Line l

$$\begin{array}{r} 3n = 6m + b \\ -\quad n = 2m + b \\ \hline 2n = 4m \end{array}$$

$$\frac{m}{n} = \frac{1}{2}$$

4. C

A system of equations has infinitely many solutions when the equations can be "matched", revealing that the equations are actually representations of the same line. Multiply the first equation by negative two-thirds to see that this system has infinitely many solutions.

$$-\frac{2}{3}(6x - 9y = 12)$$

$$-4x + 6y = -8$$

5. D

Since the equations of lines a and b are provided in standard form, understand that the slope is found in part by the ratio of the coefficients to x and y. And since these lines are perpendicular, their slopes must be opposite reciprocals. Therefore, we can simply swap both and negate one of the coefficients to create a perpendicular line. Plugging in the (x, y) point determines the constant on the other side. In general, perpendicular lines take the shape:

$$\text{A}x + \text{B}y = \text{C}_1$$

$$\text{B}x - \text{A}y = C_2$$

In the context of this problem, we get the following equations:

$$x + 2y = -7$$

$$2x - y = \text{C}_2$$

Plugging in $(1, -4)$ into the equation proves C_2 must equal six.

$$2x - y = 6$$

6. A

Substitute the known expression for y from the first equation into the second one.

$$3\left(\frac{1}{2}x - 5\right) + 15 = x - 8$$

$$\frac{3}{2}x - 15 + 15 = x - 8$$

$$\frac{3}{2}x = x - 8$$

$$\frac{1}{2}x = -8$$

$$x = -16$$

7. A

Match the equations by multiplying the first equation by 2 so that the x and y terms are the same in both equations. Then compare the constants on the right side of the equation.

$$2(x - 3y = -8)$$

$$2x - 6y = -16$$

For the equations to have no solution, the constants -16 and $4a$ cannot be equal.

$$4a \neq -16$$

$$a \neq -4$$

8. C

Set up and solve the system of equations for the number of kids under the age of 9.

Variables:

$$x\text{: \# of kids under 9}$$

$$y\text{: \# of kids aged } 9-15$$

First Equation: Total number of Kids

$$x+y=46$$

Second Equation: Total Cost of Program

$$16.24x+14.29y=\frac{5{,}106.43}{7}$$

$$16.24x+14.29=729.49$$

Solve using substitution, elimination, or graphing.

Elimination:

$$16.24x+14.29y=729.49$$

$$-\quad 14.29(x+y=46)$$

$$1.95x=72.15$$

$$x=37$$

9. C

Set up and solve system of equations to solve for the number of odd numbers rolled.

Variables:

$$x\text{: \# of evens}$$

$$y\text{: \# of odds}$$

First Equation: Total number of Rolls

$$x+y=7$$

Second Equation: Total number of Points

$$30x-6y=102$$

Solve for y through elimination or substitution.

Elimination:

$$30(x+y=7)$$

$$-\quad 30x-6y=102$$

$$36y=108$$

$$y=3$$

10. A

Find equation of line j first. Then see if the expression $h-g$ can be generated before moving on.

Slope of line j:

$$m=\frac{3-0}{2-(-1)}$$

$$m=1$$

Equations of line j:

$$y-y_1=m(x-x_1)$$

$$y-0=1\big(x-(-1)\big)$$

$$y=x+1$$

Plug in (p,g) as an (x,y) point where $x=g$ and $y=h$.

Generate the expression $h-g$.

$$h=g+1$$

$$h-g=1$$

11. B

Subtract equations and solve for $a + b$.

$$
\begin{array}{r}
5a + 2b = 10 \\
-\quad 3a = 2 \\
\hline
2a + 2b = 8
\end{array}
$$

$$a + b = 4$$

12. D

Compare equations in slope intercept form since the second equations is already almost solved for y.

First equation:

$$7y = 9x + 42$$

$$y = \frac{9}{7}x + 6$$

Second equation:

$$1.75y = 2.25x + 10.5$$

$$y = \frac{2.25}{1.75}x + 6$$

$$y = \frac{9}{7}x + 6$$

Same equation means there are infinitely many solutions, or many more than 2.

*Tip: Use Math *frac* operation in calculator to quickly simplify 2.25/1.75.

1.3 Linear Inequalities

a. Solution Sets

Solving inequalities does not have to be immensely different from solving linear equations and their systems. We can still use substitution and elimination to find points of intersection, but graphing will provide the best representation of the full solution set as it displays an entire region, or shaded area, that satisfies the given conditions. When algebraically solving for systems of inequalities, we can treat the inequality as an equal sign and then proceed with substitution or elimination. However, to shade in the solution set correctly, we must either solve the inequality for a variable, such as y, or use test points.

Q+A

System of Linear Inequalities **No Calculator**

$$y < 2x - 5$$
$$y \geq -x + 1$$

Given the inequalities above, which could be a possible value for x?

A. 0

B. 1

C. 2

D. 3

Workspace

Solution Outline

Step 1: Treat the inequalities as equal signs.

$$y = 2x - 5$$
$$y = -x + 1$$

Step 2: Substitute to find x.

$$2x - 5 = -x + 1$$
$$3x = 6$$
$$x = 2 \text{ and } y = -1$$

Step 3: Sketch a graph to determine the region of overlap. Remember to use dashed lines to represent inequalities that are not "or equal to".

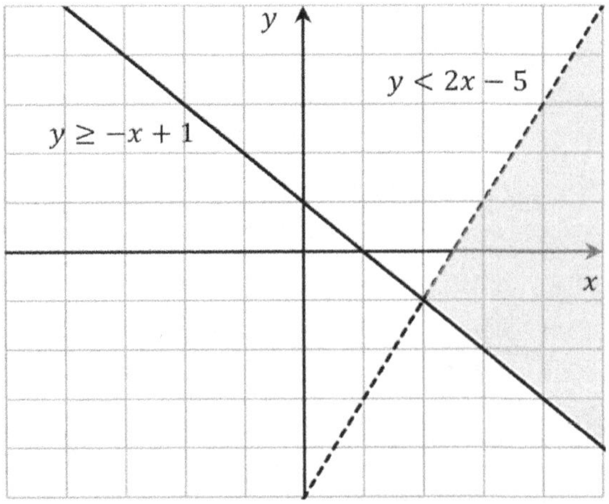

Step 4: With the shaded area drawn in, it is clear that our solution set lies to the right of the point of intersection. All x values must be greater than 2.

Answer: D

Notes

One method that can help us skip several steps when confronted with systems of linear inequalities involves working with the inequalities as they are provided to make useful substitutions or eliminations. The following example shows how to use substitution of a known constraint to quickly find a solution set.

Q+A

System of Linear Inequalities **No Calculator**

$$y > 3x - 8$$
$$3x \geq 10$$

Given the system of linear inequalities above, which could be a possible value for y?

A. -25

B. -8

C. 2

D. 13

Workspace

Solution Outline

Step 1: Based on the second inequality, we can substitute values of ten or more for $3x$ into the first inequality. This will establish a lower bound for y and reveal its range.

$y > 10 - 8$ or $y > 2$

$y > 11 - 8$ or $y > 3$

Step 2: This trend confirms that eventually y will reach a value of thirteen.

Answer: D

Notes

The next example maps out an interesting way to use elimination to quickly reach a partial solution set. An important point to recognize when performing elimination with the inequalities intact is that we can only add the "greater than" sides together and the "less than" sides together. Sometimes this is enough to get an answer; however, it may only offer a partial view of the entire solution to the system.

Q+A

System of Linear Inequalities **Diagnostic Exam | Calculator Q26**

$$x + y < -2$$
$$-x + 2y < 6$$

Given the inequalities above, which could be a possible value for y?

A. 1

B. 2

C. 3

D. 4

Workspace

Solution Outline

Step 1: Add the two inequalities together to eliminate x and establish a range for y. Since y must be less than four-thirds, the correct answer is choice A.

$$3y < 4$$
$$y < 4/3$$

Answer: A

Notes

Tip: There is a theme when answering multiple-choice questions related to linear inequalities. If asked to find a possible value for a variable in the system, the answer will only ever be the least or greatest quantity. With that alone, you can get into a 50 /50 position without any work.

Important Note: Multiplying or dividing through by a negative will flip the direction of the inequality. Same reversal applies when taking the reciprocal of both sides

b. Constructing Inequalities

Many of the inequality questions on the test are actually "set up" questions that require students to only form the system based on a description. Although these questions require a lot reading, they do not take much time because the problem does not actually need to be solved: we just need to write down the inequalities that can be used to solve the system. Consequently, there are common phrases that are used to define each inequality. Use Table 1.31 below to help distinguish exactly which inequality to use when confronted with some of these phrases.

Table 1.31 | Key Phrases for Inequalities

Greater than	Greater than or equal to	Less than	Less than or equal to
$>$	$\geq$	$<$	$\leq$
More than Over Above Larger Exceeds	At least No less than At minimum Lowest	Under Below Smaller Fewer	At most No more than Does not exceed Up to Highest

Q+A

Constructing Inequalities **Diagnostic Exam | No Calculator Q9**

A distance runner training for a 5 kilometer race wants to run at least 20 laps around a track in at most 35 minutes as a workout. The runner decides to run r laps at race pace j laps at a jog. If a lap at race pace takes the runner 80 seconds and a lap at a jog takes the runner 100 seconds, which system of inequalities represents the runner's workout?

A. $\frac{5}{3}j + \frac{4}{3}r < 35$
$r + j \geq 20$

B. $\frac{5}{3}j + \frac{4}{3}r \leq 35$
$r + j \geq 20$

C. $100j + 80r \leq 35$
$r + j > 20$

D. $80j + 100r \geq 20$
$r + j \leq 35$

Workspace

Solution Outline

Step 1: Understand that the two inequalities represented in each answer choice are designed to meet the constraints placed on the number of laps and allotted time.

Step 2: Find key phrases that define each inequality. The phrases "At most 35 minutes" and "at least 20 laps" mean less than or equal to ($\leq$) 35 and greater than or equal to ($\geq$) 20, respectively.

Step 3: Narrow down the answer choices to only those with the correct inequality.

Step 4: The total number of laps must greater than or equal to 20. Therefore, $r + j \geq 20$ and Choice B is correct. To take into account the amount of time it takes to run those laps, ensure that the units used for time match throughout the equation. Since the value of 35 minutes is used in every answer choice, 80 and 100 seconds must be converted to minutes, which requires dividing by 60. The reduced fractions are seen in Choice B.

Answer: B

Notes

We can also use inequalities to establish a range for which a particular variable can have. The following example details how we can find the lower and upper bounds of a variable so that we can incorporate them into a compound inequality statement.

Q+A

Compound Inequality **Calculator**

Thin aluminum spokes used in bicycle tires are manufactured from longer source rods. The company producing the spokes creates them by cutting longer source rods between the lengths of 493 and 567 inches, inclusive, into shorter spokes between the lengths of 21 and 29 inches, inclusive. Given these constraints, how many spokes, x, can be made from any one source rod?

A. $17 < x < 27$

B. $17 \leq x \leq 27$

C. $20 < x < 24$

D. $20 \leq x \leq 24$

Workspace

Solution Outline

Step 1: Understand that we need to find the lower and upper bounds for the value of x, which represents the total number of spokes that can be made out of the source rod.

Step 2: Find the bounds. The least number of spokes that can be made is found by taking the length of the smallest rod and dividing it by the length of the longest spoke. Conversely, the greatest number of spokes that can be made is found by taking the length of longest rod and dividing it by the length of the smallest spoke.

Step 3: Perform the division and set up the inequality. Since the values are described as inclusive, they are included using "or equal to" inequalities.

$$\frac{493}{29} = 17 \quad \text{and} \quad \frac{567}{21} = 27$$

$$17 \leq x \leq 27$$

Answer: B

Notes

Problem Set: Inequalities

10 Questions

1

$$y < 6x + 7$$

$$y > \frac{1}{2}x$$

If the system of inequalities is graphed on the xy-coordinate plane, which quadrant contains no solution to the system?

A. I

B. II

C. III

D. IV

2 Calculator Allowed

A 4,200m to 5,800m stretch of highway is to have streetlights placed between 22m and 35m apart. Which inequality best represents the possible number of streetlights N to the nearest integer.

A. $120 < N \leq 264$

B. $120 \leq N < 264$

C. $165 < N \leq 190$

D. $165 \leq N < 190$

3

$$y + x \leq 3$$

$$y - 2x > -9$$

Which value of x can satisfy the system of inequalities above?

A. 3

B. 5

C. 9

D. 11

4

$$y < 2x + 8$$

$$y < -2x + 8$$

The system of inequalities above are plotted on the xy-plane. If the point (a, b) is a solution to the system and a and b are integer constants, what is the maximum value of $a + b$?

A. 3

B. 4

C. 5

D. 6

5

Molly works as a camp counselor for \$15.75 per hour and retail clerk for \$12.50 per hour to secure extra income over summer break. Both jobs require Molly to work at least 5 hours every week. If she can spend up to 20 hours working each week, what is the most that Molly can earn over eight weeks of summer while maintaining both jobs?

A. \$300

B. \$315

C. \$2,390

D. \$2,520

6

Which system of inequalities has no solution?

A. $y < x + 4$
$y \geq x - 7$

B. $y \leq 2x - 1$
$y > 2x + 3$

C. $y \leq x + 4$
$y < x + 1$

D. $y > 2x$
$y \geq 2x + 5$

7

In her daily exercise regimen, Gwen runs 3 miles and bikes 12 miles. On average, Gwen runs at x miles per hour and bikes at y miles per hour. If her regimen takes no more than 1 hour, then which of the following inequalities best represents the range of possible average speeds she might run or bike on any given day?

A. $3x + 12y \leq 1$

B. $\frac{x}{3} + \frac{y}{12} \leq 1$

C. $\frac{3}{x} + \frac{12}{y} \leq 1$

D. $\frac{12}{x} + \frac{3}{y} \leq 1$

8

The length of a rectangular patio must be m feet and its width n feet. The area of the patio must be more than 400 square feet and its perimeter at most 100 feet. Which of the following systems matches the specifications for both area and perimeter?

A. $m + n \leq 100$
$mn > 400$

B. $m + n \geq 50$
$mn > 400$

C. $m + n \geq 100$
$mn > 400$

D. $m + n \leq 50$
$mn > 400$

9

$$y > 2x + c$$

$$y > \frac{1}{2}x + d$$

The system of inequalities above contains the solution $(4, -4)$. Given that c and d are constants, which of the following can be the value of $c + d$?

A. -20

B. -18

C. -16

D. -14

10

$$\text{I. } p^2 > q^2$$

$$\text{II. } |p| > |q|$$

$$\text{III. } p - q > 0$$

If $p > q$, which of the above is NOT necessarily true?

A. I

B. II

C. I & II only

D. I, II, & III

Explanations

Problem Set: Inequalities

1. D

Use a rough sketch of the lines on the xy-plane to see the solution set. Only quadrant IV is left out of the solution set indicated by the shaded area.

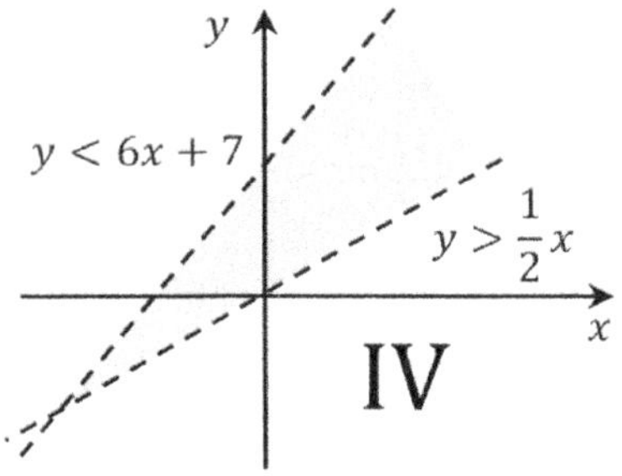

2. B

The minimum number of streetlights can be found by calculating how many times 35 goes in 4,200 and the maximum number of streetlights can be found by calculating how many times 22 goes in 5,800.

Minimum:

$$\frac{4200}{35} = 120$$

Maximum:

$$\frac{5800}{22} = 263.6$$

The lower bound includes 120 while the upper bound excludes 264.

$$120 \leq N < 264$$

3. A

Solve the system through elimination. Create equal but opposite terms, and only add less than sides together and greater than sides together.

Multiply second equation by -1.

$$-1(y - 2x > -9)$$

$$-y + 2x < 9$$

Add equations together.

$$\begin{array}{r} y + x \leq 3 \\ + \quad -y + 2x < 9 \\ \hline 3x < 12 \end{array}$$

$$x < 4$$

*Tip: Possible values to a variable in a system of linear inequalities will always be the smallest or largest answer choice. Only A or D could have been correct.

4. D

Solve by graphing. Since a and b must be integers with the greatest possible sum, we should look at the coordinates in the solution set that belong to first quadrant. The lines formed by the inequalities create a symmetric solution set relative to the y-axis that shades under both lines. In the shaded area of first quadrant, we can see distinct coordinates at $(1,5)$, $(2,3)$, and $(3,1)$. The maximum sum is 1+5 or 6.

5. C

Since Molly must maintain both jobs, she will spend five hours every week working retail at \$12.50 per hour and 15 hours every week counseling at \$15.75 per hour. Every week, she can earn $5(12.50) + 15(15.75)$ or \$298.75. Over eight weeks then, she can earn 8(298.75) or \$\$2,390.

6. B

A system of inequalities will have no solution only if there is no overlapping shaded region. This occurs in Choice B.

7. C

Use an equation to relate distance, rate, and time together. Then manipulate the equation so that it is equal to time.

$$\text{Distance} = \text{rate} \times \text{time}$$

$$\frac{\text{Distance}}{\text{rate}} = \text{time}$$

$$\text{Run time} = \frac{3}{x}$$

$$\text{Bike time} = \frac{12}{y}$$

$$\text{Total time} = \frac{3}{x} + \frac{12}{y}$$

Total time has to be less than or equal to1.

8. D

Create perimeter and area equations by recalling their formulas.

Perimeter: Sum of all sides

$$2m + 2n \leq 100$$

$$m + n \leq 50$$

Area: Product of length and width

$$mn > 400$$

9. **A**

Substitute the (x, y) point of $(4, -4)$ into each inequality and isolate the unknown constants, c and d, to one side.

First inequality:

$$y > 2x + c$$

$$-4 > 2(4) + c$$

$$-4 > 8 + c$$

$$-12 > c$$

Second inequality:

$$y > \frac{1}{2}x + d$$

$$-4 > \frac{1}{2}(4) + d$$

$$-4 > 2 + d$$

$$-6 > d$$

Add the inequalities to generate the desired expression, $c + d$:

$$\begin{array}{rc} & -12 > c \\ + & -6 > d \\ \hline & -18 > c + d \end{array}$$

In other words, the sum of c and d must be less than -18.

10. **C**

Squaring and taking the absolute value does not always maintain the integrity of an equation, let alone an inequality. However, subtracting a smaller number always leads to a positive difference. Therefore, I and II are not necessarily true, while III is. We can also come up with counterexamples by plugging in values of p and q that prove I and II false. For example, if $p = -1$ and $q = -2$, then I and II are shown not to work.

LINEAR ALGEBRA EXAM

DIRECTIONS

For questions 1-10, solve every problem and either circle or record the best answer from the answer choices provided. **For questions 11-15**, solve every problem and record your answer in the appropriate format. On the real test, you must grid in your answers according to the directions. You may use any available space in your workbook for scratch work.

NOTES

1. Calculator usage is NOT allowed in Part 1.
2. Assume real numbers unless otherwise indicated.
3. Figures may not be drawn to scale.
4. All figures are planar, unless otherwise indicated.
5. Domains represent all real numbers of for which f(x) is also a real number.

REFERENCE

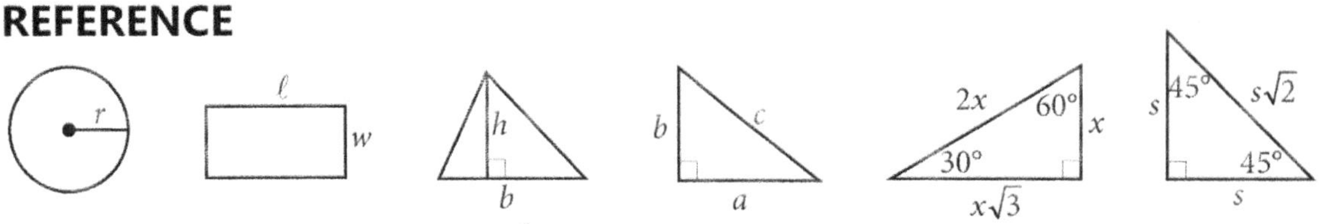

$A = \pi r^2$

$C = 2\pi r$

$A = \ell w$

$A = \frac{1}{2}bh$

$c^2 = a^2 + b^2$

Special Right Triangles

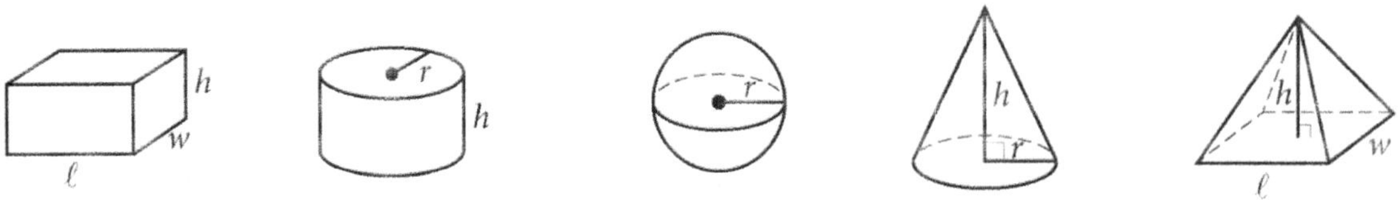

$V = \ell wh$

$V = \pi r^2 h$

$V = \frac{4}{3}\pi r^3$

$V = \frac{1}{3}\pi r^2 h$

$V = \frac{1}{3}\ell wh$

The arc of a circle contains 360 degrees.

The arc of a circle contains 2π radians.

The interior angles of a triangle add up to 180 degrees.

Part 1: No Calculator

19 Minutes, 15 Questions

1

$$F = 1.27B + 2.96$$

The number of grams F of ferric bromide required for a reaction of benzene to take place can be estimated using the equation above, where B represents grams of benzene. Which of the following statements is the best interpretation of the value 1.27 in the context of this equation?

A. The grams of benzene that will react when 2.96 grams of ferric bromide are present

B. The increase in grams of benzene required for each increase of 2.96 grams in ferric bromide

C. The increase in grams of ferric bromide required for each one-gram increase of benzene

D. The grams of benzene that will react when one gram of ferric bromide is present

2

Dora is using the equation $V_f = a + bt$ to determine the final velocity V_f in meters per second of her mousetrap car for any t seconds after $t = 0$. Given that a and b are constants, what could be the units of b?

A. seconds

B. meters

C. meters per second

D. meters per second squared

3

$$\frac{a(3x+4)}{2} = 10$$

Using the equation above, what is the value of x when $a = 5$?

A. -1

B. 0

C. 1

D. 2

4

$$y = ax + 4$$

$$y = bx - 8$$

If a and b are constants and the equations above share an x-intercept at $(c, 0)$, what positive value of c would make the lines perpendicular?

A. $\sqrt{2}$

B. 2

C. $4\sqrt{2}$

D. 6

5

$$(4x + x + x) - (x - 3x) = 18 - 2x$$

In the equation above, what is the value of $5x$?

A. 9/5

B. 9

C. 18/5

D. 18

6

For every foot that a diver descends in water, the pressure increases by 0.43 psi. If the diver observes a pressure change of a psi, which expression represents the distance that the diver descended in feet?

A. $\frac{100}{43}a$

B. $0.43a$

C. $0.43 + a$

D. $\frac{a}{43}$

7

$$\frac{3}{4}p + \frac{1}{2}q = \frac{17}{6}$$

$$3p - q = -\frac{8}{3}$$

If the solution to the system of equations above is (p, q), then what is the value of p?

A. 1/6

B. 2/3

C. 14/3

D. 3

8

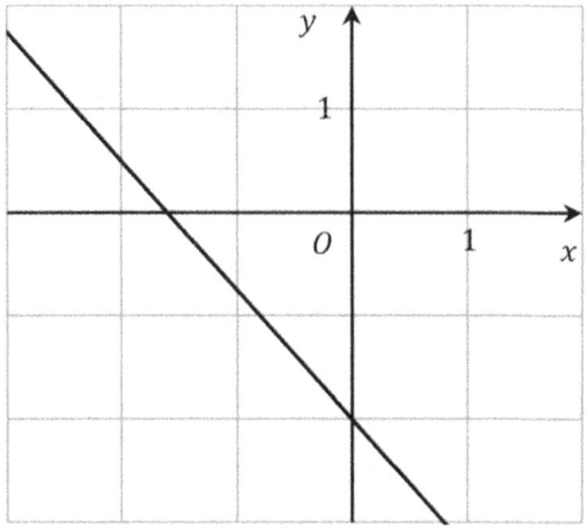

What is the equation of the line graphed on the xy-plane above?

A. $5x + 4y = -8$

B. $5x - 4y = -8$

C. $4x + 5y = 8$

D. $4x - 5y = 8$

9

$$3ax + 2ky = 8$$

$$ax + 6y = 10$$

For the system of equations given above, a and k are constants. What value of k will ensure that the system has no solution?

A. 3

B. 6

C. 9

D. 12

10

Line m crosses through the origin and contains the point $(2b, b - 3)$. If the slope of line m is $^1/_5$ then what is the value b?

A. $-1/3$

B. $5/3$

C. 5

D. Cannot be determined

11

The force F (in newtons) required to displace a spring x meters from its resting position is modeled by the equation $F = kx$, where k is a constant. If 40 newtons are required to displace a spring 0.20 meters away from its resting position, what is the value of k?

12

y	4	6	8
x	-2	$\frac{k}{2}$	5

The table above can be generated using the formula $y = ax + b$, where a and b are constants. What is the value of k in the table?

13

$$2b + c = 10$$

$$3b = 5$$

Using the equations above, what is the value of $c - b$?

14

If $4m - 2 = 3$, then what is the value of $8m - 4$?

15

$$y = ax + 250$$

Jane walks on a hiking trail that increases 1 meter in elevation for every 10 meters traveled along the trail. She starts her hike at 250 meters in elevation and travels 30 meters every minute. If the equation above models Jane's elevation y, in meters, x minutes after she begins her hike, what is the value of the constant a?

Part 2: Calculator

19 Minutes, 15 Questions

1

$$\frac{b}{2}x + 3y = c$$

$$bx - 4y = -2c$$

The solution to the system of equations above is $(-1,2)$. If b and c are constants, then what is the value of c?

A. -4

B. 1

C. 3

D. 5

2

$$6x - 2y = 5$$

When graphed on the xy-plane, lines a and b are parallel. If line a is formed by the equation above, then which of the following can represent line b?

A. $3x + y = 4$

B. $3x - y = 12$

C. $x + 3y = 2$

D. $x - 3y = 5$

3

An art supply store needs to stock up on r gallons of red paint and b gallons of blue paint. Red paint costs \$10.48 per gallon and blue paint costs \$18.36 for every two gallons. If the store needs at least 14 gallons of red and blue paint combined, but only has \$140 to spend, which of the following systems appropriately demonstrates the art supply store's needs and constraints?

A. $10.48r + 18.36b \leq 140$

$r + b \geq 14$

B. $10.48r + 9.18b \leq 140$

$r + b > 14$

C. $10.48r + 9.18b \leq 140$

$r + b \geq 14$

D. $10.48r + 18.36b > 14$

$r + b > 140$

4

$$y < -\frac{1}{2}x + 5$$

$$y > x + 2$$

Given the system of inequalities above, which CANNOT be a value of x?

A. -11

B. -9

C. 1

D. 2

5

A furniture rental company requires a deposit of \$500 from its customers. Every week, customers are charged \$8 for every piece of furniture rented. Lola wants to rent a sofa. She uses the equation $c = ax + b$ to project the total cost c (in hundreds of dollars) for renting the sofa for x weeks. If a and b are constants, what is the value of b?

A. 0.08

B. 5

C. 8

D. 500

6

Lines q and s both cross through the origin. The slope of line q *is* 3 more than double the slope of line s. If line q contains the point $(c, 5)$ and line s contains the point $(c, 2)$, what is the value of c?

A. 1/3

B. 5/3

C. 3/5

D. 3

7

A hot air balloon ascends as it travels through the air. It rises 18 meters in elevation for every 100 meters it travels along its path. If the hot air balloon is currently at 140 meters of elevation and traveling at 5 meters per second, which expression represents its elevation after t seconds?

A. $140 + 18t$

B. $140 + 5t$

C. $140 + \frac{18}{5}t$

D. $140 + \frac{9}{10}t$

8

A dining hall must seat exactly 200 occupants using two differently sized tables: small and large. Small tables seat four and large tables seat 10. If there are 6 more large tables than small ones and each table is completely filled, how many tables in total are there?

A. 20

B. 22

C. 24

D. 26

9

The coordinates $(b, 8)$ and $(10, c)$ are the same distance from the origin. If $b > 0, c > 0$, and $b - c = 4$, what is the value of $b + c$?

A. 8

B. 9

C. 10

D. 11

10

If the equation $y = bx$ contains the coordinate point $(0.5m, m)$, then what is the value of the constant b?

A. 1/2

B. 1

C. 2

D. 4

11

If $x - 4 < 5$ and $y + 2 < -4$, what is the greatest value of $x + y$, given that x and y are both integers?

12

$$y = 3x + 5$$

$$2y = \frac{1}{a}x + 10$$

What value of the constant a would ensure that the system of equations above has infinitely many solutions?

13

Line k crosses through the origin and contains the point (a, b). Given that a and b are positive integers less than five and the slope of line k is greater than one, what could be the value of the fraction b/a ?

14

A local company sells two different kinds of jam: plum and strawberry. They sell p jars of plum jam at \$4.89 each and s jars of strawberry jam at \$3.44 each. If the company made \$119.52 through selling 28 jars of jam, what is the value of p?

15

A test grants 4 points for every correct answer and takes off 1 point for every incorrect answer. If a score of 110 is received on a 40-question test where every question is answered, how many more questions were answered correctly than incorrectly?

Explanations

Linear Algebra Exam: Part 1

1. C

Recall that to answer linear equations in context correctly, three parts must be identified: Value, Units, and Key-terms. Using slope-intercept form, the value of 1.27 or 1.27/1 represent the slope. The units of slope follow rise over run, where the key-terms each, every, and per take the place of the fraction bar.

$$\text{Units: } \frac{\text{units of "}y\text{" variable}}{\text{units of "}x\text{" variable}} \text{ or } \frac{\text{1.27g ferric bromide}}{\text{1g benzene}}$$

Therefore, an additional 1.27 grams of ferric bromide are required for every 1-gram increase of benzene.

2. D

Since, a and bt are being added together to find velocity, each must have the same units as velocity. Therefore, bt must have dimensions in meters per second. Plug in the units for time to solve.

$$bt = \frac{m}{s}$$

$$b * s = \frac{m}{s}$$

$$b = \frac{m}{s^2}$$

3. B

Plug in 5 for a and then manipulate the equation to solve for x.

$$\frac{a(3x + 4)}{2} = 10$$

$$\frac{5(3x + 4)}{2} = 10$$

$$5(3x + 4) = 20$$

$$3x + 4 = 4$$

$$3x = 0$$

$$x = 0$$

4. C

Ensure that the perpendicular lines have slopes that are opposite reciprocals.

$$y = ax + 4$$

Points: $(c, 0)$ and $(0,4)$

$$\text{Slope} = \frac{0 - 4}{c - 0} \text{ or } -\frac{4}{c}$$

$$y = bx - 8$$

Points: $(c, 0)$ and $(0, -8)$

$$\text{Slope} = \frac{0 + 8}{c - 0} \text{ or } \frac{8}{c}$$

The opposite reciprocal of $-4/c$ is $c/4$.

$$\frac{c}{4} = \frac{8}{c}$$

$$c^2 = 32$$

$$c = 4\sqrt{2}$$

5. B

Simplify the left-hand side of the equation and then continue to solve for $5x$.

$$(4x + x + x) - (x - 3x) = 18 - 2x$$

$$6x - (-2x) = 18 - 2x$$

$$8x = 18 - 2x$$

$$10x = 18$$

$$5x = 9$$

6. A

Set up a proportion to relate the pressure in psi to depth in feet.

$$\frac{0.43 \text{ psi}}{1 \text{ foot}} = \frac{a \text{ psi}}{x \text{ feet}}$$

$$\frac{43 \text{ psi}}{100 \text{ feet}} = \frac{a \text{ psi}}{x \text{ feet}}$$

$$43x = 100a$$

$$x = \frac{100}{43}a$$

7. B

Solve the system of equations using elimination.

First Equation:

$$12\left(\frac{3}{4}p + \frac{1}{2}q = \frac{17}{6}\right)$$

$$9p + 6q = 34$$

Second Equation:

$$6\left(3p - q = -\frac{8}{3}\right)$$

$$18p - 6q = -16$$

Elimination:

$$\begin{array}{rr} & 9p + 6q = 34 \\ + & 18p - 6q = -16 \\ \hline & 27p = 18 \end{array}$$

$$p = \frac{18}{27}$$

$$p = \frac{2}{3}$$

8. A

The slope of the line must be more negative than -1. In standard $Ax + By = C$ form, the slope can be found using the relationship $m = -A/B$. Only choice A has slope that is both steep and negative enough to create line depicted.

9. C

The slopes must be the same for both equations. Thus, the ratios of A to B are equal in $Ax + By = C$ form.

$$\frac{2k}{3a} = \frac{6}{a}$$

$$2k = 18$$

$$k = 9$$

10. C

Ensure that the slope between the known points is equal to 1/5 .

$$m = \frac{y_2 - y_1}{x_2 - x_1}$$

$$(x_1, y_1) = (0,0)$$

$$(x_2, y_2) = (2b, b - 3)$$

$$\frac{1}{5} = \frac{b - 3 - 0}{2b - 0}$$

$$\frac{1}{5} = \frac{b - 3}{2b}$$

$$2b = 5(b - 3)$$

$$2b = 5b - 15$$

$$-3b = -15$$

$$b = 5$$

11. 200

Use the force equation provided to plug in the known force F and displacement x.

$$F = kx$$

$$40 = k(0.2)$$

$$k = 200$$

12. 3

Seeing as 6 is halfway between 4 and 8, $k/2$ must also be halfway between -2 and 5. Therefore, $k/2$ must equal 1.5, which means k must equal 3

13. 5

Subtract the equations from each other.

$$\begin{array}{r} 2b + c = 10 \\ -\quad 3b = 5 \\ \hline c - b = 5 \end{array}$$

14. 6

Double the given equation.

$$2(4m - 2 = 3)$$

$$8m - 4 = 6$$

15. 3

The ratio between the distances and rates must be the same in each case. Use a proportion to solve for the unknown rate in elevation.

Distance along trail: 10 meters

Distance in elevation: 1 meter

Rate along trail: 30 meters per minute

Rate in elevation: x meters per minute

$$\frac{\text{Distance along trail}}{\text{Distance in elevation}} = \frac{\text{Rate along trail}}{\text{Rate in elevation}}$$

$$\frac{10}{1} = \frac{30}{x}$$

$$10x = 30$$

$$x = 3$$

Explanations

Linear Algebra Exam: Part 2

1. D

Since the (x, y) solution of $(-1,2)$ is known, it can be used in both equations. Plug in $x = -1$ and $y = 2$.

First Equation:

$$\frac{b}{2}x + 3y = c \quad \text{or} \quad bx + 6y = 2c$$

$$b(-1) + 6(2) = 2c$$

$$-b + 12 = 2c$$

Second Equation:

$$bx - 4y = -2c$$

$$b(-1) - 4(2) = -2c$$

$$-b - 8 = -2c$$

Elimination:

$$\begin{array}{rl} & -b + 12 = 2c \\ - & -b - 8 = -2c \\ \hline & 20 = 4c \\ & 5 = c \end{array}$$

2. B

Ensure that the ratio between the coefficients of x and y is the same in both equations so that the lines are parallel. This means that we can multiply or divide the coefficients of these terms in line a by some number, and the lines will still be parallel. The constant C does not affect the slope, so we can ignore it in this problem. For example, we can take half of both coefficients to find a new version of $Ax + By$ that would still represent a parallel line.

$$6x - 2y = C_1$$

$$3x - y = C_2$$

Only answer choice B has the proper ratio between the coefficients of x and y, so it must be correct. Parallel lines have same slope, whereas perpendicular lines have slopes that are opposite reciprocals.

3. C

Break down each variable and its associated cost to write the system of inequalities.

Gallons of red paint: r

Unit cost of red paint: \$10.48/1 gallon

Total cost of red paint: $10.48r$

Gallons of blue paint: b

Unit cost of blue paint: \$18.36/2 gallons

Total cost of blue paint: $9.18b$

$$10.48r + 9.18b \leq 140$$

$$r + b \geq 14$$

4. **D**

Quickly sketch the inequalities on the xy-plane and shade in the solution set. Since the inequalities are already in slope-intercept form, we can begin at the y-intercept then follow the slope to generate the line, which should be dashed in both cases. Remember to shade above the line when $y > mx + b$ and shade below it when $y < mx + b$. We can see from the sketch that the overlapping shaded area lies to the left of the point of intersection, bounded by the two lines. Because of that fact, we know that the maximum value in the answer choices must be excluded from the solution set, and is therefore the correct answer. As it turns out, if we solve the system algebraically, treating the inequalities as equal signs, we can see that the lines intersect at $x = 2$. However, because the inequalities are NOT "or equal to" inequalities, $x = 2$ is restricted from the solution set.

5. **B**

The constant b represents the y-intercept, which is used for any one-time or initial fees. Since the cost is represented in hundreds, b must equal 5.

6. **A**

Find the slope of each line, then relate them together according to the details of the question.

Slope of line q

Points: $(0,0)$ and $(c, 5)$

$$\text{Slope} = \frac{5-0}{c-0} \text{ or } \frac{5}{c}$$

Slope of line s

Points: $(0,0)$ and $(c, 2)$

$$\text{Slope} = \frac{2-0}{c-0} \text{ or } \frac{2}{c}$$

Relationship between the slopes

$$m_q = 2m_s + 3$$

$$\frac{5}{c} = 2\left(\frac{2}{c}\right) + 3$$

$$\frac{5}{c} = \frac{4}{c} + 3$$

$$5 = 4 + 3c$$

$$c = 1/3$$

7. **D**

Since all answer choices have an equation with the same y-intercept, we need to determine the slope that represents the rate of change in elevation. We can use a proportion to solve fo this unknown.

$$\frac{\text{Distance along Path}}{\text{Distance in elevation}} = \frac{\text{Rate along Path}}{\text{Rate in elevation}}$$

$$\frac{100}{18} = \frac{5}{m}$$

$$m = \frac{9}{10}$$

8. **D**

Create and solve a system of equations.

Variables: s = small tables, l = large tables

Equation 1: Relationship between small and large tables

$$l = s + 6$$

Equation 2: Number of Occupants

$$4s + 10l = 200$$

Substitute the expression for l found in equation 1 into equation 2.

$$4s + 10(s + 6) = 200$$

$$14s + 60 = 200$$

$$14s = 140$$

$$s = 10, l = 16$$

Total number of tables: $s + l = 26$

9. **B**

Use the Pythagorean Theorem to relate the distances to each other. Remember that these distances are relative to the origin (0,0). We can then factor to solve for the desired expression.

$$(\text{Distance})^2 = (\text{Distance})^2$$

$$b^2 + 8^2 = 10^2 + c^2$$

$$b^2 + 64 = 100 + c^2$$

$$b^2 - c^2 = 36$$

$$(b + c)(b - c) = 36$$

$$(b + c) * 4 = 36$$

$$b + c = 9$$

10. **C**

Use the equation and the given (x, y) point to solve for the constant b.

$$y = bx$$

$$m = b * 0.5m$$

$$b = \frac{m}{0.5m}$$

$$b = 2$$

11. **1**

Find the greatest integers of x and y from the given inequalities.

$$x - 4 < 5$$

$$x < 9$$

Greatest integer: 8

$$y + 2 < -4$$

$$y < -6$$

Greatest integer: -7

Sum of greatest integers: 1

12. 1/6

Match equations by multiplying the top equation by two. For these equations to have infinitely many solutions, all corresponding terms must match. Therefore, the coefficients of the x-terms must be equal.

First Equation:

$$2(y = 3x + 5)$$

$$2y = 6x + 10$$

Second Equation:

$$2y = \frac{1}{a}x + 10$$

Matching:

$$1/a = 6$$

$$a = 1/6$$

13. 4/3, 3/2 or 1.5, 2, 3, 4

Since the line crosses the goes through the origin, the slope would be equivalent to b/a. Use the integer values below to ensure the fraction is greater than one.

Possible values of b: $1, 2, 3, 4$

Possible values of a: $1, 2, 3, 4$

14. 16

Solve a system of equations using elimination.

Total Cost: $4.89p + 3.44s = 119.52$

Number of Jars: $p + s = 28$ or $3.44p + 3.44s = 96.32$

Eliminate s.

$$\begin{array}{r} 4.89p + 3.44s = 119.52 \\ - \quad 3.44p + 3.44s = 96.32 \\ \hline 1.45p = 23.20 \end{array}$$

$$p = 16$$

15. 20

Solve a system of equations by elimination based on the given information, then find the desired quantity.

$$x = \text{number of correct answers}$$

$$y = \text{number of incorrect answers}$$

$$x + y = 40$$

$$4x - y = 110$$

Eliminate y, then find x to find the difference between x and y.

$$\begin{array}{r} x + y = 40 \\ + \quad 4x - y = 110 \\ \hline 5x = 150 \end{array}$$

$$x = 30 \text{ and } y = 10$$

$$x - y = 20$$

II

ADVANCED MATHEMATICS

Exponential Properties and Models

- Properties of Exponents
- Comparing Relationships
- Exponential Models

Expressions and Equations

- Polynomial Expressions
- Rational Expressions
- General Algebra
- Factors of Change

Function Properties

- Function Notation
- Quadratic Functions
- Polynomial Functions
- Transformations & Theorems

Advanced Functions and Equations

- Domain and Range
- Equations and Systems

Introduction

The Advanced Mathematics section delves into higher-level math operations and functions, such as exponentials and quadratics. Test-takers are expected to demonstrate their math fluency by employing more involved solving techniques that require a solid grasp of algebraic operations and their properties. Combined, the topics in Advanced Mathematics cover 28% of the math questions on the test. We will introduce all of the content necessary to answer questions in this section as quickly as possible by focusing on methods that require the least number of steps. The less steps taken, the less room there is for error.

2.1 EXPONENTIAL PROPERTIES AND MODELS

a. Properties of Exponents

Recall that exponents indicate the number of times a number, the base, is to be multiplied to itself. For example, the expression a^4 equals $a \cdot a \cdot a \cdot a$. We can see in this example that the base a is multiplied to itself four times because its exponent is four. While this is a simple idea in and of itself, it leads to very distinct properties. Among the properties that students must master to score high on the test involve multiplying and dividing by a common base, relating rational exponents to radicals, and understanding the implication of negative exponents. Table 2.11 below summarizes the most important properties of exponents.

Table 2.11 | Properties of Exponent Summary

$x^0 = 1$	$(xy)^a = x^a y^a$
$x^1 = x$	$\left(\frac{x}{y}\right)^a = \frac{x^a}{y^a}$
$x^a = \frac{1}{x^{-a}}$	$(x^a)^b = x^{ab}$
$x^a x^b = x^{a+b}$	$\sqrt[b]{x^a} = \sqrt[b]{x}^a = x^{\frac{a}{b}}$
$\frac{x^a}{x^b} = x^{a-b} = \frac{1}{x^{b-a}}$	$\sqrt[b]{\frac{x}{y}} = \frac{\sqrt[b]{x}}{\sqrt[b]{y}}$

Note: $(x+y)^a$ does NOT equal $x^a + y^a$

Q+A

Properties of Exponents **Diagnostic Exam | Calculator Q28**

$$2^{3x-1} = 5w$$

What is the value of 8^x in terms of the positive constant w using the equation above?

A. $4w$ B. $6w$ C. $8w$ D. $10w$

Workspace

Solution Outline

Step 1: Rewrite the left-hand side of the equation by following the subtraction rule for exponents. That is, we subtract exponents when dividing by a common base.

$$\frac{2^{3x}}{2^1} = 5w$$

Step 2: Multiply both sides by two and continue using properties of exponents to reach the desired expression: 8^x.

$$2^{3x} = 10w$$

$$(2^3)^x = 10w$$

$$8^x = 10w$$

Answer: D

Notes

Simplifying expression and solving equations that involve exponents often require us to rewrite current bases using common ones. For example, the numbers 8 and 16 can be written as 2^3 and 2^4. The following question depicts this idea and expands on additional properties of exponents. We will see that many questions on the test often involve expressions. After creating common bases to apply our rules for multiplying or dividing by a common base, we might generate an expression as the new exponent. In many cases, the problem will supply information that can be used in tandem with this new exponent.

Q+A

Properties of Exponents **Diagnostic Exam | Calculator Q8**

$$\frac{27^{2x}}{9^y}$$

If $3x - y = 2$, what is the value of the expression above?

A. 9 B. 27 C. 81 D. 243

Workspace

Solution Outline

Step 1: Rewrite the expression so that the numerator and denominator share the same base.

$$\frac{27^{2x}}{9^y} = \frac{(3^3)^{2x}}{(3^2)^y} = \frac{3^{6x}}{3^{2y}} = 3^{6x-2y} = 3^{2(3x-y)}$$

Step 2: Substitute the known value of $(3x - y)$ into the expression.

$$3^{2(2)} = 3^4 = 81$$

Answer: C

Notes

When solving for a variable in an equation, understand that variable has an exponent of one. For example, when we isolate x to one side of the equal sign, we have actually isolated x^1, or x to the first power. This is important to consider because when the variable has an irregular exponent, we must quickly bring it to one. One way we can do this is by multiplying that exponent by its reciprocal. What this mean in terms of solving an equation is that we can to raise both sides by the reciprocal of the current exponent to the variable we are asked to find, which amounts to multiplying its exponent by that amount, thereby bringing it to one. The following question summarizes this idea and expands on how we can quickly evaluate expressions with tricky, fractional exponents.

Q+A

Properties of Exponents **No Calculator**

$$x^{-\frac{2}{5}} = 4$$

What is the value of x in the equation above?

A. 1/32 B. 1/16 C. 16 D. 32

Workspace

Solution Outline

Step 1: Raise both sides of the equation by $-5/2$. The goal here is to find x, or x^1. So raising both sides of the equation by the reciprocal of the current exponent isolates x immediately. In addition, it is sometimes helpful to see other exponents as well. For example, $4 = 2^2$. Remember: Raising exponents to other exponents requires multiplication.

$$\left(x^{-\frac{2}{5}}\right)^{-\frac{5}{2}} = (2^2)^{-\frac{5}{2}}$$

$$x = 2^{-5}$$

Step 2: Use property of negative exponents to find the exact value of x.

$$2^{-5} = \frac{1}{2^5} = \frac{1}{32}$$

Answer: A

Notes

In addition to exponents, radicals have also earned a prominent spot on the exam. Note that rational exponents (exponent that are fractions) can be rewritten in radical form, and vice versa. Recall that the denominator of the exponent determines the index of the radical and the numerator stays, but may be written inside or outside the radical. In other words, the denominator determines the type of root. Table 2.12 summarizes general properties of radicals.

Exponents and Radicals	$x^{\frac{a}{b}} = \sqrt[b]{x^a}$ or $\sqrt[b]{x}^a$

Table 2.11 | General Properties of Radicals

	Example	Notes
Add and Subtract	$2\sqrt{3} + 5\sqrt{3} = 7\sqrt{3}$	Terms with the same radical are like-terms. Think of how $2x + 5x = 7x$
Multiply and Divide	$\frac{10\sqrt[3]{15}}{2\sqrt[3]{5}} = 5\sqrt[3]{3}$	Multiply or divide the coefficients to each radical, then multiply or divide the radicands. Think of "outside" and "inside".
Simplify	$\sqrt[3]{81x^8} = \sqrt[3]{3^4x^8}$ $\sqrt[3]{3^3 * 3^1 * x^6 * x^2}$ $3x^2\sqrt[3]{3x^2}$	Take out factors that occur in groups the size of the index. A quick way to do this is to find exponents of each factor that are divisible by the index. In this example, we should look for exponents that are divisible by the index of 3.
Rationalize	$\frac{1}{\sqrt{3x}} * \frac{\sqrt{3x}}{\sqrt{3x}} = \frac{\sqrt{3x}}{3x}$ $\frac{1}{1+\sqrt{x}} * \frac{1-\sqrt{x}}{1-\sqrt{x}} = \frac{1-\sqrt{x}}{1-x}$	The aim is to create an equivalent fraction that does not have a radical in the denominator. This technique is often used with expressions involving complex numbers. Visit 4.1 Complex Numbers to learn more.

b. Comparing Relationships

Comparing exponential and linear relationships is another new testing point. Understand that exponential relationships occur through the *multiplication* of a constant where linear relationships occur through the *addition* of a constant. Follow Table 2.13 below to see these relationships in practice.

Table 2.13 | Exponential and Linear Relationships

Exponential Relationship		Linear Relationship	
Form	$y = ab^x$	Form	$y = mx + b$
Example	$y = 2(3)^x$	Example	$y = 2x + 1$

x	0	1	2	3
y	2	6	18	54

x	0	1	2	3
y	1	3	5	7

Q+A

Exponential Relationship **Diagnostic Exam | No Calculator Q10**

The function f is graphed on the xy-plane below. If the function g is the reflection of f across the y-axis, which of the following can represent the equation of $g(x)$?

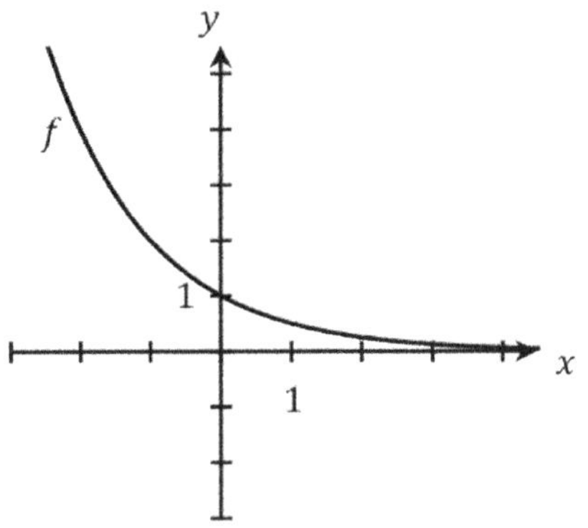

A. $g(x) = 2x$

B. $g(x) = 2x + 1$

C. $g(x) = 2^x$

D. $g(x) = 2^x + 1$

Workspace

Solution Outline

Step 1: Since the graph is curved, it cannot represent the linear equations described in A and B. The reflection across the y-axis will not change the y-intercept where $x = 0$. Therefore, C must be the correct answer since it holds that $f(0) = 1$, whereas D does not.

Answer: C

Notes

Since exponentials form curved functions when graphed on the xy-plane, they can intersect lines at up to two distinct points. If the graphed line and exponential do in fact cross at two points, then one function will be above and the other will be below at all other points. The test-makers like to challenge students on this idea by describing two points of intersection to then ask for the interval in which one function is greater than the other. The following example demonstrates this type of question students could expect on the test.

Q+A

Exponential Relationship

Calculator

The linear function f and exponential function g are graphed on the xy-plane over the closed interval $[0, 8]$. The two functions intersect at the points $(m, 1)$ and $(n, 5)$, where m and n are positive constants between 0 and 8 and $m < n$. Which interval ensures that $f(x) > g(x)$ given that $g(x) > 0$ for all values of x?

A. $(0, m)$

B. $(0, n)$

C. (m, n)

D. $(n, 8)$

Workspace

Solution Outline

Step 1: Make a quick sketch of the two functions on the xy-plane. Start by placing the two points $(m, 1)$ and $(n, 5)$ relative to each other in the interval $[0, 8]$. Then draw a line through the points for the linear function f and a curved exponential function g that will always remain above the x-axis. We will see that the line f is above the exponential g in the interval (m, n).

Answer: C

Notes

c. Exponential Models

Exponential growth and decay functions are formed when the output is repeatedly multiplied by a constant positive factor other than one. This multiplication creates functions that are increasing at an increasing rate (growth) or decreasing at a decreasing rate (decay). Table 2.14 below depicts exponential growth and decay functions on the xy-plane for the general function f.

Table 2.14 | Exponential Growth and Decay Graphs

Exponential Growth		Exponential Decay	
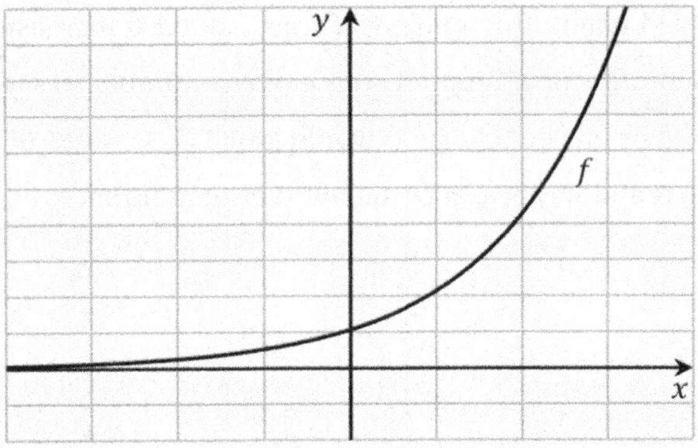		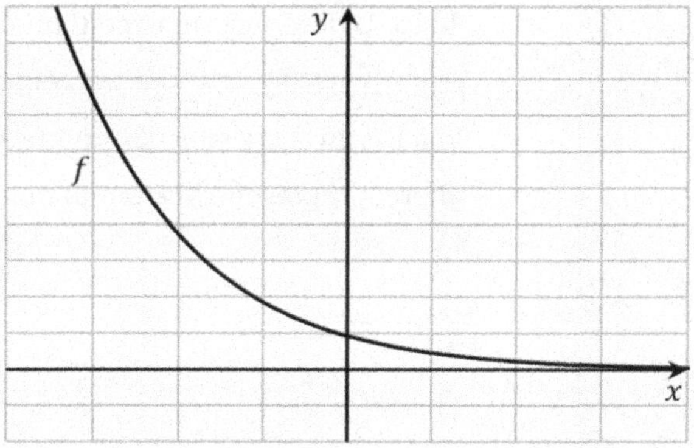	
Form	Condition	Form	Condition
$f(x) = b^x$	$b > 1$	$f(x) = b^x$	$0 < b < 1$

An important application to this is idea of growth and decay is that any trend that involves a percent increase or decrease for every additional unit of x forms an exponential relationship. The base of the model will be $(1 \pm r)$ where r is the percent of increase or decrease written in decimal form. In the long run, exponential growth outpaces linear.

Q+A

Exponential vs Linear Growth **Diagnostic Exam | No Calculator Q11**

An environmental committee wants to restore the greatest number of trees possible in a given acreage of forest over the long run. Which plan fits the needs of this committee?

A. Every month, 20 trees are added
B. Every month, 10% of the original number of trees is added
C. Every month, 15% of the original number of trees is added
D. Every month, 5% of the current number of trees is added

Workspace

Solution Outline

Step 1: Find the scenario where the number of tress added to the forest is increasing since this indicates exponential growth, which outpaces linear growth over time.

Discussion: Choice A describes a linear growth plan since the number of trees being added every month is exactly the same: only 20 trees are added each month. Choices B and C also describe plans where the number of trees added does not change. The reason for this is the percent of trees added is based off the original number of trees present. If there are 150 trees initially in the specified acreage, then 10% is 15 trees. This means that 15 trees will be added every month. Only choice D describes an exponential growth plan where the number of trees added is increasing. This occurs because the percentage is based off a new amount: the *current* number of trees, which is always rising. The model represented by answer choice D would look like $y = \mathrm{A}(1.05)^n$, where A represents the initial number of trees and n represents the number of months.

Answer: D

Notes

We may also need to find a relationship between percent change and a defined time interval based off a description or equation. In these cases, we will have to take into account both the base and its exponent. How the base deviates from one and what it takes for the exponent to reach consecutive integers will both go into determining this relationship. Recall that the percent change can be determined be finding how much the base differs from one, then multiplying by one-hundred to create a percent. Additionally, the interval required to ensure the exponent reaches the next integer will of often describe the time it takes for the aforementioned percent change to take place. The following example details exactly how to find these two characteristics of exponential models so that we can analyze these situations with efficiency and precision.

Q+A

Interpreting Percent Change

Calculator

Using satellite-based data cataloguing the canopy biomass of a kelp forest off the coast of southern California, a group of researchers discovered that the adult kelp population had changed at an exponential rate over the course of a ten-year period. The researchers found that the adult kelp population $a(t)$ could be modeled by the equation $a(t) = a(0.98)^{4t}$, where t is the number of years since the start of the study. Which of the following accurately describes the change in the adult kelp population that the researchers observed during their study?

A. The adult kelp population declined 2% every year

B. The adult kelp population declined 2% every three months

C. The adult kelp population declined 98% every year

D. The adult kelp population declined 98% every three months

Workspace

Solution Outline

Step 1: Determine how the base of the exponential model differs by one. Since 0.98 is 0.02 lower than one, the model indicates a 2% decline for a certain time period.

Step 2: Determine what values of t are required so that the exponent reaches integer values (e.g. 1, 2, and 3), This is the time interval for which the 2% decrease takes place. It takes multiples of one-fourth to reach integer values, and since t is in years, one-fourth of a year is three months. Therefore, there is a 2% decrease every three months.

Answer: B

Notes

Problem Set: Exponentials

12 Questions

1

$$\frac{\sqrt{300} - \sqrt{48}}{\sqrt{27}}$$

Which is equivalent to the expression above?

A. 1/2

B. $\sqrt{3}/2$

C. $\sqrt{3}$

D. 2

2

$$b^{\frac{5}{3}}$$

Which is equivalent to the expression above, for all values of b?

A. $\sqrt[5]{b^3}$

B. $\sqrt[3]{b^5}$

C. $\sqrt{b^5}$

D. $\sqrt[5]{b^{\frac{1}{3}}}$

3

A researcher is studying the growth of a certain strain of bacteria on an agar plate. She determines that the number of bacteria, B, on the agar plate after t minutes can be modeled by the function $B = 125(1.03)^t$. What is the percent increase in the number of bacteria every minute as demonstrated by this equation?

A. 0.03

B. 1.03

C. 3

D. 103

4 Calculator Allowed

x	0	1	2	3
y	128	160	200	250

From the data above, which equation best relates x and y?

A. $y = 1.25x + 128$

B. $y = 128x + 1.25$

C. $y = 1.25(128)^x$

D. $y = 128(1.25)^x$

5

$$7^x = \frac{25}{49}$$

Using the equation above, what is the value of 7^{x+2}?

A. 25/7

B. 25

C. 175

D. 1225

6

$$a^{2x} = \frac{1}{9}$$

Using the equation above, what is the value of a^x?

A. 1/18

B. 1/3

C. 3

D. 18

7

A ball is dropped from a height of 128 feet above the ground. If the ball only rises to a third of its previous maximum height after each bounce off the ground, what is the maximum height (in feet) the ball can reach after its fourth bounce?

A. 128/81

B. 128/27

C. 128/9

D. 128/3

8 Calculator Allowed

Manny determines that the value of his car is modeled by the equation below, where A is the value of his car in dollars, m is a constant, and t is the number of years after 2010.

$$A = 25{,}000(m)^t$$

If the value of Manny's car is $10,000 in 2015, what is the value of m rounded to the nearest hundredth?

A. 0.74

B. 0.83

C. 0.96

D. 1.07

9

The half-life of an isotope is the amount of time required for its mass to reach half of its original value. A scientist studying half-life determined that the mass of a certain isotope fell to an eighth of its original value at the end of one trial period of 17 days.

If the decrease in mass can be solely attributed to the effects of the isotope's half-life, what is the length in days of one half-life for this particular isotope?

A. 17/8

B. 17/3

C. 51

D. 136

10

$$C = 200(1.05)^{\frac{t}{4}}$$

The average cost in thousands of dollars of a house in a certain suburb can be modeled using the function above, where t represents years after 2008. In what year will the cost have increased by 5%?

A. 2010

B. 2011

C. 2012

D. 2013

11

In the equation $y = b^{kx}$, b and k are constants and $0 < b < 1$. Which of the following must be true for this equation to represent exponential growth?

I. $|k| > 1$

II. $k < 0$

III. $k > 0$

A. I only

B. II only

C. III only

D. I and II

12

$$x^a = 4$$

$$x^b = 3$$

Given the values of x^a and x^b above, which of the following is equivalent to x^{2a-b} ?

A. $\frac{3}{16}$

B. $\frac{4}{9}$

C. $\frac{9}{4}$

D. $\frac{16}{3}$

Explanations

Problem Set: Exponentials

1. D

Simplify each radical by taking out perfect squares.

$$\sqrt{300} = \sqrt{3 * 100} = 10\sqrt{3}$$

$$\sqrt{48} = \sqrt{3 * 16} = 4\sqrt{3}$$

$$\sqrt{27} = \sqrt{3 * 9} = 3\sqrt{3}$$

Simplify:

$$\frac{10\sqrt{3} - 4\sqrt{3}}{3\sqrt{3}} = \frac{6\cancel{\sqrt{3}}}{3\cancel{\sqrt{3}}} = 2$$

2. B

Rational exponents can be written into radical form by keeping the numerator as the exponent to the base and swinging out the denominator as the index.

$$b^{\frac{5}{3}} = \sqrt[3]{b^5}$$

3. C

Find how much the base of the exponent deviates from 1, then convert to a percent. The base is 1.03, which is 0.03 above 1. The decimal 0.03 translates to 3%

4. D

Plug in x-values into the answer choices to see which equation matches for every x and y pair. Note the gap between each consecutive y-value is increasing, so it cannot be a linear relationship.

5. B

Rewrite the expression, keeping in mind exponent rules, like adding exponents when multiplying by common bases.

$$7^x = \frac{25}{49}$$

$$7^x * 49 = 25$$

$$7^x * 7^2 = 25$$

$$7^{x+2} = 25$$

6. B

Remember: A square root can be written as to the ½ power. This means dividing the exponent by 2. To square root fractions, take the square root of the numerator and denominator independently.

$$a^{2x} = \frac{1}{9}$$

$$\sqrt{a^{2x}} = \sqrt{\frac{1}{9}}$$

$$a^x = \frac{1}{3}$$

7. A

Multiply by 1/3 after every bounce. Since there are four bounces, multiply 128 by 1/3 four times.

$$128\left(\frac{1}{3}\right)^4 = \frac{128}{81}$$

8. B

After 5 years, the value of the car is $10,000. Plug in $t = 5$ and $A = 10{,}000$ to solve for m.

$$10{,}000 = 25{,}000m^5$$

$$m^5 = \frac{10{,}000}{25{,}000}$$

$$m^5 = 0.4$$

$$(m^5)^{\frac{1}{5}} = (0.4)^{\frac{1}{5}}$$

$$m = 0.83$$

9. B

One half-life: 1/2 of original value

Two half-lives: 1/4 of original value

Three half-lives: 1/8 of original value

Three half-lives occurred in a 17-day period. Divide 17 by 3 to get the amount of time of just one half-life.

10. C

The t value must be 4 to make the exponent go to one. Four years after 2008 is 2012.

11. B

If b is between 0 and 1, then k must be negative to create an exponential growth model.

12. D

Use properties of exponents to breakdown the desired expression and relate it back to the given values of x^a and x^b.

$$x^{2a-b} = \frac{x^{2a}}{x^b}$$

$$\frac{(x^a)^2}{x^b} = \frac{(4)^2}{3}.$$

$$\frac{16}{3}$$

2.2 EXPRESSIONS AND EQUATIONS

a. Polynomial Expressions

To score well on the test, it is crucial for us to master the algebra of simplifying, expanding, and factoring expressions in the context of larger algebraic problems. This begins with adding and subtracting algebraic expressions, like polynomials. To do this, we need to find like-terms, then add or subtract their coefficients. Recall that like-terms are terms that have the same variable component. Realize that the exponents to variables of polynomials are whole numbers; they are never fractions or negatives, nor are the variables contained in operations like trig functions or roots.

Like-Terms	Unlike-Terms
$4x$ and $9x$	$4x$ and $9y$
$2xy^2$ and $-3xy^2$	$2x^2y$ and $-3xy^2$
$3\sqrt{x}$ and $22\sqrt{x}$	$3\sqrt{x}$ and $22\sqrt[3]{x}$

Q+A

Operations of Polynomials **Diagnostic Exam | No Calculator Q7**

Which of the following is equivalent to the difference between the expressions $b^2 + 2b$ and $3b + 2$?

A. $b^2 - b + 2$ B. $b^2 - b - 2$ C. $b^2 + 5b + 2$ D. $b^2 + b - 2$

Workspace

Solution Outline

Step 1: Follow order of operations and combine like-terms.

$(b^2 + 2b) - (3b + 2)$

$b^2 + 2b - 3b - 2$

$b^2 - b - 2$

Answer: B

Notes

Multiplying and factoring polynomials are also important skills needed for the test. In particular, perfect square trinomials and difference of squares have unique factoring patterns that are worth being familiar with. It is not necessary to memorize these patterns since they are subject to normal rules of factoring and multiplying polynomials; however, when these expressions show up on the test, they will often take on more complex structures that will benefit students who are familiar with them. Table 2.21 summarizes the factoring patterns of perfect square trinomials and difference of squares with an example for each case. Following the discussion on these special cases, there will be a further analysis on factoring trinomials that will help us answer questions on polynomial expressions and functions, namely quadratics.

Table 2.21 | Special Factoring Patterns

	Rule	Example
Perfect Square Trinomial (+)	$a^2 + 2ab + b^2 = (a + b)^2$	$9y^2 + 12xy + 4x^2 = (3y + 2x)^2$
Perfect Square Trinomial (−)	$a^2 - 2ab + b^2 = (a - b)^2$	$25x^4 - 40x^2 + 16 = (5x^2 - 4)^2$
Difference of Squares	$a^2 - b^2 = (a + b)(a - b)$	$4x^6 - 25 = (2x^3 + 5)(2x^3 - 5)$

Q+A

Factoring Quadratics **Diagnostic Exam | No Calculator Q3**

If the expression $5x^2 - 10$ is rewritten in the form $5(x - a)(x + a)$, what is the value of the positive constant a?

A. $\frac{\sqrt{2}}{2}$ B. $\sqrt{2}$ C. 2 D. 5

Workspace

Solution Outline

Step 1: Recognize the factoring format in the rewritten form. The expression $5x^2 - 10$ is being factored as a difference of squares after taking out the greatest common factor, 5.

$$5x^2 - 10 \quad \text{or} \quad 5(x^2 - 2) \quad \text{or} \quad 5(x - \sqrt{2})(x + \sqrt{2})$$

Answer: B

Notes

Factoring Methods

There are several factoring techniques, but generally the first step is to take out the greatest common factor. Then from there, use the ac method and factor by grouping. The discussion that follows will primarily center around quadratics, or a polynomial that can be written in the form: $ax^2 + bx + c$. The ac method gets its name from this form. The idea here is to take the product of a and c, then add the factors of this product to reach b.

Q+A

Factoring Quadratics **Diagnostic Exam | Calculator Q3**

Which of the following is equivalent to $(5x^2 + 1) - (2x^2 - 8x - 3)$?

A. $(3x + 2)(x + 2)$ B. $(3x + 2)(x - 2)$ C. $(3x - 2)(x + 2)$ D. $(3x - 2)(x - 2)$

Workspace

Solution Outline

Step 1: Expand and simplify the given expression.

$5x^2 + 1 - 2x^2 + 8x + 3$

$3x^2 + 8x + 4$

Step 2: Factor the trinomial using the ac method and factor by grouping. Since $a = 3$ and $c = 4$, $ac = 12$. The The factors of 12 that add to 8 are 2 and 6. Use these to expand the middle term.

Expand: $3x^2 + 2x + 6x + 4$

Group: $(3x^2 + 2x) + (6x + 4)$

Factor each group: $x(3x + 2) + 2(3x + 2)$

Factor again: $(3x + 2)(x + 2)$

Answer: A

Notes

b. Rational Expressions

As much as working with factors to condense or expand polynomial expressions is an important tool, using them to your advantage when solving problems involving *rational* expressions is just as crucial. The main point to remember when working with rational expressions is to identify factors, especially those of each denominator. Table 2.22 below summarizes the properties of rational expressions as they relate to the most common operations.

Table 2.22 | Properties of Rational Expressions

Addition and Subtraction	Multiplication
Denominators must be identical before you can add or subtract the numerators. Create equivalent fractions to reach common denominators.	Factor, then multiply. Remember to cross simplify and multiply straight across when multiplying two or more fractions together
Given: $\frac{2}{x}+\frac{1}{x^2}$	Given: $\frac{2x}{x+2}*\frac{x^2+3x+2}{x-1}$
Step 1: $\frac{2}{x}*\frac{x}{x}+\frac{1}{x^2}$	Step 1: $\frac{2x}{(x+2)}*\frac{(x+2)(x+1)}{x-1}$
Step 2: $\frac{2x}{x^2}+\frac{1}{x^2}$	Step 2: $\frac{2x}{1}*\frac{x+1}{x-1}$
Final: $\frac{2x+1}{x^2}$	Final: $\frac{2x(x+1)}{x-1}$
Division	**Exponents**
To divide by a fraction, simply multiply by the reciprocal of that fraction and use the aforementioned properties of multiplication.	To raise a fraction to a power, apply that power to both the numerator and denominator. Always simplify where possible.
Given: $\frac{x+1}{x+2}\div\frac{x^2-1}{x^2-4}$	Given: $\left(\frac{x+1}{x+4}\right)^2$
Step 1: $\frac{x+1}{x+2}*\frac{x^2-4}{x^2-1}$	Step 1: $\frac{(x+1)^2}{(x+4)^2}$
Step 2: $\frac{(x+1)}{(x+2)}*\frac{(x+2)(x-2)}{(x+1)(x-1)}$	Step 2: $\frac{(x+1)(x+1)}{(x+4)(x+4)}$
Final: $\frac{x-2}{x-1}$	Final: $\frac{x^2+2x+1}{x^2+8x+16}$

Advanced Rational Expressions

The test also contains more complex rational expressions. These often involve several different fractions and operations. Generally, there a couple strategies that can help you answer these questions quickly and accurately. Often, the goal is to simplify the rational expression as much as possible, ultimately getting to a simplified fraction with a polynomial expression in the numerator and another in the denominator. To see this in practice, attempt the example that follows, then confirm your answer and strategy to the one provided. There are many methods for simplifying these complex rational expressions, so committing to one is just a matter of preference.

Q+A

Rational Expressions **Diagnostic Exam | Calculator Q4**

$$\frac{2m}{\frac{2}{m-2}+\frac{1}{m+1}}$$

Which of the following is equivalent to the expression above for all values of m, given $m > 2$?

A. $\frac{m^3-m^2-2m}{3}$ B. $\frac{3}{2m^2-2m-4}$ C. $\frac{m^2-m-2}{2m}$ D. $\frac{2m^2-2m-4}{3}$

Workspace

Solution Outline

Strategy 1

Step 1: Add the fractions in the denominator by creating common denominators.

$$\frac{2m}{\frac{2}{m-2}*\frac{m+1}{m+1}+\frac{1}{m+1}*\frac{m-2}{m-2}}$$

$$\frac{2m}{\left[\frac{3m}{(m-2)(m+1)}\right]}$$

Step 2: Multiply by reciprocal and simplify.

$$\frac{2m}{1}*\frac{(m-2)(m-1)}{3m}$$

$$\frac{2m^2-2m-4}{3}$$

Answer: D

Strategy 2

Step 1: Create an equivalent fraction by using factors that cancel with each denominator. Notice that as we distribute the factors into the denominator, one factor cancels out in each term.

$$\frac{2m}{\frac{2}{m-2}+\frac{1}{m+1}}*\frac{(m-2)(m+1)}{(m-2)(m+1)}$$

$$\frac{2m(m-2)(m+1)}{2(m+1)+1(m-2)}$$

$$\frac{2m(m-2)(m+1)}{3m}=\frac{2m^2-2m-4}{3}$$

Notes

c. General Algebra

Now that there has been a discussion on the most essential algebra skills – from solving linear equations to working with more advanced expressions – it will be easier to master another topic that will undoubtedly find its way on the test: manipulating an equation to solve for unknowns in terms of other variables and constants. Though this is a specific question type, the process of solving for a variable or expression in algebraic equations spills over into almost every aspect of math. Often, these questions will give an equation, usually made up of many variables, and then ask to solve for one of those variables. The following example will give us a good idea of what this looks like on the test.

Q+A

Algebraic Manipulation **No Calculator**

$$E = \frac{1}{2}kx^2$$

Springs store potential energy when they are stretched or compressed as a result of an applied force. The equation above shows how to calculate this stored energy, where E is the elastic potential energy of the spring, k is the spring constant, and x is the spring's displacement from its equilibrium position. Which of the following accurately calculates the spring constant?

A. $k = \frac{1}{2}Ex^2$ B. $k = \frac{\sqrt{E}}{2x}$ C. $k = \frac{2E}{x^2}$ D. $k = 2Ex^2$

Workspace

Solution Outline

Step 1: Isolate k to one side by working with the given equation.

$$E = \frac{1}{2}kx^2 \quad \text{or} \quad 2E = kx^2 \quad \text{or} \quad \frac{2E}{x^2} = k$$

Answer: C

Notes

Some questions that involve algebraic manipulation require students solve for a variable that appears in the equation in more than one place, as we will see in the following example. The strategy for this type of question is to expand, group, factor, and solve.

Q+A

Algebraic Manipulation **No Calculator**

$$M = \frac{x}{1 - x}$$

Economists use the tax multiplier to determine how tax cuts and increases can affect Gross Domestic Product. The equation above shows how the magnitude of the tax multiplier M is calculated based on an individual's marginal propensity to consume, x. Which correctly solves for the marginal propensity to consume in terms of the magnitude of the tax multiplier?

A. $x = \frac{M}{1 + M}$

B. $x = \frac{M}{1 - M}$

C. $x = \frac{1 + M}{M}$

D. $x = \frac{1 - M}{M}$

Workspace

Solution Outline

Step 1: Expand the equation to group all x terms on one side of the equation.

$$M(1 - x) = x$$

$$M - Mx = x$$

$$M = x + Mx$$

Step 2: Factor and solve for x.

$$M = x(1 + M)$$

$$x = \frac{M}{1 + M}$$

Answer: A

Notes

d. Factors of Change

Another new testing mainstay includes questions that ask us to observe the relationship among variables in an equation and determine a factor of change. A factor of change is the multiplicative conversion between one set of conditions and another. In this section, we will rely on semi-quantitative reasoning methods to calculate the factor of change produced when one set of variables changes in a given relationship. Essentially, these questions will ask students to rationalize the effect that changing one or more variables by a certain factor has on another variable. Students can employ several strategies here, but the two most helpful ones will be either to take a ratio between the "new" and "given" equations or to reduce the given equation with the appropriate substitutions to "see" the factor by which another variable changes.

Q+A

Quantitative Reasoning **Diagnostic Exam | No Calculator Q6**

$$T = 2\pi\sqrt{\frac{L}{g}}$$

The period of a pendulum can be measured using the equation above, where T is the period in seconds, L is the length of the pendulum in meters, and g is the acceleration due to gravity in meters per second squared. If the length of the pendulum increases by a factor of 4, by what factor does the period increase?

A. 2

B. 4

C. 8

D. 16

Workspace

Solution Outline

Strategy 1

Find the ratio between "new" and given equations. The new equations should include the variables and their factor of change.

$$T_{new} = 2\pi\sqrt{\frac{4L}{g}} \quad \text{or} \quad 4\pi\sqrt{\frac{L}{g}}$$

$$\frac{T_{new}}{T_{old}} = \frac{4\cancel{\pi}\sqrt{\frac{\cancel{L}}{\cancel{g}}}}{2\cancel{\pi}\sqrt{\frac{\cancel{L}}{\cancel{g}}}} \quad \text{or} \quad 2$$

Answer: A

Strategy 2

Reduce, substitute, and simplify to observe the factor of change. Reduce: Ignore all variables and constants that do not change, but maintain the structure of the equation, by including placeholders of 1 if necessary.

$$T \propto \sqrt{L}$$

Substitute: Apply the change in the variable that the question specifies. The factor seen out front, represents the factor of change

$$T_{new} \propto \sqrt{4L} \quad \text{or} \quad 2\sqrt{L}$$

Notes

Problem Set: Expressions and Equations

18 Questions

1

$$(x-1)(x^5+x^4+x^3+x^2)$$

Which of the following is equivalent to the expression above?

A. $x^2(x-1)(x+1)(x^2+1)$

B. $x^2(x^2+1)^2$

C. $(x^5-1)(x+1)$

D. $x(x^4+1)(x-1)$

2

$$\frac{2p^3}{\frac{1}{p-3}+\frac{1}{p+3}}$$

Which of the following is equivalent to the expression above for all values $p > 5$?

A. $\frac{2p^3}{p^2-9}$

B. $2p^2(p+3)^2$

C. $p^2(p^2-9)$

D. $\frac{p^2-9}{2p^3}$

3

$$(2g-3s)\left(\frac{s}{6g^2}\right)$$

Which of the following is equivalent to the expression above for all nonzero values of the variables s and g?

A. $12g^3-3s^2$

B. $\frac{s}{3g}-\frac{s^2}{2g^2}$

C. $\frac{gs}{3}-\frac{3s^2}{g^2}$

D. $\frac{12g^3}{s}-18g^2$

4

The area of a square is represented by the expression $4(x^2y^2+6xy+9)$. Which of the following expressions represent the side length of this square?

A. $xy+3$

B. $2xy+3$

C. $2xy+6$

D. $4xy+12$

5

$$y = 4b - 49x^2$$

In the equation above, b is a positive constant and the graph of the equation in the xy-plane is a parabola. Which of the following is an equivalent form of the equation?

A. $y = (2\sqrt{b} + 7x)(2\sqrt{b} - 7x)$

B. $y = (2b + 7x)(2b - 7x)$

C. $y = (\sqrt{b} + 7x)(4\sqrt{b} - 7x)$

D. $y = (2b - 7x)^2$

6

$$z = 36x^2 + 25y^2$$

$$c = xy$$

$$q = z - 60c$$

Given the system of equations above where $c, q,$ and z are constants, which expression is equivalent to q?

A. $(6x - 5y)(6x + 5y)$

B. $(6x + 5y)^2$

C. $(6x - 5y)^2$

D. $-60xy(36x^2 + 25y^2)$

7

$$\frac{\frac{w}{w-6} + \frac{2w}{w+1}}{\frac{w^2}{w+1}}$$

Which of the following is equivalent to the expression above for all values $w > 7$?

A. $\dfrac{3w - 11}{w^2 - 6w}$

B. $\dfrac{w + 2w}{w^2 - 6w}$

C. $\dfrac{w^2 + 6w}{w + 1}$

D. $\dfrac{w^2}{w^2 - 5w - 6}$

8

$$\frac{5^{-3} + 5^{-5}}{5^{-4}}$$

Which is equivalent to the expression above?

A. $\sqrt{5}$

B. 1/5

C. 26/5

D. 1/25

9

$$a^2 - b^2 = 64$$

The equation above defines a relationship between constants a and b. If the value of $a - b$ is 32, what is the value of $a^2 + 2ab + b^2$?

A. 2

B. 4

C. 6

D. 8

10

$$(u^2 - 2qu^2 - w^3) - (-2u^2 - 2qu^2 + w^2)$$

Which of the following is equivalent to the expression above?

A. $3u^2 - w^2 - w^3$

B. $3u^2 - 4qu^2 - w^5$

C. $2u^4 - 4q^2u^4 + w^5$

D. $-u^2 - 4qu^2 - w^3 + w^2$

11

$$V = \frac{V_{max}[s]}{k_m + [s]}$$

The Michaelis-Menten equation above is used to determine the reaction velocity V established by an enzyme given the substrate concentration $[s]$, maximum reaction velocity V_{max}, and Michaelis constant k_m. What is the substrate concentration equal to in terms of V, V_{max}, and k_m?

A. $[s] = \dfrac{Vk_m}{V_{max} - V}$

B. $[s] = \dfrac{V - V_{max}}{Vk_m}$

C. $[s] = \dfrac{Vk_m}{V_{max}} - 1$

D. $[s] = \dfrac{V_{max}}{Vk_m} - 1$

12

$$V_f^2 = V_i^2 + 2ad$$

In the kinematic equation above, what is the value of d in terms of V_i, V_f, and a?

A. $d = \dfrac{V_f^2}{2a} - V_i^2$

B. $d = \dfrac{\sqrt{V_f^2 - V_i^2}}{2a}$

C. $d = \dfrac{V_f^2 + V_i^2}{2a}$

D. $d = \dfrac{V_f^2 - V_i^2}{2a}$

13

$$F = \frac{Gm_1m_2}{r^2}$$

The force of gravity F between two masses, m_1 and m_2, r meters apart can be found using the equation above where G is a constant. If m_1 is doubled and the distance between the two masses is quadrupled, by what factor will the force of gravity change?

A. 1/16

B. 1/8

C. 1/2

D. 2

14

$$R = PQ$$

Shannon can determine the total revenue R after selling Q magazines at P dollars each by using the equation above. If she triples her sales by halving the price per magazine, by what factor will her revenue change?

A. 2/3

B. 3/2

C. 5

D. 6

15

$$5h = \sqrt[3]{1250a^4b^3}$$

Which of the following is equivalent to h, given the equation above?

A. $h = \dfrac{ab\sqrt[3]{10a}}{5}$

B. $h = 5\sqrt[3]{10a}$

C. $h = ab\sqrt[3]{10a}$

D. $h = \sqrt[3]{10a}$

16

$$6^{x^2-y^2} = 6^{48}$$

If in the equation above $x + y = 4$, what is the value of $x - y$?

A. 8

B. 12

C. 16

D. 48

17

$$\frac{I_1}{I_2} = 10^{0.1(dB_1 - dB_2)}$$

A sound wave can be characterized in terms of both intensity I and decibels dB. The ratio between the intensities of two sound waves is related using the equation above. If the difference of $dB_1 - dB_2$ doubles from 40 to 80, by what factor will the ratio of I_1 to I_2 change?

A. $10^{0.2}$

B. $10^{0.4}$

C. 10^2

D. 10^4

18

$$K = \frac{1}{2}mv^2$$

The kinetic energy K of an object can be found using the equation above, where m is the mass in kilograms and v is the velocity in meters per second. By what factor does the velocity need to change to double the kinetic energy of a particular object?

A. 1/2

B. $\sqrt{2}/2$

C. $\sqrt{2}$

D. 2

Explanations

Problem Set: Expressions and Equations

1. A

Continue to factor the polynomial by taking out the greatest common factor, x^2, then through grouping.

$$(x-1)(x^5+x^4+x^3+x^2)$$

$$x^2(x-1)(x^3+x^2+x+1)$$

$$x^2(x-1)[(x^3+x^2)+(x+1)]$$

$$x^2(x-1)[x^2(x+1)+1(x+1)]$$

$$x^2(x-1)(x+1)(x^2+1)$$

2. C

Create an equivalent fraction through multiplying the numerator and denominator by factors that cancel out the denominators of the small fractions.

$$\frac{2p^3}{\frac{1}{p-3}+\frac{1}{p+3}} * \frac{(p-3)(p+3)}{(p-3)(p+3)}$$

$$\frac{2p^3(p-3)(p+3)}{\frac{\cancel{(p-3)}(p+3)}{\cancel{(p-3)}}+\frac{(p-3)\cancel{(p+3)}}{\cancel{(p+3)}}}$$

$$\frac{2p^3(p-3)(p+3)}{(p+3)+(p-3)}$$

$$\frac{\cancel{2}p^3(p-3)(p+3)}{\cancel{2}p}$$

$$p^2(p-3)(p+3)$$

$$p^2(p^2-9)$$

3. B

Distribute and simplify to reach the final answer.

$$(2g-3s)\left(\frac{s}{6g^2}\right)$$

$$2g*\frac{s}{6g^2}-3s*\frac{s}{6g^2}$$

$$\frac{s}{3g}-\frac{s^2}{2g^2}$$

4. C

Recall that the area of a square is its side length squared.

$$\text{Area} = (\text{side length})^2$$

$$\text{side length} = \sqrt{\text{Area}}$$

$$\text{side length} = \sqrt{4(x^2y^2+6xy+9)}$$

$$\text{side length} = \sqrt{4(xy+3)^2} \text{ or } 2(xy+3)$$

5. A

Factor as a difference of squares:

$$\sqrt{4b} = 2\sqrt{b}$$

$$\sqrt{49x^2} = 7x$$

$$y = 4b - 49x^2$$

$$y = (2\sqrt{b}+7x)(2\sqrt{b}-7x)$$

6. C

Plug in z and c into the equation of q.

$$q = z - 60c$$

$$q = 36x^2 + 25y^2 - 60xy$$

$$q = 36x^2 - 60xy + 25y^2$$

Factor as a perfect square trinomial.

$$q = (6x - 5y)^2$$

7. A

To divide by a fraction, multiply by its reciprocal.

$$\frac{\frac{w}{w-6} + \frac{2w}{w+1}}{\frac{w^2}{w+1}}$$

$$\left(\frac{w}{w-6} + \frac{2w}{w+1}\right) * \frac{w+1}{w^2}$$

$$\frac{\cancel{w}}{w-6} * \frac{w+1}{w^{\cancel{2}}} + \frac{2\cancel{w}}{\cancel{(w+1)}} * \frac{\cancel{(w+1)}}{w^{\cancel{2}}}$$

$$\frac{w+1}{w(w-6)} + \frac{2}{w}$$

$$\frac{w+1}{w(w-6)} + \frac{2(w-6)}{w(w-6)}$$

$$\frac{w+1+2(w-6)}{w(w-6)}$$

$$\frac{3w+11}{w^2-6w}$$

8. C

$$\frac{5^{-3}+5^{-5}}{5^{-4}} \text{ or } \frac{\frac{1}{5^3}+\frac{1}{5^5}}{\frac{1}{5^4}}$$

$$\frac{\frac{1}{5^3}+\frac{1}{5^5}}{\frac{1}{5^4}} * \frac{5^5}{5^5} = \frac{5^2+1}{5} \text{ or } \frac{26}{5}$$

9. B

Factor both the given and desired expressions to see the exact quantities involved. Since we know the value of $a - b$ is 32, we can see that $a + b$ is 2 because only 32 times 2 is 64.

$$a^2 - b^2 = 64$$

$$(a+b)(a-b) = 64$$

$$a^2 + 2ab + b^2 = (a+b)^2 = 2^2 = 4$$

10. A

Expand and simplify to reach final answer.

$$(u^2 - 2qu^2 - w^3) - (-2u^2 - 2qu^2 + w^2)$$

$$u^2 - \cancel{2qu^2} - w^3 + 2u^2 + \cancel{2qu^2} - w^2$$

$$3u^2 - w^2 - w^3$$

11. A

Manipulate the equation to solve for the desired variable.

$$V = \frac{V_{max}[s]}{k_m + [s]}$$

$$V(k_m + [s]) = V_{max}[s]$$

$$Vk_m + V[s] = V_{max}[s]$$

$$Vk_m = V_{max}[s] - V[s]$$

$$Vk_m = [s](V_{max} - V)$$

$$[s] = \frac{Vk_m}{(V_{max} - V)}$$

12. D

Manipulate the equation to solve for the desired variable.

$$V_f^2 = V_i^2 + 2ad$$

$$2ad = V_f^2 - V_i^2$$

$$d = \frac{V_f^2 - V_i^2}{2a}$$

13. B

Use the template of the equation to keep only the variables that change.

$$F = \frac{Gm_1m_2}{r^2}$$

$$F \propto \frac{m_1}{r^2}$$

$$F' \propto \frac{2m_1}{(4r)^2}$$

$$F' \propto \frac{2m_1}{16r^2}$$

$$F' \propto \frac{1}{8} * \frac{m_1}{r^2}$$

14. B

Employ the factors of change for each variable.

$$R' = \frac{1}{2}P * 3Q$$

$$R' = \frac{3}{2} * PQ$$

15. C

$$5h = \sqrt[3]{125 * 10 * a^3 * a * b^3}$$

$$5h = 5ab\sqrt[3]{10a}$$

$$h = ab\sqrt[3]{10a}$$

16. B

Since both sides of the given equation have the same base, their exponents must be equal. Set the exponents equal to each other, factor, the solve for the desired quantity.

$$x^2 - y^2 = 48$$

$$(x + y)(x - y) = 48$$

$$(4)(x - y) = 48$$

$$(x - y) = 12$$

17. D

$$\text{Factor of Change} = \frac{\text{New Value}}{\text{Old Value}}$$

$$\text{New Value} = 10^{0.1(80)} \text{ or } 10^8$$

$$\text{Old Value} = 10^{0.1(40)} \text{ or } 10^4$$

$$\text{Factor of Change} = \frac{10^8}{10^4} = 10^4$$

18. C

Relate v and K to each other to determine the overall factor of change.

$$K = \frac{1}{2}mv^2$$

$$K \propto v^2$$

$$v \propto \sqrt{K}$$

$$v' \propto \sqrt{2K}$$

$$v' \propto \sqrt{2} * \sqrt{K}$$

2.3 FUNCTION PROPERTIES

a. Function Notation

This section is all about functions. Remember that the textbook definition of a function is a relationship whereby each input only has one output. That is to say, for example, that on the xy-plane any given x-value can only have one distinct y-value. Linear equations, therefore, represent linear functions because they follow this principle. The most important functions we should know for the test are linear, quadratic, higher-degree polynomial, square root, absolute value, and exponential.

Recall that $f(x)$ is a more elaborate way to write the output y. Additionally, it is helpful to write this statement out as an equation: $f(x) = y$. So, when given the statement $f(-2) = 4$, we then know that $x = -2$ and $y = 4$. To get a better sense of this idea, analyze the placement of each number in the equation.

$$f(x) = y$$

$$f(-2) = 4$$

$$x = -2 \text{ and } y = 4$$

Function notation provides both the input and output in the equation. The exam now tests students on their ability to plug in numbers or expressions as inputs into an equation that utilizes function notation.

Q+A

Function Notation **Calculator**

$$f(x) = x^2 - 3x - 10$$

Using the function $f(x)$ above, which of the following is equivalent to $f(2ab)$?

A. $2(2ab - 5)(ab + 1)$

B. $(4ab + 5)(ab + 1)$

C. $(4a^2b^2 + 5)(ab - 2)$

D. $2(2ab + 5)(ab - 1)$

Workspace

Solution Outline

Step 1: Since all answer choices are in factored form, factor the function first.

$$f(x) = (x - 5)(x + 2)$$

Step 2: Plug in $2ab$ for x.

$$f(2ab) = (2ab - 5)(2ab + 2)$$

Step 3: Take out the common factor of 2 in the second group.

$$f(2ab) = 2(2ab - 5)(ab + 1)$$

Answer: A

Notes

Functions and Charts

Chart A to the right defines x-values for three different functions: $f(x), q(x)$, and $z(x)$. Before we really get into this with an example, do this small exercise. List out each function value when $x = 1$. The function values are as follows:

$$f(1) = -5 \qquad q(1) = 5 \qquad z(1) = 2$$

To get these values, go to the row where x is one, then move over to each function column. The information from Chart A is required to answer the following examples that come from questions 19 to 21 in the diagnostic exam.

Chart A

x	$f(x)$	$q(x)$	$z(x)$
-2	8	-2	-9
-1	1	0	7
0	2	4	-1
1	-5	5	2
2	4	1	6

Q+A

Function Values and Factors

Diagnostic Exam | Calculator Q19

For which of the following is $x + 1$ a factor? (Refer to Chart A)

I. $f(x)$

II. $q(x)$

III. $z(x)$

A. I only

B. II only

C. I and II

D. I and III

Workspace

Solution Outline

Step 1: Set factor equal to zero to identify the x-intercept. Factors tell us where x-intercepts are located. Recall that the x-intercepts occur when $y = 0$.

$$x + 1 = 0 \quad \text{or} \quad x = -1$$

Step 2: Find $x = -1$ in the chart and see which functions have a value of zero.

Answer: B

Notes

Q+A

Functions and Charts **Diagnostic Exam | Calculator Q20**

What is the value of $z\left(\frac{1}{4}f(2)\right)$? (Refer to Chart A)

A. -1 B. 2 C. 6 D. 7

Workspace

Solution Outline

Step 1: Evaluate the innermost function using the chart: $f(2) = 4$

Step 2: Substitute in the value of $f(2)$ and continue to evaluate.

$$z\left(\frac{1}{4} * 4\right) \quad \text{or} \quad z(1)$$

Step 3: Find the value of $z(1)$ using the chart.

Answer: B

Notes

Q+A

Function Intervals **Diagnostic Exam | Calculator Q21**

Given that $f(x)$ is a continuous function, in what interval would it must have an x-intercept? (Refer to Chart A)

A. $(-0.5,1.5)$

B. $(-2,0)$

C. $(1.5,2.5)$

D. None of the above

Workspace

Solution Outline

Step 1: Continuous functions will have an x-intercept when the value of the function changes sign. The interval itself refers to the span of its x-values.

Step 2: Find values from the chart using values within the given intervals to determine if a sign change takes place. Remember only to use x-values from the chart that are within the interval, or *above* the minimum and *below* the maximum. For example, in answer choice A, the lowest x-value we can use in the chart is zero and the highest is 1 Use the values to see if a sign change occurs.

$$f(0) = 2$$

$$f(1) = -5$$

Since the value of the continuous function changed sign within the interval, the function must have crossed the x-axis.

Answer: A

Notes

Some questions may give a table of values for one function that is to be related to another function. The following example details this idea as it gives values for the function, $f(x)$, and its relationship to the function $g(x)$.

Q+A

Functions and Charts

Calculator

$f(x)$	32	−2	5	9	−12
x	−8	−3	4	7	9

The table above provides information about the function f. If $g(x) = -f(2x + 6) + 4$, what value of the constant c would make $g(c) = -5$?

A. -1

B. 0

C. $1/2$

D. $3/2$

Workspace

Solution Outline

Step 1: Substitute c in for x into the equation that defines $g(x)$.

$$g(x) = -f(2x + 6) + 4$$

$$g(c) = -f(2c + 6) + 4$$

Step 2: Substitute -5 for $g(c)$.

$$-5 = -f(2c + 6) + 4$$

Step 3: Isolate the function f to one side.

$$f(2c + 6) = 9$$

Step 4: Use the chart to find when the function f has a value of 9.

$$f(7) = 9$$

Therefore: $f(2c + 6) = f(7)$

Step 5: Set $2c + 6 = 7$ and solve for c.

$$2c + 6 = 7$$

$$c = 1/2$$

Answer: C

Notes

Function Operations

Additional function operations are summarized in Table 2.31 below. These operations include general arithmetic and composite functions. All algebraic operations can be applied to functions and may be expressed in condensed or expanded form. Pay close attention to the composite function operation where one function becomes an input into the other.

Table 2.31 | Summary of Function Operations

Operation	Condensed Notation	Expanded Notation
Add	$(f + g)(x)$	$f(x) + g(x)$
Subtract	$(f - g)(x)$	$f(x) - g(x)$
Multiply	$(f \cdot g)(x)$	$f(x) \cdot g(x)$
Divide	$(f/g)(x)$	$f(x)/g(x)$
Composite	$(f \circ g)(x)$	$f(g(x))$

Q+A

Composite Functions **Calculator**

$$f(x) = x^2 + x + 1$$
$$g(x) = x + 1$$

Using the functions f and g above, what expression is equivalent to the composite function $f \circ g$ for all values of x?

A. $x^2 + 2x + 2$

B. $x^2 + 3x + 3$

C. $x^3 + 2x^2 + 2x + 1$

D. $x^3 + 3x^2 + 3x + 1$

Workspace

Solution Outline

Step 1: Rewrite $f \circ g$ in expanded notation to plug in the new input, $g(x)$, into $f(x)$.

$$f \circ g = f(g(x))$$

$$f(x) = (x)^2 + (x) + 1$$

$$f(g(x)) = (g(x))^2 + (g(x)) + 1$$

Step 2: Substitute the value of $g(x)$.

$$f(g(x)) = (x + 1)^2 + (x + 1) + 1$$

Step 3: Expand and simplify.

$$f \circ g = (x^2 + 2x + 1) + (x + 1) + 1$$

$$f \circ g = x^2 + 3x + 3$$

Answer: B

Notes

b. Quadratics

Quadratic functions are a specific type of polynomial function that can be represented in the form $f(x) = ax^2 + bx + c$ and form parabolas when graphed on the coordinate plane. Though all polynomial functions share general characteristics that will be discussed in the next section, quadratics have several key features that have earned a prominent spot on the redesigned test. For example, quadratics contain a vertex, or turning point, that may be the function's minimum or maximum value through which a line of symmetry can be drawn. Quadratics always contain a y-intercept and up to two x-intercepts. With so many unique features, quadratics can be represented in several different forms: General, Vertex, and Factored. In this section, we will discuss when to use one form over another and how to create our own quadratic functions that satisfy any number of criteria.

Key Features of Quadratic Functions

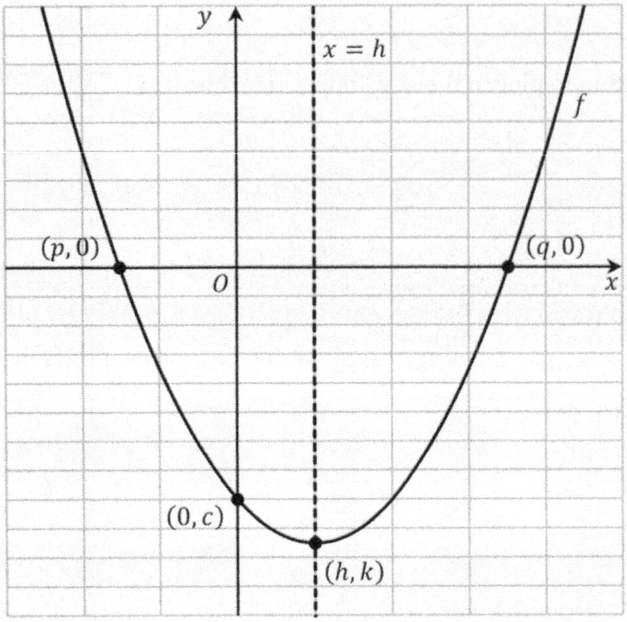

Function Forms	
General	$f(x) = ax^2 + bx + c$
Vertex	$f(x) = a(x-h)^2 + k$
Factored	$f(x) = a(x-p)(x-q)$
Characteristics	
Vertex	(h, k)
Intercepts	$(p, 0), (q, 0), (0, c)$
Axis of Symmetry	$x = h$

Among the most important characteristics that we will need to locate on the parabola is the vertex. As mentioned in the opening, the vertex to a parabola is the turning point that is represented by the coordinates (h, k). Since parabola can open up or down, the vertex can represent either minimum or maximum value. Thus, Vertex Form is the equation of a parabola that shows the vertex – either a minimum or maximum – as constants. We should note that since the parabola is symmetric, the x-coordinate of the vertex, h, is located right between the x-intercepts, p and q. To find the middle of any two numbers, we can simply take an average. Therefore, one way we can locate h is by taking an average of the quadratics x-intercepts. If the equation of the quadratic is in general, we can also find h using the constants a and b.

Formula for h	$\dfrac{p+q}{2}$ or $-\dfrac{b}{2a}$

The next example covers how this property – the fact that the vertex lies precisely between the two x-intercepts – may be applied on the test.

Q+A

Vertex of Quadratic Functions **No Calculator**

$$f(x) = (x + 2)(x - b)$$

The function f provided above is graphed on the xy-plane, where $y = f(x)$. The vertex of function f is located at the coordinate point $(1, a)$ and constants a and b are integers. What is the value of a?

A. -9

B. -4

C. 4

D. 9

Workspace

Solution Outline

Step 1: Find b using the known x-intercept and vertex. The zero to the first factor is -2, making it our first known x-intercept. The x-coordinate of the vertex is 1, which is three units to the right of this x-intercept. Going another three units to the right brings us to 4, which must be that value of b, the second x-intercept, as that value ensures the vertex sits squarely in the middle.

Step 2: Find a using known value of b and the function equation. Since a occurs when $x = 1$, we just need to plug $x = 1$ into the equation to find its value. In other words, $a = f(1)$.

$$f(x) = (x + 2)(x - 4)$$

$$f(1) = (3)(-3) \quad \text{or} \quad -9$$

Answer: A

Notes

The previous example also shows how we can find the value of k, the y-coordinate to a quadratic function's vertex. All we need to do is plug h back into the function.

Formula for k	$k = f(h)$

It is very important that we recognize when to use a particular form to represent the equation of a quadratic function. These forms are useful because each one contains constants that locate particular points. For example, general form contains the constant c, which is the y-intercept. Additionally, vertex form contains the constants h and k, which are x- and y-coordinates of the vertex, and thus locates the function's min or max. And finally, factored form contains the constants p and q, which are the quadratic's x-intercepts, or roots. Table 2.32 below summarizes these three forms and the constants they contain.

Table 2.32 | Summary of Quadratic Functions and Constants

Form	Equation	Constants
General	$f(x) = ax^2 + bx + c$	y-intercept
Vertex	$f(x) = a(x - h)^2 + k$	Vertex (Min or Max)
Factored	$f(x) = a(x - p)(x - q)$	x-intercepts

Sometimes, knowing how a form is structured is enough to answer a question about a quadratic function, or at least eliminate some wrong answer choices. The test often challenges students to determine when to use vertex or factored form by describing the constants that should be seen in the equation. The next example explains this process in detail. We will follow up on this example to see exactly how we can reformat the structure of quadratic function into vertex form, even when answer choices are not present, which will be useful in the grid-in component of the exam.

Q+A

Quadratic Functions **Diagnostic Exam | No Calculator Q13**

$$f(x) = -2x^2 - 4x + 8$$

The function f provided above is graphed on the xy-plane, where $y = f(x)$. Which of the following is an equivalent form of the function f that contains constants that locate its maximum?

A. $f(x) = -2x(x + 2) + 8$

B. $f(x) = -2x(x + 2) + 10$

C. $f(x) = -2(x + 1)^2 + 8$

D. $f(x) = -2(x + 1)^2 + 10$

Workspace

Solution Outline

Step 1: Identify answer choices in vertex form. Only choices C and D are in vertex form, which will, in this case, locate the maximum of the quadratic function.

Step 2: Determine the coordinates of the vertex. Since we know that $h = -1$ from the choices we are left with, we can find k. Since $k = f(h)$, we just need to plug -1 into the function to determine the value of k. We will see that $k = 10$, so choice D is correct.

$$f(-1) = -2(-1)^2 - 4(-1) + 8$$

$$f(-1) = 10$$

Answer: D

Notes

In the previous example, we determined how to identify the correct answer choice when asked to bring a quadratic function written in general form into vertex form. Vertex form requires us to input three constants: $a, h,$ and k. Table 2.33a below summarizes how to find these three constants when starting out in general form. Recall that general form is the expanded form where $f(x) = ax^2 + bx + c$ and vertex form is where $f(x) = a(x - h)^2 + k$.

Table 2.33a | General to Vertex Form

Step 1: Find a	Step 2: Find h	Step 3: Find k
$f(x) = \boldsymbol{a}x^2 + bx + c$	$h = -\frac{b}{2a}$	$k = f(h)$

If we look back at the function $f(x) = -2x^2 - 4x + 8$, it is clear that $a = -2$ and $b = -4$. Plugging these values into the formula in step two above tells us that $h = -1$. Now all we need to do is plug this value of h back into the function to find k, which we already did in the last step of the previous example: $k = 10$. Since we know $a, h,$ and k, we can plug them into the template of vertex form, $f(x) = a(x - h)^2 + k$; doing so brings us to $f(x) = -2(x + 1)^2 + 10$.

In using these three steps to find $a, h,$ and k, we can quickly get the needed information to bring a quadratic function written in general form into vertex form; however, it is not the only method. We can also complete the square. Completing the square is an important technique that students can use to not only to reformat a quadratic equation into vertex form, but also to find rational, irrational, and complex roots without the quadratic formula. It is also a critical process to finding the standard equation of a circle on the coordinate plane, which will be discussed later on. The process comes down to the five steps listed in Table 2.33b. For comparison, we will stick with the same function used in the previous example, $f(x) = -2x^2 - 4x + 8$.

Table 2.33b | Complete the Square

Steps	Description	Example
Step 1	Group variable terms to one side	$f(x) - 8 = -2x^2 - 4x$
Step 2	Divide both sides by a	$\frac{f(x) - 8}{-2} = x^2 + 2x$
Step 3	Add $\left(\frac{b}{2}\right)^2$ to both sides	$\frac{f(x) - 8}{-2} + 1 = x^2 + 2x + 1$
Step 4	Factor the perfect square trinomial	$\frac{f(x) - 8}{-2} + 1 = (x + 1)^2$
Step 5	Solve for $f(x)$	$f(x) = -2(x + 1)^2 + 10$

Solving Quadratic Equations

The usefulness of having a quadratic function in vertex form is that it indicates the positioning of the parabola on the xy-plane, making it easier to visualize and, therefore, sketch. Though the vertex form of a quadratic function highlights the coordinates of the vertex in its equation, it also helps in determining the number and types of x-intercepts. There will be a more detailed discussion on this later in this section, but the takeaway right now is that another major testing point is finding the x-intercepts of quadratic functions. It is important to realize that x-intercepts are also called zeros and roots, which might also be referred to as solutions when solving quadratic equations. The x-intercepts of quadratic functions can be found through several different methods, the most common of which rely on factoring, general algebra, and the quadratic formula. The fastest method will often depend on the nature of the quadratic, as discussed in the examples that follow. Be sure to know all three methods as they all frequently show up on the test.

1. **Factor**: Use when in expanded form.

    Step 1: Set the polynomial equal to zero

    Step 2: Reduce and Factor.

    Step 3: Set each factor equal to zero and solve for x.

    Example: Find the x-intercepts to the function $f(x) = x^2 - 5x + 6$

    Solution: $0 = x^2 - 5x + 6$

    $0 = (x - 3)(x - 2)$

    $x - 3 = 0$ or $x - 2 = 0$

    $x = 3$ or $x = 2$

2. **Algebra**: Use when variable only occurs in one place. Typically avoids expansion.

Step 1: Plug in any necessary information

Step 2: Isolate x to one side.

Example: Find the roots of the function $f(x) = 2(x+1)^2 - 14$

Solution: $0 = 2(x+1)^2 - 14$

$14 = 2(x+1)^2$

$7 = (x+1)^2$

$\pm\sqrt{7} = x + 1$

$x = -1 \pm \sqrt{7}$

3. **Quadratic Formula**: Use when in expanded form and the quadratic cannot be factored.

Step 1: Set the quadratic equal to zero to identify $a, b,$ and c.

Step 2: Use the quadratic formula

Quadratic Formula	$x = \dfrac{-b \pm \sqrt{b^2 - 4ac}}{2a}$

Example: Find the solutions of x that satisfy the equation: $-2 = 3x^2 + 11x$

Solution: $0 = 3x^2 + 11x + 2$

$a = 3 \quad b = 11 \quad c = 2$

$$x = \frac{-11 \pm \sqrt{(11)^2 - 4(3)(2)}}{2(3)}$$

$$x = \frac{-11 \pm \sqrt{97}}{6}$$

Sum and Product Formulas

If the x-intercepts, roots, zeros, or solutions are difficult to find, the following formulas can save us considerable time on questions that ask for their sum or product. The quadratic must be in general form and set equal to zero when we use sum and product formulas.

Sum of the Roots	$-\dfrac{b}{a}$
Product of the Roots	$\dfrac{c}{a}$

Q+A

Sum of the Roots **Diagnostic Exam | Calculator Q10**

$$f(x) = x + 2$$

$$g(x) = x^2 - 3$$

The functions $f(x)$ and $g(x)$ provided above are graphed on the xy-plane and intersect at the points (m, n) and (p, q). What is the value of $m + p$?

A. -1

B. 0

C. 1

D. 2

Workspace

Solution Outline

Step 1: Understand that the question asks for the sum of the x-coordinates to the solutions, so we can use the Root Sum Formula: $-b/a$.

Step 2: Set $f(x) = g(x)$ and bring all terms to one side.

$$x + 2 = x^2 - 3$$

$$x^2 - x - 5 = 0$$

Step 3: Use Root Sum formula, $-b/a$

$a = 1$ and $b = -1$

$$-\frac{b}{a} = 1$$

Answer: C

Notes

The Discriminant

The value of the discriminant informs us as to how many roots a quadratic function contains or how many solutions a quadratic equation has. Recall that roots are also called x-intercepts or zeros, which we can often find from the function's factors; similarly, solutions may be referenced to as intersections because they can be found when two curves cross on the coordinate plane. Since the discriminant is the numerical value found using the formula $b^2 - 4ac$, it can be positive, zero, or negative.

Discriminant Formula	$b^2 - 4ac$

When the discriminant is positive, the quadratic function contains two real roots. When the discriminant is zero, the quadratic function contains one real root, and when the discriminant is negative, the quadratic function contains no real roots. Likewise, a quadratic equation has two, one, and zero solutions when the discriminant is positive, zero, and negative, respectively. Since the discriminant actually comes from the quadratic formula, we can only use it when the quadratic is set equal to zero.

Q+A

Discriminant **Diagnostic Exam | No Calculator Q5**

$$2x^2 - qx = -4$$

In the equation above, x is a variable and q is a positive constant. What value of q ensures that the equations has exactly one solution?

A. 0　　B. $2\sqrt{2}$　　C. $4\sqrt{2}$　　D. 16

Workspace

Solution Outline

Step 1: Bring the equation into $ax^2 + bx + c = 0$ form.

$$2x^2 - qx + 4 = 0$$

Step 2: Set discriminant equal to zero and solve for q.

$$b^2 - 4ac = 0$$

$$(-q)^2 - 4(2)(4) = 0$$

$$q^2 - 32 = 0$$

$$q^2 = 32$$

$$q = 4\sqrt{2}$$

Answer: C

Notes

By mastering the application of the discriminant, we can often skip unnecessary math, which will save us time and energy when taking the test. Use Table 2.34 to visualize the significance of the discriminant in each of its three cases by identifying the number times the parabola crosses the x-axis and the number of real roots.

Table 2.34 | Discriminant Summary

Discriminant	Description	Graph
Positive	Two Real Roots $b^2 - 4ac > 0$	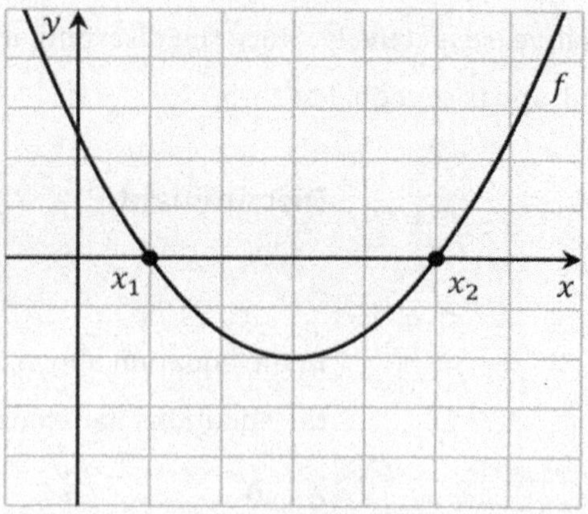
Zero	One Real Root $b^2 - 4ac = 0$	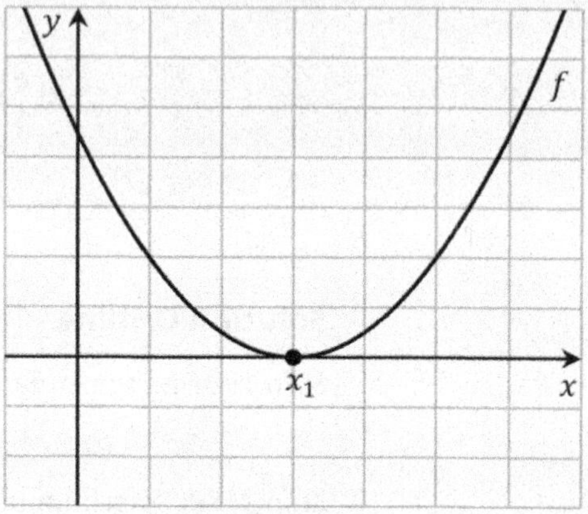
Negative	No Real Roots $b^2 - 4ac < 0$	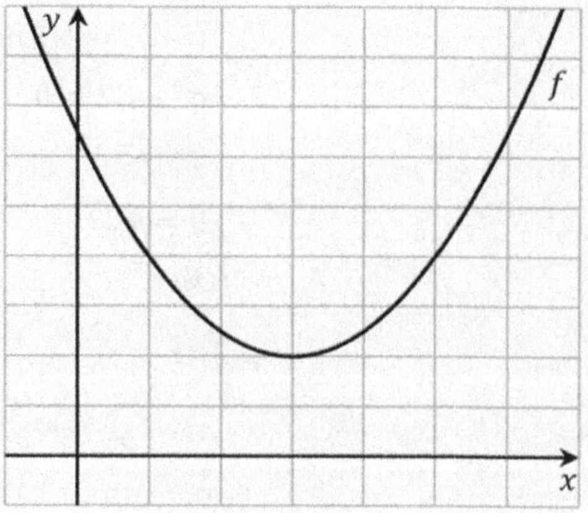

c. Polynomial Functions

Though linear and quadratic relationships fall under the umbrella of polynomial functions, they have unique properties that distinguish them from other polynomial functions. More generally, all polynomial functions share some notable characteristics – namely, end behavior and zero multiplicities – that are determined by the function's degree, leading coefficient, and factors. Before we delve into further detail, some basics: here are examples of what polynomial functions are and are *not*.

Polynomial Functions	Not Polynomial Functions
$f(x) = x^3 - 2x + 1$	$f(x) = 2^x - 9$
$f(x) = -x^4 + 8$	$f(x) = \frac{1}{x^2} + 5$
$f(x) = \frac{1}{2}x - 5$	$f(x) = \sqrt{x^2 - 16}$
$f(x) = 3x^2 + \frac{4}{5}x - 9$	$f(x) = x^{-\frac{3}{2}} + 13$
$f(x) = 5x^7 - 4x^2 + 2$	$f(x) = -x^{-1} + 7$

Degree and Leading Coefficient

Polynomial functions have variables that have positive, whole number exponents. There are no fractional exponents or variables occurring in denominators or as exponents. Polynomial functions can be grouped into different categories by taking into account their degree and leading coefficient. In fully expanded form, the degree of the polynomial is the largest exponent on x and the leading coefficient is the coefficient to this same x-term. This means that the function $f(x) = -x^4 + 8$ has a degree of 4 and leading coefficient of -1. The function $f(x) = 5x^7 - 4x^2 + 2$ has a degree of 7 and leading coefficient of 5. To abbreviate, the degree can be odd or even and the leading coefficient can be positive or negative.

End Behavior

End behavior describes what happens to the function as it moves left and right on the coordinate plane. Polynomial functions with even degrees have matching end behaviors, whereas polynomials with odd degrees do not. Specifically, for polynomial functions with an even degree, both sides will go up when the leading coefficient is positive and both sides will go down when it's negative. And for polynomial functions with an odd degree, the left side goes down and the right side goes up when the leading coefficient is positive and vice versa when the leading coefficient is negative. Use Table 2.35 to check out the end behavior of polynomial functions that have different degrees and leading coefficients. The table contains the graphs of quadratic and cubic functions that can be used to generalize and visualize the end behavior for any polynomial function with an odd or even degree and a positive or negative leading coefficient. The table uses arrow notation to describe what happens to the function $f(x)$ on the left- and right-hand sides of the coordinate plane. For example, $x \to \infty$ is translated to "as x approaches infinity", but in practical terms refers to the right side of the graph. Likewise, $f(x) \to -\infty$ means the function value approaches negative infinity, or goes down.

Table 2.35 | End Behavior Summary

Leading Coefficient		Degree: Even		Degree: Odd	
Positive		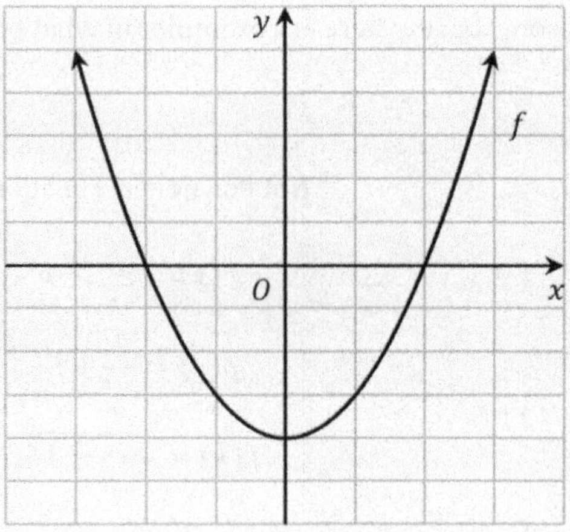		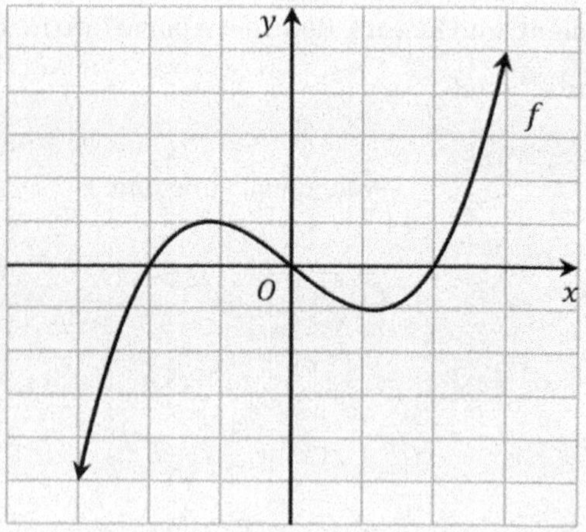	
		$x \to -\infty, f(x) \to \infty$	$x \to \infty, f(x) \to \infty$	$x \to -\infty, f(x) \to -\infty$	$x \to \infty, f(x) \to \infty$
Negative		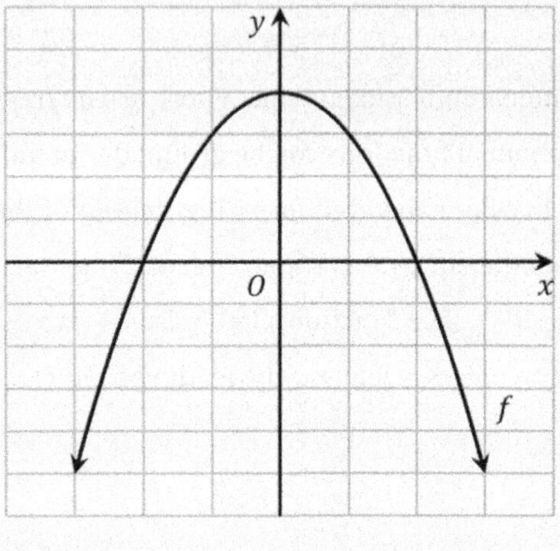		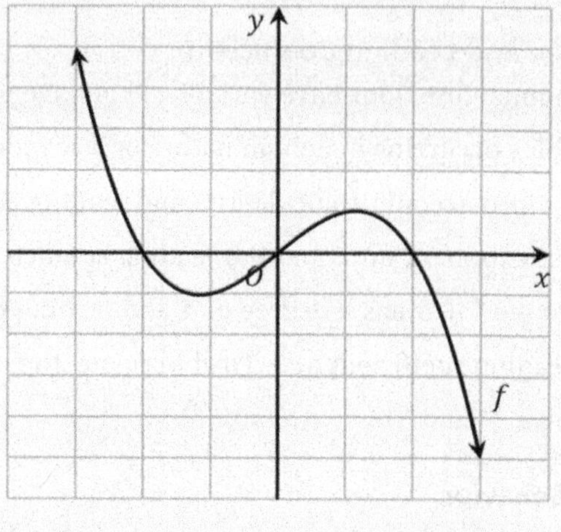	
		$x \to -\infty, f(x) \to -\infty$	$x \to \infty, f(x) \to -\infty$	$x \to -\infty, f(x) \to \infty$	$x \to \infty, f(x) \to -\infty$

Zero Multiplicity

In addition to end behavior, intercepts are also central to our ability to accurately sketch a function on the coordinate plane. A quick reminder: a function's x-intercepts occur when y is zero and y-intercepts occur when x is zero. More pertinent to this section are a function's x-intercepts, or zeros, which can be found given its factors. The exponents to those factors determine the multiplicity of each zero, and will allow us to determine whether the function will cross the x-axis or "bounce" off it. Factors that have odd exponents show where the function will cross the x-axis and factors that have even exponents show where the function will bounce off the x-axis. Important Note: Even and odd degree polynomial functions are NOT the same as even and odd functions. Even functions are symmetric about the y-axis and odd functions are symmetric about the origin. The implication here is that Even Functions follow the rule $f(x) = f(-x)$ and Odd Functions follow the rule $-f(x) = f(-x)$.

Q+A

Zero Multiplicity **Diagnostic Exam | Calculator Q35**

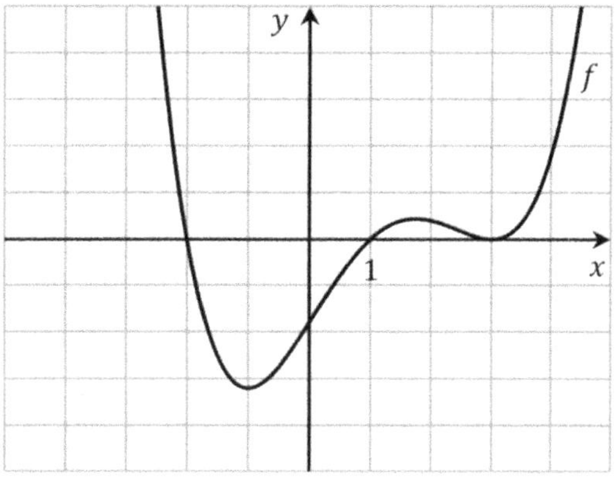

The function f is graphed on the xy-plane above where $y = f(x)$. It can be modeled in its factored form using the equation $f(x) = (x + a)(x - b)^2(x - c)$. Provided that $a, b,$ and c are positive integer constants, what is the value of b?

Workspace

Solution Outline

Step 1: Identify x-intercepts: $-2, 1,$ and 3.

Step 2: Write out each factor using the x-intercepts. Factors will take the shape $(x - x_{int})$.

$(x + 2), (x - 1),$ and $(x - 3)$

Step 3: Determine each factor's multiplicity. Since the graph smoothly crosses through the x-axis at -2 and 1, those factors will have multiplicities of one. In contrast, since the graph of f only touches the x-axis at 3, that factor must have an even multiplicity - in this case, of 2.

$f(x) = (x + 2)^1(x - 3)^2(x - 1)^1$

Answer: 3

Notes

In the previous example, both sides of the function go up, which is a feature of even degree polynomial functions. Does the function in question meet this standard? To find the degree, focus on just multiplying the x-term of each factor. Therefore, to find the degree of $f(x) = (x + a)(x - b)^2(x - c)$, multiply $(x)(x)^2(x) = x^4$. The function f is in fact a polynomial function with an even degree.

Degree and Extrema

The degree indicates the maximum number of real zeros (or x-intercepts) that the function can have. Odd degree polynomial functions always have at least one real zero, whereas even degree polynomial functions may not have any. Keep in mind the implication here: Odd-degree polynomial functions necessarily cross the x-axis and even-degree polynomial functions do not. Another characteristic we can observe is the number of turning points that a polynomial function has. The rule is that one less than the degree (n) indicates the maximum number of turning points that the polynomial function can have.

Turning Points	$n - 1$

Turning points are local maxima and minima where the function changes from increasing to decreasing or decreasing to increasing. On paper, these extrema look like the function's hill tops and valley lows. Even degree polynomial functions always have at least one turning point that would then also act as the global maximum or minimum, whereas odd degree polynomial functions may not have any turning points, such as in the case of $f(x) = x^3$.

d. Transformations and Theorems

The concepts that we will discuss in this section include transforming parent functions, finding equivalent polynomial functions, and dividing polynomial functions. Grouping these topics together allows us to better distinguish the higher-level skills the test-makers expect of the test-takers. Though they may not be tested directly on every exam, the ideas that are discussed in this section can still be applied in variety of ways, as well as rounds out our understanding of functions and their properties.

Transformation of Parent Functions

Transformations, as they are often represented on the test, include changes that can be made to functions on the coordinate plane. Such changes include translations, dilations, and reflections. Recall that translations are simply movements to a curve that retain its shape while shifting along the x- or y-axes. On the other hand, dilations stretch or compress curves, thereby distorting a curve's general features. Reflections flip the curve over a particular line or axis. To better apply such transformations, we should become familiar with the parent functions that dominate the test.

A **parent function** is the most basic function of a family of functions. The most important parent functions we need to know are linear, quadratic, cubic, square root, cube root, absolute value, exponential, and rational functions. By knowing the shape and features of these functions, we will be more adept at transforming them and answering questions surrounding their specific characteristics, like domain and range. Table 2.36 displays the graphs of the most relevant parent functions. There will be a further discussion on domain and range in section 2.4 Advanced Functions and Equations.

For now, just acknowledge the general shape of each function, as well as the presence of unique points or asymptotic behavior. For example, it is relevant to note which parent functions cross through the origin and which ones do not. It is also worthwhile to identify the asymptotes of the exponential and rational functions. Asymptotes are lines that the function approaches as the function nears it. From its graph we can see that the general exponential function $f(x) = a^x$ contains a horizontal asymptote at $y = 0$ – which overlaps with the x-axis – because as the function gets closer to it on the left side of the coordinate axis.

Table 2.36 | Parent Functions

Linear

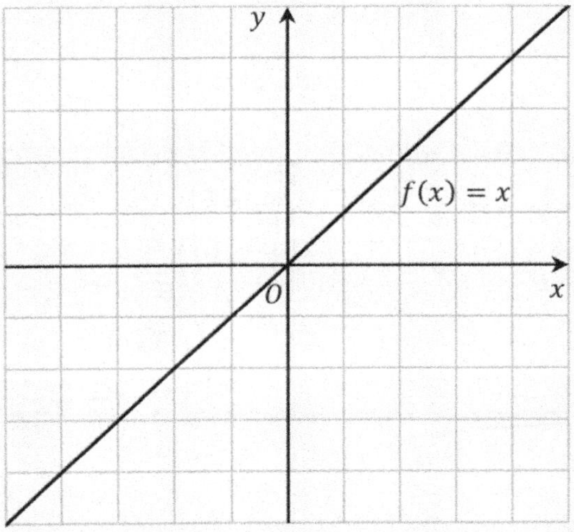

Absolute Value

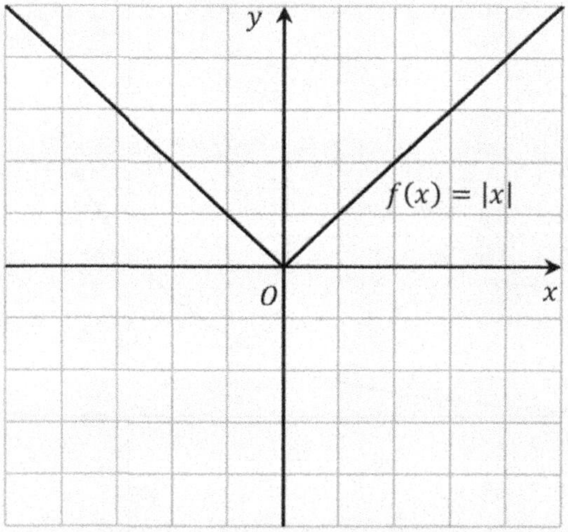

Quadratic

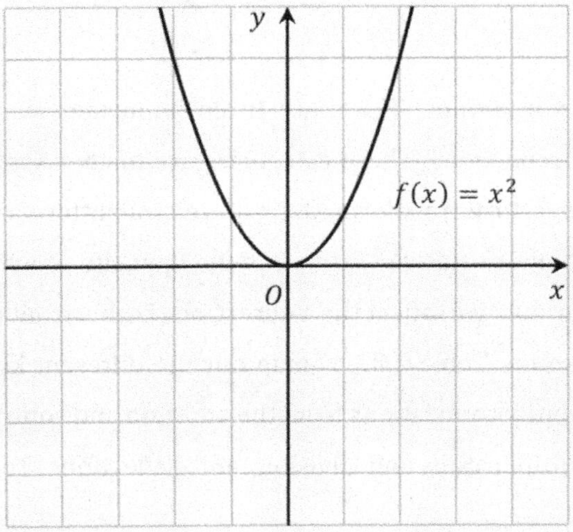

Cubic

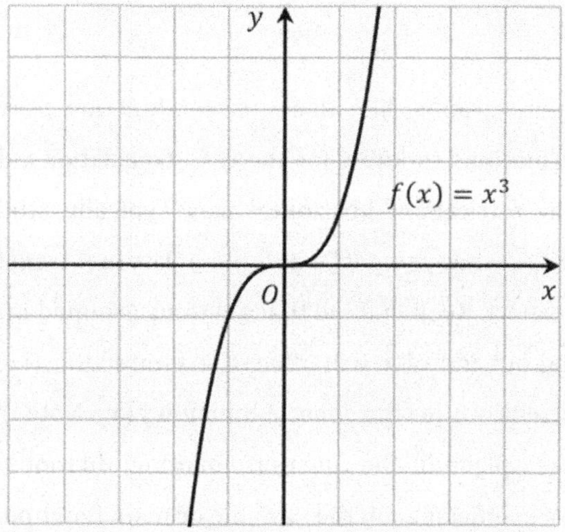

Square Root

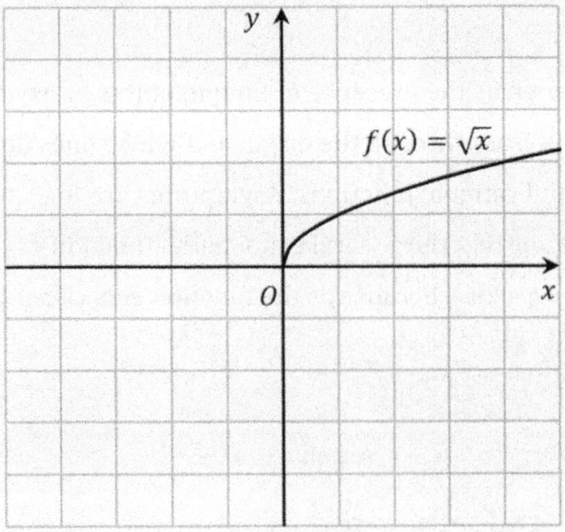

Cube Root

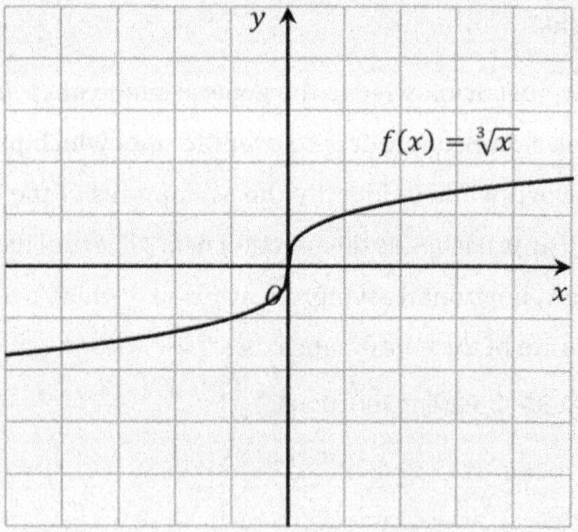

Exponential

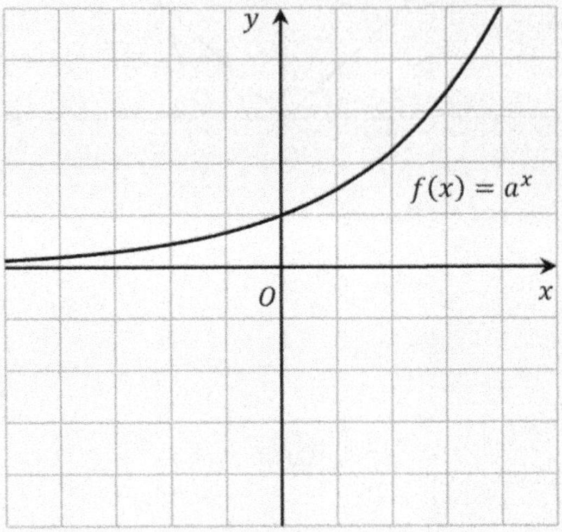

Rational

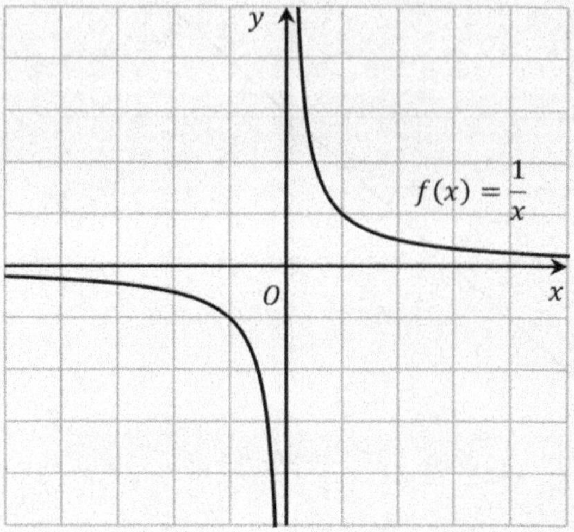

Changes to graphs, like these parent functions, are called transformations. There are three important classes of transformations to know for the test: **translations, dilations,** and **reflections.** These categories are further separated along the vertical and horizontal axes. Typically, students find the vertical transformations more straightforward – as in "you see what you get" – and the horizontal transformations counterintuitive – as in it is the opposite of what you would expect. Keep in mind that anything grouped in with the x-variable will affect the x-direction (i.e. horizontal) and anything outside of x will affect the y-direction (i.e. vertical). Use the Table 2.37 to map out the different kinds of transformations for the general function $f(x)$. Note that we can identify translations from the addition and subtraction that takes place in the function equation, dilations from the multiplication and division, and reflections from the negative coefficients on the variable or main function operation.

Table 2.37 | Transformation Rules

Transformation	Vertical		Horizontal	
Translations	$f(x)+b$	shift up b units	$f(x-b)$	shift right b units
	$f(x)-b$	shift down b units	$f(x+b)$	shift left b units
Dilations	$af(x)$		$f(ax)$	
	$\lvert a\rvert > 1$	vertical stretch by a factor of a	$0 < \lvert a\rvert < 1$	horizontal stretch by a factor of $1/a$
	$0 < \lvert a\rvert < 1$	vertical compression by a factor of a	$\lvert a\rvert > 1$	horizontal compression by a factor of $1/a$
Reflections	$-f(x)$	reflection over the x-axis	$f(-x)$	reflection over the y-axis

Q+A

Distance of Translation **Diagnostic Exam | Calculator Q38**

$$f(x) = (x+2)^2 - 6$$

$$g(x) = (x-4)^2 + 2$$

The functions f and g are defined above and graphed on the xy-plane. Function g is the image formed after each point on function f has been translated several units along the vertical and horizontal axes. What is the overall displacement that each point on f must undergo to form g?

Workspace

Solution Outline

Step 1: Observe the differences between the two functions. The only parts that change between the two equations are the addition and subtraction. In other words, only translations are taking place: movements left or right and up or down. Look first at what was being added or subtract from x. We went from adding 2 to subtracting 4. This is an overall change of -6, or shift right 6 units. Now let's move to the addition and subtraction taking place outside of x. We went from a subtracting 6 to adding 2. The overall change is $+8$, or a shift up 8 units. So, each point moved 6 units to the right and 8 units up.

Step 2: Find the overall displacement using the Pythagorean Theorem, where the legs to the right triangle are the shifts of 6 and 8 found in Step 1 and the hypotenuse is the magnitude of the displacement. Pythagorean Theorem: $a^2 + b^2 = c^2$

$$6^2 + 8^2 = \text{distance}^2$$

$$100 = \text{distance}^2$$

$$10 = \text{distance}$$

Answer: 10

Notes

While the previous example covered translations, the next will go over dilations. Simply put, dilations encompass the various stretches and compressions that can be made to a function; necessarily, there is a certain factor of change that results in the transformed function.

Q+A

Functions and Dilations

No Calculator

$$g(x) = f(ax)$$

Function g is related to function f for all values of x by using the equation above. When both are graphed on the xy-plane, function g is the image formed after function f has been horizontally stretched by a factor of 3, what could be the value of the constant a?

A. 0

B. 1/3

C. 1

D. 3

Workspace

Solution Outline

Step 1: The factor of change used to perform a horizontal stretch is the reciprocal of the coefficient on the input variable. Therefore, a and 3 have a reciprocal relationship: $a = 1/3$.

Answer: B

Notes

Equivalent Polynomial Functions

Equivalent polynomials have corresponding coefficients that are equal. When equivalent polynomials or expressions are set equal to each other, they form equations that have infinitely many solutions. In this scenario, the prevailing strategy is to expand and match coefficients of like terms between the two sides. The test often asks students to equate linear functions or quadratics, but this is true for any polynomial function. The general rule below for equivalent quadratics can be applied to any polynomial equation – the idea being students should be able to identify and match corresponding terms to ensure both sides represent the same quantity.

Equivalent Quadratics	If $\boldsymbol{ax^2 + bx + c = fx^2 + gx + h}$ for all values of x, then $\boldsymbol{a = f}, \boldsymbol{b = g}$, and $\boldsymbol{c = h}$

Before we discuss equivalent quadratics, let's work through an example that has with less terms: linear equations. And since we have already seen this idea at work in Section 1.2b, we can compare it to our new matching strategy, which can be used for all equivalent polynomials.

Q+A

Equivalent Polynomials — Diagnostic Exam | Calculator Q7

$$5x - a(x - b) = 2x + 5$$

In the equation above, a and b are constants. If the equation has infinitely many solutions, what is the value of b?

A. 0　　B. 3/5　　C. 5/3　　D. 5

Workspace

Solution Outline

Step 1: Expand and group like-terms of the x-variable

$$5x - ax + ab = 2x + 5$$

$$(5 - a)x + ab = 2x + 5$$

Step 2: Match the coefficients and constants of the corresponding terms between the two sides.

Match x coefficients: $5 - a = 2$ so $a = 3$

Match constants: $ab = 5$ so $3b = 5$

Solve for b: $b = 5/3$

Answer: C

Notes

Since higher-degree polynomials have more like-terms, even questions involving equivalent quadratics can get a little messy. Still, we will rely on matching the coefficients of corresponding terms, and abiding by any additional parameters set in place. In many cases, the process will involve a system of equations, often with easily identifiable integer solutions. Use the following example from the Diagnostic Exam to see this process in full.

Q+A

Equivalent Polynomials **Diagnostic Exam | Calculator Q37**

$$(mx + 2)(nx - 1) = 28x^2 + cx - 2$$

In the equation to the above, $c, m,$ and n are constants. If the equation is true for all values of x and $m - n = 3$, then what could be a positive integer value for c?

Workspace

Solution Outline

Step 1: Expand and Group

$$(mx + 2)(nx - 1) = 28x^2 + cx - 2$$

$$mnx^2 + 2nx - mx - 2 = 28x^2 + cx - 2$$

$$mnx^2 + (2n - m)x - 2 = 28x^2 + cx - 2$$

Step 2: Match Coefficients

Match x^2 coefficients: $mn = 28$

Match x coefficients: $2n - m = c$

Step 3: Use given information to find correct values of m and n. We need factors of 28 that have a difference of 3.

Case One: $m = 7$ and $n = 4$

Case Two: $m = -4$ and $n = -7$

Step 4: Use each case to find correct values of c.

Case One: $2(4) - 7 = 1$

Case Two: $2(-7) - (-4) = -10$

Answer: 1

Notes

Dividing Polynomials

Dividing polynomials can be easy if they lend themselves to canceling out factors; however, on the test this sometimes not the case. In this scenario, there are two methods available: long division and synthetic division. Long division of polynomials works very much like long division of numbers. As a result, it is both a familiar and robust tool to perform a new task. Synthetic division is a very different way of finding a quotient, but it provides a quick means to divide polynomials by a linear factor. Herein lies the problem with synthetic division: It only works when we are dividing by the linear factor $x - c$, where c is a constant.

Q+A

Division of Polynomials | **Diagnostic Exam | No Calculator Q14**

$$\frac{2x^3 + 5x^2 + 3x + 4}{x + 1}$$

Provided that $x \neq 0$, which of the following is equivalent to the expression above for all included values of x?

A. $2x^2 + 7x + \dfrac{14}{x+1}$

B. $2x^2 + 7x + \dfrac{6}{x+1}$

C. $2x^2 + 3x + \dfrac{4}{x+1}$

D. $2x^2 + 3x - \dfrac{4}{x+1}$

Workspace

Solution Outline

Strategy 1

Long Division requires figuring out what to multiply the divisor by to "match" the first term of the polynomial (dividend). Multiply through, then subtract and repeat with the difference until the remainder is reached.

Quotient: $2x^2 + 3x + \dfrac{4}{x+1}$

$$x + 1 \;\big)\; 2x^3 + 5x^2 + 3x + 4$$

$2x^2(x + 1)$ $\quad -(2x^3 + 2x^2)$

$$3x^2 + 3x + 4$$

$3x(x + 1)$ $\quad -(3x^2 + 3x)$

$$R = 4$$

Strategy 2

Synthetic division requires using the "zero" to the divisor and the coefficients to each term. Note that in synthetic division we always bring down the first coefficient, multiply across by the "zero" and add down as demonstrated in the example that follows.

$(x + 1)$	$2x^3$	$5x^2$	$3x$	4
-1	2	5	3	4
		+	+	+
		-2	-3	0
	2	3	0	$R = 4$

Quotient: $2x^2 + 3x + \dfrac{4}{x+1}$

Answer: C

Notes

Though it was nice to go through long division and synthetic division in this last example, it would actually be easier to solve the problem using the Remainder Theorem, which states that the value of a polynomial function at $x = c$ is equal to the remainder when it is divided by the linear factor $x - c$.

Remainder Theorem	When $p(x)$ is divided by $x - c$, then the remainder equals $p(c)$

The idea here is that when we divide by a linear factor, like $x + 1$ in our last example, then the remainder is actually the value of the polynomial function (the dividend) at the zero to that factor. In the context of the last example, this means we can just plug in $x = -1$ into $2x^3 + 5x^2 + 3x + 4$ to get the remainder. Since the choices in the last example all had different remainders, this is enough for us to reach an answer.

Plug in $x = -1$ into $2x^3 + 5x^2 + 3x + 4$:

$2(-1)^3 + 5(-1)^2 + 3(-1) + 4$

$2(-1) + 5(1) + 3(-1) + 4$

$(-2) + (5) + (-3) + 4$

$R = 4$

Answer: C (Remainder must be $+4$)

Note: A remainder of zero indicates that the divisor is actually a factor of the dividend and can therefore be used to locate x-interepts. In other words, factors indicate x-intercepts. In reference to the Remainder Theorem, since the remainder is zero, so too is the value of the function at that point.

Q+A

Remainder Theorem **Diagnostic Exam | Calculator Q22**

$$f(x) = 2ax^3 + 8bx^2 + 3x - 10$$

If $x + 1$ is a factor of the function f given above, what is the value of $4b - a$?

A. -1

B. 0

C. 1/2

D. 13/2

Workspace

Solution Outline

Step 1: Find the zero to the factor to find the x-value that we will plug into the function. The value to the function at this x-value must also be zero. Remember that factors indicate x-intercepts, which always have a y-value of zero; according to the remainder theorem, since factors have remainders of zero, the function also must have a value of zero at this point.

$$x + 1 = 0$$
$$x = -1$$
$$f(-1) = 0$$

Step 2: Plug in $x = -1$ and $y = 0$ to then reach the desired expression through algebraic manipulation.

$$f(x) = 2ax^3 + 8bx^2 + 3x - 10$$
$$0 = 2a(-1)^3 + 8b(-1)^2 + 3(-1) - 10$$
$$0 = -2a + 8b - 13$$
$$13 = 8b - 2a$$
$$\frac{13}{2} = 4b - a$$

Answer: D

Notes

Problem Set: Function Properties

14 Questions

1

$$\frac{3x^2 - x + 2}{x - 1}$$

Which of the following is equivalent to the expression above?

A. $x - 4 + \dfrac{6}{x - 1}$

B. $x - 4 + \dfrac{8}{x - 1}$

C. $3x + 2 + \dfrac{2}{x - 1}$

D. $3x + 2 + \dfrac{4}{x - 1}$

2

$$f(x) = (2x + 1)(x - b)$$

The function f is defined above, where b is a constant. If f has a minimum at $x = 4$, what is the value of b?

A. 5/2

B. 11/2

C. 17/2

D. 23/2

3

$$(ax + 7)(bx - 2) = 6x^2 + kx - 14$$

If the equation above is true for all values of x and $a + b = 5$, what are the two possible values for k?

A. 2 and 3

B. 8 and 17

C. 17 and 21

D. −5 and 40

4

$f(x)$	12	7	22	10	−8
x	−2	−1	0	1	2

$$g(x) = \frac{1}{2}f(3x) + 4$$

Information about functions f and g are provided above. If c is a constant such that $g(c) = 9$, which of the following can be the value of c?

A. 1/3

B. 1

C. 3

D. 10

5 Calculator Allowed

$$p(x) = -3x^2 + 480x - 18000$$

A local coffee roasting company uses the function p to model its weekly profit for selling x kilograms of coffee per week. Which of the following is an equivalent form of $p(x)$ that shows the company's maximum weekly profit as a constant?

A. $p(x) = -3(x + 60)(x + 100)$

B. $p(x) = -3(x - 80)^2 + 1200$

C. $p(x) = -3(x - 60)(x - 100)$

D. $p(x) = -3(x - 80)^2 - 18000$

6

$$x^2 - 4x - 5$$

$$x^2 - 3x + b$$

What is the sum of all possible values of b for which the polynomials above have a root in common?

A. -14

B. -10

C. -4

D. -1

7

$$f(x) = 2(5)^x + 1$$

$$g(x) = 2(5)^x - 6$$

Given that the y-intercept of f occurs at $(0, r)$ and the y-intercept of g occurs at $(0, s)$, what is the value of $r - s$?

A. 1

B. 3

C. 5

D. 7

8

$$f(x) = -\frac{1}{2}x^2 - 4x - k$$

If the function f above has two x-intercepts, which of the following can be the value of the constant k?

A. 7

B. 8

C. 9

D. 10

9 Calculator Allowed

$$h(t) = 30t - 5t^2$$

The height of a projectile in meters t seconds after it is launched can be modeled by the function h above. For how many seconds is the projectile beyond 10 meters in height?

A. $\sqrt{7}$

B. $2\sqrt{7}$

C. 3

D. 6

10

$g(x)$	0	-8	-3	0	-2
x	-2	-1	0	1	2

When $g(x)$ is divided by $x - 2$, what is the remainder?

A. -8

B. -3

C. -2

D. 0

11

$$y = x^2 + 3$$

If the equation above is graphed on the xy-plane, how many units up must it be translated to cross through the point $(1,6)$?

A. 2

B. 4

C. 6

D. 8

12

$$5(cx^4 + 1) = 4x^4 + 5$$

If the equation above is true for all values of x, then what must be the value of the constant c?

A. 1/5

B. 4/5

C. 1

D. 4

13

$$x^2 - x = 3x - 4$$

How many solutions does the equation above contain?

A. Zero

B. One

C. Two

D. More than two

14

A parabola with the equation $y = bx - x^2$, where b is a constant, is graphed on the xy-plane and contains the point $(m, 0)$ located on the negative x-axis. Which of the following can be the value of b?

A. -3

B. 0

C. 2

D. 3

Explanations

Problem Set: Function Properties

1. D

All remainders are different. Use Remainder Theorem by plugging in the zero of the denominator into the numerator.

Zero to the denominator:

$$x - 1 = 0$$

$$x = 1$$

Plug $x = 1$ into the numerator:

$$3(1)^2 - (1) + 2$$

$$3 - 1 + 2$$

$$4$$

2. C

The min or max of a parabola always occurs in the middle of its zeros. The zeros can be found by setting its factors equal to zero.

$$2x + 1 = 0$$

$$x = -\frac{1}{2}$$

$$x - b = 0$$

$$x = b$$

Average the zeros to find the middle. In this case the middle is 4.

$$\frac{-\frac{1}{2} + b}{2} = 4$$

$$-\frac{1}{2} + b = 8$$

$$b = 8.5 \text{ or } 17/2$$

3. B

Expand the left-hand side of the equation to match corresponding coefficients. Then pug in values of a and b that work with the given parameters.

$$(ax + 7)(bx - 2) = 6x^2 + kx - 14$$

$$abx^2 + (7b - 2a)x - 14 = 6x^2 + kx - 14$$

Therefore:

$$ab = 6 \text{ and } a + b = 5$$

$$a = 3. b = 2 \text{ or } a = 2, b = 3$$

Since $k = 7b - 2a$:

$$k = 7(2) - 2(3) = 8$$

$$k = 7(3) - 2(2) = 17$$

4. A

Use the equation of $g(x)$ and the fact that $g(c) = 9$ to begin solving for c.

$$g(c) = \frac{1}{2}f(3c) + 4$$

$$9 = \frac{1}{2}f(3c) + 4$$

$$5 = \frac{1}{2}f(3c)$$

$$10 = f(3c)$$

The function f is 10 only at $f(1)$

$$f(1) = f(3c)$$

$$1 = 3c$$

$$c = 1/3$$

5. B

To identify a max or min in the equation of a quadratic function, vertex form must be used. Only choices B and D are in vertex form. Since this quadratic has a maximum value that occurs off the y-axis, the max (k) must be greater than the y-int of -18000. This leads to choice B.

To find the values without using the answer choices as a clutch, find a, h, and k independently. The value of a is still -3 in vertex form. The value of h is $-b/2a$, which is $-480/(2(-3))$ or 80. Plugging 80 back in to the original equation means that k is 1200.

$$y = a(x-h)^2 + k$$

$$y = -3(x-80)^2 + 1200$$

6. A

Find the roots of $x^2 - 4x - 5$ by factoring.

$$x^2 - 4x - 5 = 0$$

$$(x-5)(x+1) = 0$$

$$x = 5 \text{ or } x = -1$$

Plug these x-values, or roots, into the second quadratic and set equal to zero to find the possible values of the constant b.

Plug in $x = 5$

$$5^2 - 3(5) + b = 0$$

$$b = -10$$

Plug in $x = -1$

$$(-1)^2 - 3(-1) + b = 0$$

$$b = -4$$

Add the two values of b: $-10 + (-4) = -14$

7. D

Since the functions have the same template, we can identify the vertical translation that takes place. In effect, we can see the gap, or difference, between the functions at any given x value.

The difference between 1 and -6 is 7.

*Note: $r = 3$ and $s = -4$

8. A

The discriminant must positive (greater than zero) to satisfy the given condition.

$$b^2 - 4ac > 0$$

$$(-4)^2 - 4\left(-\frac{1}{2}\right)(-k) > 0$$

$$16 - 2k > 0$$

$$16 > 2k$$

$$k < 8$$

9. B

Set up the equation and move all terms to one side to find the difference between zeros from the quadratic formula. Or use a graphing calculator to find the two points of intersection between the quadratic and the line $y = 10$.

$$10 = 30t - 5t^2$$

$$5t^2 - 30t + 10 = 0$$

$$t^2 - 6t + 2 = 0$$

The quadratic formula yields the two solutions as the irrational expressions $3 + \sqrt{7}$ and $3 - \sqrt{7}$. The difference between them is $2\sqrt{7}$.

10. C

Find the zero of the factor to get the x-value to use in the table. The value of the function is equal to the remainder at that point.

$$x - 2 = 0$$

$$x = 2$$

$$\text{Remainder: } g(2) = -2$$

11. A

Apply vertical shift by adding a constant to the current equation and ensuring it works for the desired point (1,6) where $x = 1$ and $y = 6$.

$$y = x^2 + 3 + a$$

$$6 = 1 + 3 + a$$

$$6 = 4 + a$$

$$a = 2$$

Another way to think about this is that, in the given equation, the y-value is 4 when $x = 1$, leading to the coordinate pair (1,4). This point must be shifted up 2 units to get to (1,6).

12. B

Expand the left-hand side, then match corresponding coefficients.

$$5(cx^4 + 1) = 4x^4 + 5$$

$$5cx^4 + 5 = 4x^4 + 5$$

$$5c = 4$$

$$c = 4/5$$

13. B

Set the equation equal to zero, then solve or use the discriminant.

$$x^2 - x = 3x - 4$$

$$x^2 - 4x + 4 = 0$$

$$(x - 2)^2 = 0$$

$$\text{One factor } = \text{ one solution}$$

14. A

The equation $y = bx - x^2$ can be factored to reach the form $y = x(b - x)$ that displays the factor x, which means there is an x-intercept at the origin. Since the other x-intercept is along the negative x-axis, then the x-coordinate of the vertex h must also be negative because it occurs between the two x-intercepts. The formula $h = -b/(2a)$ can be used to locate the x-coordinate of the vertex. From the given equation of the parabola $y = bx - x^2$ we can see that $a = -1$ because that is the coefficient of the x^2 term. To keep the overall value of h negative, b must also be negative. Therefore, $b = -3$ could work for the described parabola.

We can also use the factor $(b - x)$ to determine the sign of the constant b. This factor gives the second x-intercept at $(m, 0)$. When we plug in m, the factor equals zero. Solve for b in that equation and keep in mind that m is on the negative x-axis. We will see that $b = m$, and since m is negative, so is b.

$$b - x = 0$$

$$b - m = 0$$

$$b = m$$

2.4 ADVANCED FUNCTIONS AND EQUATIONS

a. Domain and Range

The domain of a function is the complete set of values of the independent variable and the range is the complete set of values of the dependent variable. In other words, domain and range refer to all possible values of x and y, respectively. In addition to knowing what values may be included in the domain and range of a particular function, it is equally important to understand when values must be restricted from these sets. In this section, we will review operations that contain restricted values, leading to functions with limited domains and ranges.

Domain

As noted in the introduction, domain refers to all possible x-values of a function. For example, polynomial functions have a domain of all real numbers: there are no necessary restrictions to the x-variable. So, when would a function have restrictions? The most important times when x would have to be restricted is when it appears in a denominator, under an even root, or inside any other restrictive operation, like logarithms and some trigonometric functions. However, the current test mainly focuses on the denominators and even roots when testing students on domain and range. Use Table 2.41 to review these two important operations that have restrictions. In short, denominators cannot be zero and radicands of even roots cannot be negative.

Table 2.41 | Operations with Restrictions

Denominators		Even Roots	
$\frac{1}{f(x)}$	Restriction $f(x) \neq 0$	$\sqrt{f(x)}$	Restriction $f(x) \geq 0$

Range

Range refers to all possible y-values of a function. Odd degree polynomials have a range of all real numbers. Even degree polynomials will have either a minimum or maximum, as is the case when working with quadratics. Many polynomial functions will also have relative maxima and minima, which look like the hills and valleys of the function; however, several operations have a minimum value of – or approaching – zero. Those are even degree, even roots, absolute values, and exponentials. Therefore, the smallest output of an even degree, even root, absolute value operation, and exponential operation is or approaching zero. This idea is best conveyed by reviewing the graphs of the parent functions to each of these operations and identifying each function's range of values. Note the shape of each function in Table 2.42 and the lowest y-value it either reaches or approaches. In short, each function is either on or above the x-axis at any given point. We can see this clearly with the quadratic, square root, and absolute value functions because each contains the origin at its lowest point. The exponential function approaches the x-axis as it moves to the left of the coordinate plane.

Table 2.42 | Lowest Output Operations

Even Root

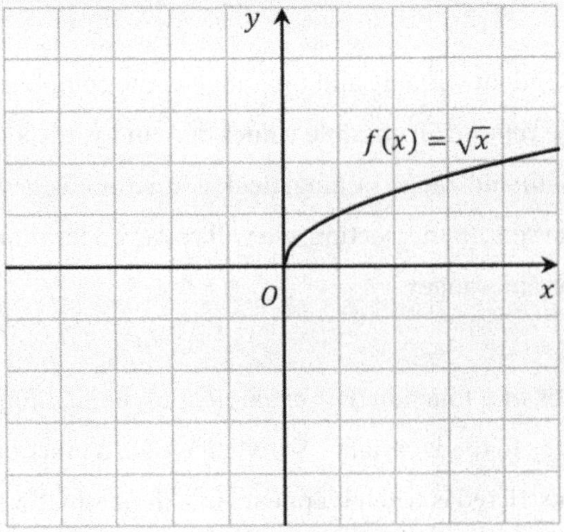

Domain	Range
$[0, \infty)$	$[0, \infty)$

Absolute Value

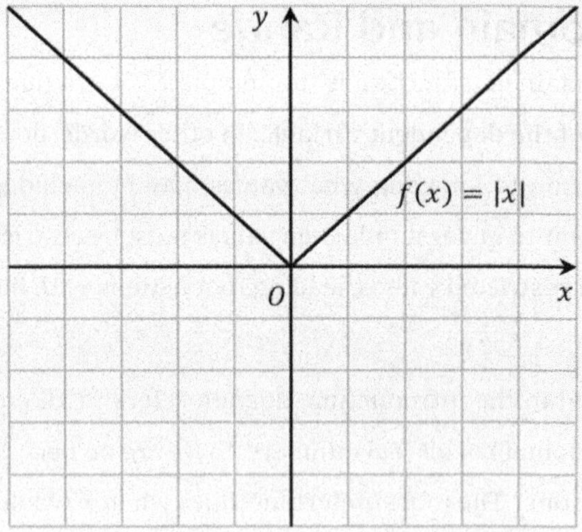

Domain	Range
$(-\infty, \infty)$	$[0, \infty)$

Even Degree

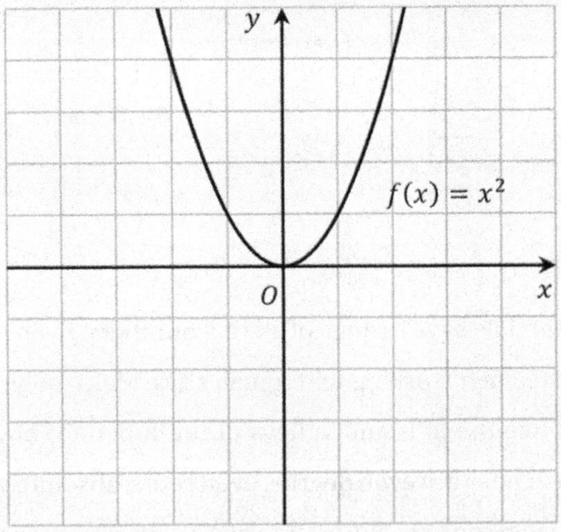

Domain	Range
$(-\infty, \infty)$	$[0, \infty)$

Exponential

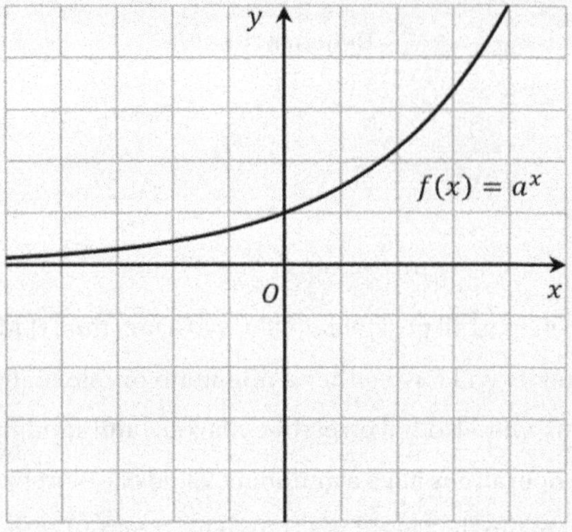

Domain	Range
$(-\infty, \infty)$	$(0, \infty)$

In the example that follows, we will see how the test challenges us to determine an element of the domain for a function that contains the two main operations we highlighted in our discussion on the domain and the restrictions to it. The example comes from the Diagnostic Exam, Calculator Question 16.

Q+A

Restrictions to Domain **Diagnostic Exam | Calculator Q16**

$$f(x) = \frac{1}{\sqrt{x-3}-2}$$

The function above is graphed on the xy-plane, where $y = f(x)$. The curve that forms function f contains the coordinate pair (m, n). If m and n are constants, which of the following can be a value of m?

A. 0

B. 2

C. 6

D. 7

Workspace

Solution Outline

Step 1: Understand that the question asks about the domain. Since the question asks for a possible value of m, which is acting as an x-coordinate, then m must be in the domain. The given function contains both instances where x values must be restricted, since it occurs in the denominator AND under an even root. Therefore, not only does the value inside the radical must be greater than zero, but the denominator as a whole cannot be zero.

Step 2: Find the restrictions. From the restrictions found, m can be any number greater than or equal to 3, except 7.

Radical: $x - 3 \geq 0$

$\therefore x \geq 3$

Denominator: $\sqrt{x-3} - 2 \neq 0$

$\therefore x \neq 7$

Answer: C

Notes

b. Equations and Systems

This section will take us through in more detail the equations and systems of advanced operations. To start, we will analyze the graphs of functions to highlight the fact that solutions occur at points of intersection. Then we will review the solving strategies of absolute value and square root operations, as well as advanced systems of equations. In the example that follows, we will determine the number of solutions to a system when a graph is provided with the question.

Q+A

Solutions Occur at Intersections **Diagnostic Exam | Calculator Q33**

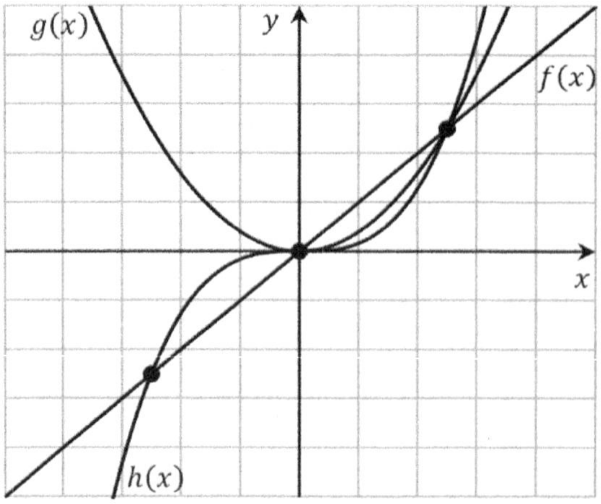

The functions f, g, and h are graphed on the xy-plane above. Given that all points of intersection are marked, how many possible values of the constant a ensure that $f(a) = g(a) = h(a)$?

Workspace

Solution Outline

Step 1: Find the number of intersections that include all three functions. Though there are three intersections labeled in the graph, only two of those intersections include all three functions.

Answer: 2

Notes

Absolute Value

The absolute value operation is the mathematical representation of a real number's magnitude, or distance from zero on the number line. Because of this, the output of an absolute value is always non-negative. That is to say that the output may be zero or positive. Consequently, univariate equations where the variable is contained in an absolute value operation can have up to two solutions. Let's consider the equation $|x| = 6$. This implies that x can equal positive or negative six because both values are six units away from zero on the number line. In practical terms, this means that to solve these types of equations, we will have to set what is inside the absolute value equal to both the positive and negative value of the output, as long as the output is non-negative. If the output of the absolute value is negative, for example in the case of $|x| = -3$, there is no solution because the rule of absolute values has been violated.

Q+A

Absolute Value **Diagnostic Exam | Calculator Q34**

$$f(x) = |x - 2|$$

$$g(x) = \frac{1}{2}x$$

Functions f and g are provided above and each contain the domain of all real numbers. When the functions are graphed on the xy-plane, they both contain the coordinate points (a, b) and (c, d) where $a, b, c,$ and d are rational constants. If b is less than d, what is the value of b?

Workspace

Solution Outline

Strategy 1

Step 1: Solve algebraically by setting the functions equal to each other to find the x-values at which both functions have the same output.

$$f(x) = g(x)$$

$$|x - 2| = \frac{1}{2}x$$

Step 2: To solve equations containing absolute values, first isolate the absolute value operation to one side of the equation – it already is – then set what is inside the absolute value equal to the output and its negative. Check solutions to ensure the equation works for that value.

$$x - 2 = \frac{1}{2}x \quad \text{and} \quad x - 2 = -\frac{1}{2}x$$

$$x = 4 \quad \text{and} \quad x = \frac{4}{3}$$

Step 3: The x-values found in the previous step correspond to the x-coordinates a and c. Plugging these values into either function yields their outputs: b and d. When $x = 4$, the functions equal 2. However, when $x = 4/3$, the functions equal $2/3$. Since $b < d$, b must equal $2/3$.

Answer: $2/3$

Strategy 2

Step 1: Use graphing calculator to plug $f(x)$ into $Y1$ and $g(x)$ into $Y2$.

Step 2: Use the calculator to find the x-value of each intersection. In a TI-83/84 graphing calculator with the graph displayed, press 2nd, Trace, 5) Intersect and follow calculator instructions to find the lower intersection. This x-value is $4/3$ and has a y-coordinate of $2/3$.

Answer: $2/3$

Notes

Square Root

The square root is another mathematical operation that we are expected to have fluency with for this test. Just like the absolute value operation, the output to a square root is always non-negative; however, the square root operation also has the restriction that its contents be non-negative. That is the minimum input of the square root is zero, just like its minimum output. In solving equations where the input variable is contained inside a square root, we should isolate the square root to one side if it is not already, ensure that the output is non-negative if possible, square both sides, then continue solving for the variable. There may be extraneous solutions when solving these types of equations, so we must check our answers at the end. Although, this is sometimes an extensive process, we are in luck if we have answer choices to choose from. We can simply check to see which answer choice has the full solution set by plugging the listed values into the equation one at a time to verify if it is valid or not.

Q+A

Square Roots **No Calculator**

$$\sqrt{2x+9} = x+5$$

The equation above is used to find the solution set A for the discrete variable x. Which of the following represents A, provided that it contains all values of x that satisfy the given equation?

A. $\{-6\}$

B. $\{-4\}$

C. $\{-4,1\}$

D. $\{-4,0,1\}$

Workspace

Solution Outline

Strategy 1

Step 1: Use answer choices, understanding that the output of the square root operation is always positive and the correct answers often create perfect squares inside the square root.

Step 2: Plug in. Choice A does not work because -6 makes the right side of the equation negative. Choices C and D do not work because 1 does not create a perfect square inside the radical.

Answer: B

Strategy 2

Step 1: Square both sides to get rid of the radical. Note, only do this once the radical is isolated to one side. Then solve the equation.

$$2x + 9 = x^2 + 10x + 25$$

$$0 = x^2 + 8x + 16$$

$$0 = (x + 4)^2$$

$$x = -4$$

Step 2: Verify the $x = -4$ is a solution by plugging it back into the given equation. Since it satisfies the equation, it is part of the solution set.

Answer B

Notes

Advanced Systems of Equations

To algebraically solve advanced systems of equations, we will use two strategies: substitution and elimination. These are the same strategies we observed and practiced in Section 1.2 Linear Systems – though, now the given equations will contain more advanced operations and terms. Recall that substitution involves solving for a variable or expression in one equation and using its value in the other to solve a univariate equation; elimination requires adding or subtraction two equations together to cancel out one of the variable terms. With these strategies in place, we can determine both the number of solutions and the exact solution sets for any system in question on the test. The subsequent example shows how we can find the number of solutions from an advanced system of equations.

Q+A

Advanced System of Equations

No Calculator

$$y^2 = x + 1$$

$$x^2 + y^2 = 13$$

The equations of the system above are graphed on the xy-plane. How many unique points of intersection are made between the two curves?

A. 0

B. 1

C. 2

D. 3

Workspace

Solution Outline

Step 1: Since one of the variable terms is already solved for, use substitution. We can write in the expression $x + 1$ instead of y^2 in the second equation. The result is a quadratic equation that can be solved through factoring.

$x^2 + x + 1 = 13$

$x^2 + x - 12 = 0$

$(x + 4)(x - 3) = 0$

$x = -4$ and 3

Step 2: Find the number of coordinate pairs that can be made with these two x-values by plugging them into the first equation. When we plug in $x = -4$, we see that $y^2 = -3$, which is not possible because squared values are always positive. Therefore, no point of intersection occurs at $x = -4$. However, when we plug in $x = 3$, we find that $y^2 = 4$, so y can equal 2 or -2. This means there are two unique points of intersection: $(3, -2)$and $(3,2)$.

Answer: C

Notes

Problem Set: Advanced Functions and Equations

11 Questions

1

The function f has a domain that contains some real number c such that $f(c) = -1$. Which of the following functions of f satisfies the outlined constraint?

A. $f(x) = |x - 2|$

B. $f(x) = (x - 2)^4$

C. $f(x) = 2^{x-2}$

D. $f(x) = \sqrt{x} - 2$

2

$$f(x) = \sqrt{x - 4}$$

The function f is described above and contains only real numbers in its domain and range. If the domain of f contains the integer a, which of the following CANNOT be the value of a?

A. 3

B. 4

C. 5

D. 6

3

$$f(x) = x^3$$

$$g(x) = x$$

The functions f and g are provided above and contain the domain of all real numbers. For how many values of c will $f(c) = g(c)$?

A. Zero

B. One

C. Two

D. More than two

4

$$x^2 + y^2 = 5$$

$$x^2 - y^2 = 3$$

The equations of the system above are graphed on the xy-plane. How many unique points of intersection are made between the two curves?

A. 2

B. 3

C. 4

D. 5

5

$$f(x) = \frac{1}{(x-3)^2} + 2$$

The function f above contains the point (a, b). Which of the following CANNOT be a value of $f(a)$?

A. 2

B. 3

C. 4

D. 5

6

Function f contains the point (a, b) and function g contains the point (c, d). Provided that $f(a) = g(c)$, then which of the following must be true?

A. $a = b$

B. $b = d$

C. $a = c$

D. $c = d$

7

$$f(x) = 2x + k$$

$$g(x) = (x-1)^2$$

For what value of k will the functions f and g share an x-intercept?

A. −2

B. −1

C. 0

D. 1

8

$$f(x) = x^3 + ax - 1$$

$$g(x) = x^4 + x^3 + bx + 4$$

Assume a and b are constants. If $f(2) = g(-1)$, what is the value of $2a + b$?

A. −5

B. −3

C. 3

D. 5

9

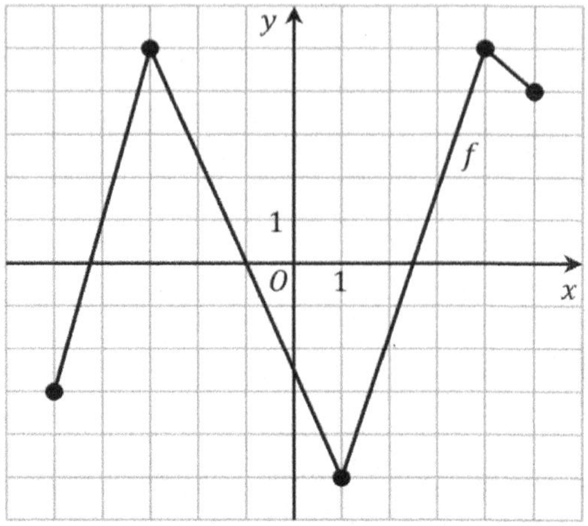

The function f is shown in the xy-plane above. For how many values of c does $f(c) = 4$?

A. 1

B. 2

C. 3

D. 4

Questions 10 and 11 refer to the following information.

Dennis throws a baseball upward with an initial velocity of 20 meters per second (m/s). The equations below describe the motion of the baseball under constant acceleration, where v_0 is the initial velocity of the baseball, v_f is the final velocity of the baseball after t seconds, y is the height of the baseball above the ground, and a is the acceleration due to gravity ($-10\ m/s^2$) .

$$v_f = v_0 + at$$

$$y = v_0 t + \frac{1}{2}at^2$$

$$v_f^2 = v_0^2 + 2ay$$

10 Calculator Allowed

To the nearest integer, what speed in meters per second will the baseball have the moment it reaches it maximum height?

A. 0

B. 10

C. 15

D. 20

11 Calculator Allowed

To the nearest tenth of a second, how long will it take the baseball that Dennis threw to reach the ground the after it reaches its maximum height?

A. 2.0

B. 2.5

C. 4.0

D. 4.5

Explanations

Problem Set: Advanced Functions and Equations

1. D

Absolute values, even degrees, even roots, and exponentials have lowest outputs of or approaching zero. Plug in zero for these operations and simplify. Choose answer choice that leaves -1 or lower.

$$\sqrt{x} - 2$$

$$0 - 2$$

$$-2$$

2. A

Even roots often have restricted values since the radicand cannot be negative. To find acceptable values, set what is under the radical greater than or equal to four.

$$x - 4 \geq 0$$

$$x \geq 4$$

Three is not in the domain because it creates a negative under the radical.

3. D

Set functions equal to each other, then bring all terms to one side of the equation and factor to determine the number of solutions.

$$x^3 = x$$

$$x^3 - x = 0$$

$$x(x^2 - 1) = 0$$

$$x(x + 1)(x - 1) = 0$$

$$3 \text{ factors} = 3 \text{ solutions}$$

4. C

Solve the system by elimination to determine the number of unique solutions.

$$x^2 + y^2 = 5$$

$$+ \quad x^2 - y^2 = 3$$

$$2x^2 = 8$$

$$x^2 = 4$$

$$x = \pm 2$$

$$y = \pm 1$$

Four solutions: $\{(2,1), (-2,1), (2,-1), (-2,-1)\}$

5. A

The fraction cannot equal zero because the numerator is always one. The value of a fraction can only be zero if the numerator is zero. This means that:

$$f(a) \neq 0 + 2$$

$$f(a) \neq 2$$

*The restriction to the domain is $x \neq 3$. Therefore, $a \neq 3$. Also note that $f(a) = b$

6. B

The outputs of each function must be equal. The output of $f(a)$ is b and the output of $g(c)$ is d. Thus, $b = d$.

7. A

To find x-intercepts, set the function equal to zero.

$$g(x) = (x - 1)^2$$

$$0 = (x - 1)^2$$

$$0 = x - 1$$

$$x = 1$$

Use the point (1,0) in $f(x)$ to solve for k.

$$f(x) = 2x + k$$

$$0 = 2(1) + k$$

$$k = -2$$

8. B

Use each function's equation as a template to plug 2 into f and -1 into g. Set the functions equal to each other and find the desired expression $2a + b$.

$$f(2) = g(-1)$$

$$(2)^3 + 2a - 1 = (-1)^4 + (-1)^3 - b + 4$$

$$8 + 2a - 1 = -b + 4$$

$$2a + 7 = -b + 4$$

$$2a + b = -3$$

9. D

Sketch the horizontal line $y = 4$. The function and the line touch four times.

10. A

Before the maximum height, the ball is on its way up, so velocity is positive. After the maximum height, the ball is on its way down, so its velocity is negative. Therefore, at its maximum height, the ball must have a velocity of ZERO.

*Note: 20 is the maximum height

11. A

Note that the time required for the baseball to go from ground level to the top of the trajectory is the same time required for it to reach the ground from the apex. The equations show that the ball has a height of zero when it is released, so it must be starting at ground level. Use the first equation listed and plug in all known information. The final velocity v_f is zero from the previous problem, the initial velocity v_0 is 20 and the acceleration, a, is -10.

$$v_f = v_0 + at$$

$$0 = 20 - 10t$$

$$10t = 20$$

$$t = 2.0$$

ADVANCED MATHEMATICS EXAM

DIRECTIONS

For questions 1-15, solve every problem and either circle or record the best answer from the answer choices provided. **For questions 16-20**,solve every problem and record your answer in the appropriate format. On the real test, you must grid in your answers according to the directions. You may use any available space in your workbook for scratch work.

NOTES

1. Calculator usage is NOT allowed in Part 1.
2. Assume real numbers unless otherwise indicated.
3. Figures may not be drawn to scale.
4. All figures are planar, unless otherwise indicated.
5. Domains represent all real numbers of for which f(x) is also a real number.

REFERENCE

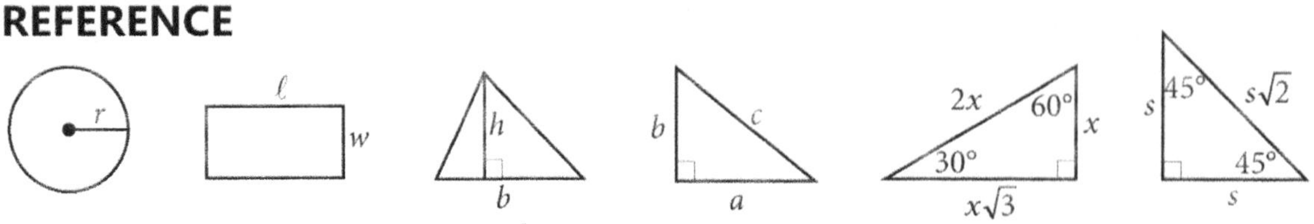

$A = \pi r^2$ $C = 2\pi r$ | $A = \ell w$ | $A = \frac{1}{2}bh$ | $c^2 = a^2 + b^2$ | Special Right Triangles

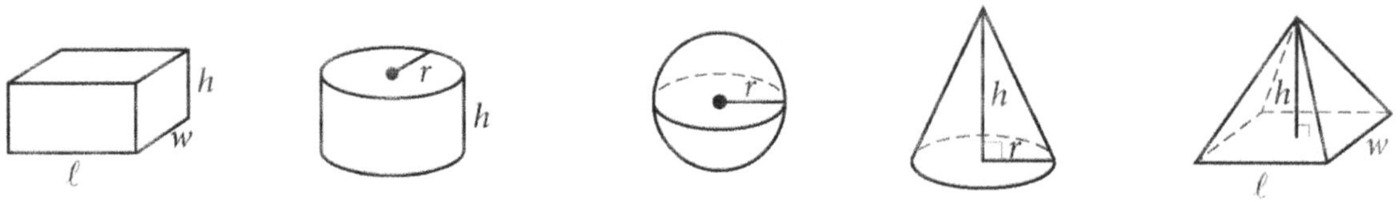

$V = \ell wh$ | $V = \pi r^2 h$ | $V = \frac{4}{3}\pi r^3$ | $V = \frac{1}{3}\pi r^2 h$ | $V = \frac{1}{3}\ell wh$

The arc of a circle contains 360 degrees.

The arc of a circle contains 2π radians.

The interior angles of a triangle add up to 180 degrees.

Part 1: No Calculator

19 Minutes, 15 Questions

1

Newton's Law of Cooling can be used to model the temperature of an object as it cools due to the cooler temperature of its surroundings.

$$T(t) = A + Be^{-kt}$$

$T(t)$ is the temperature of the object after t seconds, A is the temperature of the surroundings, and B is the difference in temperature between the warmer object and its cooler surroundings. Which of the following must be true?

A. The initial temperature is equal to $A + B$

B. The final temperature $A - B$

C. A must be larger than B

D. A must be less than B

2

$$\sqrt{x-2} = 2x - 5$$

What is the solution set to the equation above?

A. $\left\{\frac{9}{4}\right\}$

B. $\{3\}$

C. $\left\{\frac{9}{4}, 3\right\}$

D. $\left\{0, \frac{9}{4}, 3\right\}$

3

$$\frac{2^{m^2}}{2^{n^2}} = 8^4$$

$$m + n = 6$$

Provided the equations above, what is the difference between m and n?

A. 2

B. 4

C. 6

D. 8

4

$$f(x) = x^2 - 64$$

$$g(x) = ax^2 + 64$$

For what value of the constant a will $f(x)$ and $g(x)$ have the same roots?

A. -2

B. -1

C. 1

D. 2

5

$$p(x) = a(x-3)^2(x+1)$$

When $p(x)$ is divided by $x - 2$, the remainder is -15. What is the value of the constant, a?

A. -5

B. 5

C. $-3/4$

D. $3/4$

6

$$f(x) = 2(x^2 - 2x + 8) - 4(x + k)$$

In the function f above, k is a constant. If $f(x)$ is divisible by x, what is the value of k?

A. 0

B. 2

C. 4

D. 6

7

Which of the following is an equivalent form of the function $f(x) = x^2 - 6x + 5$, from which the coordinates of its vertex can be identified as constants in the equation?

A. $f(x) = (x-5)(x-1)$

B. $f(x) = x(x-6) + 5$

C. $f(x) = (x-3)^2 - 4$

D. $f(x) = (x+3)^2 - 4$

8

$$\rho = mv$$

Momentum ρ can be calculated by multiplying an object's mass m in kilograms by its velocity v in meters per second, as described by the equation above. The graph below represents the momentum of an object with constant mass given changes to its velocity.

(1 kilogram = 1000 grams)

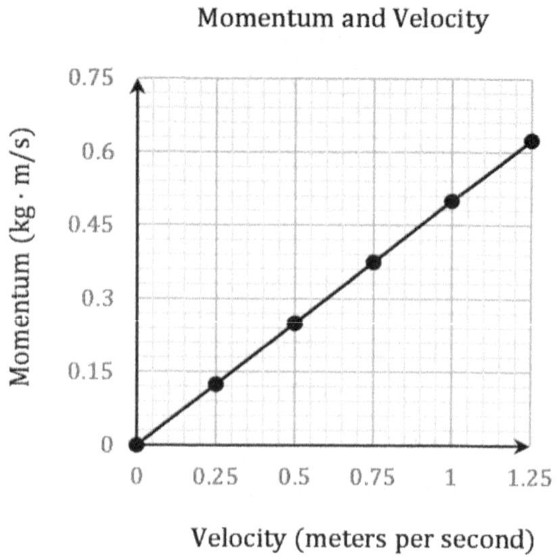

From the graph, which of the following can be inferred?

I. Momentum is directly proportional to velocity

II. The slope of the line above represents the object's mass

III. The mass of the object represented in the graph is 500 grams

A. I only

B. III only

C. II and III

D. I, II, and III

9

$$(7x-1)^2-(7x-1)$$

Which of the following is an equivalent form of the expression above?

A. $(7x-1)(x-2)$

B. $(7x-1)(x+1)$

C. $(7x-1)(7x-2)$

D. $(7x-1)(7x+1)$

10

The polynomial function p has zeros at $x=-1$ and $x=2$. If $p(x) \geq 0$ for all values of $x<2$, which of the following can be $p(x)$?

A. $p(x)=(x+1)(x-2)$

B. $p(x)=(x+1)^2(x-2)$

C. $p(x)=-(x+1)(x-2)$

D. $p(x)=-(x+1)^2(x-2)$

11

$$\frac{3}{2}x^2-8x=-5$$

What is the sum of the solutions to the equation above?

12

$$5x^2+(p+q)x-2=ax^2+9x-2$$

The equation above is true for all values of x. What is the value of $a(p+q)$?

13

$$g(x) = f(cx)$$

If the graph of function g is the graph of function f stretched horizontally by a factor of 2, what is the value of the constant c?

14

$$f(g(x)) = x + 4$$

$$g(x) = 4x - 1$$

Using the functions above, what is the value of $f(3)$?

15

$$f(x) = (x - 1)^3 + 6$$

$$g(x) = (x - a)^3 + b$$

If the function g is obtained by shifting the function f left 4 units and down 1 unit, what is the value of $a + b$?

Part 2: Calculator

19 Minutes, 15 Questions

Questions 1-3 refer to the graph below

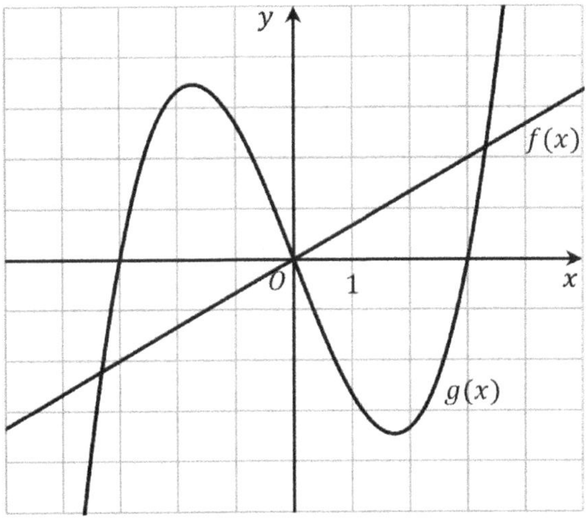

1

For how many values of x does $f(x) = g(x)$?

A. 0

B. 1

C. 2

D. 3

2

$$f(g(c)) = 0$$

What value of the constant c would satisfy the equation above?

A. −2

B. −1

C. 2

D. 3

3

$$g(x) > f(x)$$

Which interval satisfies the inequality above?

A. $(-4, -2)$

B. $(-3, -1)$

C. $(0, 2)$

D. $(2, 4)$

4

The polynomial expression $3x^2 - 1$ is rewritten in the form $3(x + b)(x - b)$. Which of the following is the value of the positive constant b?

A. 1

B. $\sqrt{3}/3$

C. $\sqrt{3}$

D. 3

5

$$x = 3^a$$

$$y = 2^b$$

Using the equations above, which of the following is 72^{ab} equivalent to?

A. x^2y^3

B. x^ay^b

C. $x^{3a}y^{2b}$

D. $x^{2b}y^{3a}$

6

The price P in dollars for good A depends on the number of units, x, sold according to the equation below.

$$P = 52 - 2x$$

Total revenue is calculated by multiplying the price of the good by the number of units sold. For how many dollars should good A be sold to maximize total revenue?

A. 13

B. 26

C. 34

D. 40

7

$$P(t) = 1000(2)^{\frac{t}{20}}$$

The population size P of a species is described above as a function of time t years after 2005. Which of the following statements is true?

A. The population size increases by 100% from 2005 to 2025

B. The population size increases by 200% from 2005 to 2025

C. The population size increases by 100% every year

D. The population size increases by 200% every year

8

$$f(x) = 2x^2 - 3x - 9$$

$$g(x) = x^2 + bx - 10$$

In the functions above, b is a constant. If f and g share an x-intercept, which of the following is a possible value of b

A. $-3/2$

B. $1/6$

C. $1/3$

D. 3

9

$$\frac{H - 2C}{5} = kC + 2$$

Using the equation above, what is C equal to in terms of H and k?

A. $C = \frac{H}{5k} - \frac{2}{3}$

B. $C = \frac{H - 10}{5k + 2}$

C. $C = \frac{H}{5} - 2$

D. $C = \frac{H - 2}{5k}$

10

$$f(x) = x + 17$$

$$g(x) = \frac{1}{2}(x - 4)^2 + 17$$

What is one possible value of the constant a such that $f(a) = g(a)$?

A. 3

B. 5

C. 6

D. 8

11

$$y = x^2 - 4x + 5$$

$$y = 5 - x$$

If the ordered pair (x, y) satisfies the system above, what is a possible positive value of y?

12

$$f(x) = x^2 + 11$$

$$g(x) = -x^2 + 9x + 1$$

Given that $f(a) = g(a)$ and $f(b) = g(b)$ in the system above, what is the magnitude of the product of a and b?

13

$$f(x) = 2\sqrt{x-3} + 9$$

Function f is given above. If its range is $[a, \infty)$, what is the smallest value of the constant a?

14

$$x = \sqrt[5]{b}$$

$$2b = 64$$

Considering the equations above, what is the value of x?

15

$$\frac{f(x)}{x-a} = p(x) + \frac{r}{x-a}$$

If $f(x) = x^3 - 3x^2 - 5x + 15$ and $r = 0$, what is one possible positive and rational value of a?

Explanations

Advanced Mathematics Exam: Part 1

1. **A**

The expression for the initial temperature can be found by plugging in $t = 0$.

$$T(0) = A + Be^{-k(0)}$$

$$T(0) = A + Be^0$$

$$T(0) = A + B$$

The final temperature would be A not $A - B$, and there is no relationship between the surrounding temperature A and the difference in temperatures B to justify choices C and D.

2. **B**

The fastest method to solve this question is by plugging in the values of x listed in the answer choices to verify if they actually work for the given equation. Remember, negative numbers cannot be square rooted and square roots cannot give negative answers. The value, $9/4$, is not correct because it leaves the right side of the equation negative. This means choices A, C, and D are not correct. Only choice B satisfies the equation.

3. **A**

Use properties of exponents to rewrite each side of the equation so they both have the same base.

$$\frac{2^{m^2}}{2^{n^2}} = 8^4$$

$$2^{m^2-n^2} = (2^3)^4$$

$$2^{m^2-n^2} = 2^{12}$$

$$m^2 - n^2 = 12$$

$$(m+n)(m-n) = 12$$

$$6(m-n) = 12$$

$$m - n = 2$$

4. **B**

For the two functions to share the same roots, there equations must both equal zero at the same x value.

$$f(x) = x^2 - 64$$

$$0 = x^2 - 64$$

$$x^2 = 64$$

$$x = \pm 8$$

$$g(x) = ax^2 + 64$$

$$0 = a(\pm 8)^2 + 64$$

$$0 = 64a + 64$$

$$-64a = 64$$

$$a = -1$$

5. A

The Remainder Theorem describes that the value of a function is equal to the remainder at the given point. The x-value that makes the factor $x - 2$ zero is 2. Therefore, $p(2) = -15$.

$$p(x) = a(x-3)^2(x+1)$$

$$-15 = a(2-3)^2(2+1)$$

$$-15 = 3a$$

$$a = -5$$

6. C

The Remainder Theorem describes that the value of a function is equal to the remainder at the given point. If a function is divisible by a factor, the remainder is zero. The value that makes the factor of x zero is, in this case, also 0. Therefore, $f(0) = 0$

$$f(x) = 2(x^2 - 2x + 8) - 4(x + k)$$

$$0 = 2((0)^2 - 2(0) + 8) - 4(0 + k)$$

$$0 = 2(8) - 4k$$

$$4k = 16$$

$$k = 4$$

7. C

Since the coordinates of the vertex must be identified as constants in the final equation, only equations in vertex form can be correct. Therefore, only choices C and D are possible.

General Form: $f(x) = ax^2 + bx + c$

Vertex Form: $f(x) = a(x-h)^2 + k$

$a = a$	$h = -b/2a$	$k = f(h)$
$a = 1$	$h = 3$	$k = -4$

$$f(x) = (x-3)^2 - 4$$

8. D

Momentum is proportional to velocity because the graph displays a linear relationship that goes through the origin. That is, it is an equation that takes the form $y = kx$. The slope of the velocity vs momentum graph represents the mass because rise over run translates to momentum over velocity. And since the graph has a slope of 0.5, the mass of the object is 0.5 kilograms, or 500 grams.

9. C

Factor the expression by first seeing that $7x - 1$ is a repeated expression. To see the factoring more clearly, make the substitution $u = 7x - 1$.

$$(7x-1)^2 - (7x-1)$$

$$u^2 - u$$

$$u(u-1)$$

$$(7x-1)\big((7x-1)-1\big)$$

$$(7x-1)(7x-2)$$

10. D

One way to approach this question is to plug in values before, between, and after each zero to find the intervals where the function is positive and negative. Only choice D contains an equation for $p(x)$ that is positive at all values of x that are less than 2. Easy values of x to plug into the listed functions include 0 and -2.

Choice A: $p(0)$ is negative

Choice B: $p(0)$ is negative

Choice C: $p(-2)$ is negative

Choice D: $p(-2)$ is positive and $p(0)$ is positive

11. 16/3

Sum of the solutions to a quadratic is equal to $-b/a$ when $ax^2 + bx + c = 0$.

$$\frac{3}{2}x^2 - 8x = -5$$

$$\frac{3}{2}x^2 - 8x + 5 = 0$$

$$3x^2 - 16x + 10 = 0$$

$$a = 3 \text{ and } b = -16$$

$$-\frac{b}{a} = -\frac{-16}{3} \text{ or } \frac{16}{3}$$

12. 45

Since the equation is true for all values of x, corresponding terms must be identical. For the x^2 terms on each side of the equation to match, a must equal 5 and, for the x-terms to match, $(a + b)$ must equal 9. Therefore, $a(p + q) = 5(9) = 45$.

13. 1/2

The coefficient c should be the reciprocal to the factor of the horizonal stretch. The reciprocal of 2 is $1/2$.

14. 5

Since f is shown as a function of $g(x)$, first find the x value that would make $g(x)$ equal 3.

$$g(x) = 4x - 1$$

$$3 = 4x - 1$$

$$4 = 4x$$

$$x = 1$$

Therefore, to find $f(3)$, plug in $x = 1$.

$$f(3) = 1 + 4 \text{ or } 5$$

15. 2

Horizontal shifts are affected by addition and subtraction directly to x whereas vertical shifts are affected by addition and subtraction outside of the primary operation. Recall that horizontal shifts to the left are found through addition and vertical shifts down are found through subtraction. Therefore, to shift the function f leftward by adding 4 to -1 and downward by subtracting 1 from 6.

$$g(x) = f(x + 4) - 1 = (x + 4 - 1)^3 + 6 - 1$$

$$g(x) = (x + 3)^3 + 5$$

$$a = -3 \text{ and } b = 5$$

$$a + b = 2$$

Explanations

Advanced Mathematics Exam: Part 2

1. D

The curves intersect at three distinct points, and, therefore, the equation $f(x) = g(x)$ has three solutions.

2. D

The function f is only equal to zero when $x = 0$. Therefore $g(c)$ must equal zero. The function g is equal to zero at $x = -3, 0, 3$.

3. B

The function g is above the function f for all x values between -3 and -1.

4. B

Factor the expression by first taking out a three from both terms, then treating the result as a difference of squares.

$$3x^2 - 1$$

$$3\left(x^2 - \frac{1}{3}\right)$$

$$3\left(x^2 - \frac{3}{9}\right)$$

$$3\left(x + \frac{\sqrt{3}}{3}\right)\left(x - \frac{\sqrt{3}}{3}\right)$$

5. D

Use properties of exponents to rewrite the given expression using the given values of x and y.

$$(72)^{ab}$$

$$(9 * 8)^{ab}$$

$$(3^2 * 2^3)^{ab}$$

$$3^{2ab} * 2^{3ab}$$

$$(3^a)^{2b}(2^b)^{3a}$$

$$x^{2b}y^{3a}$$

6. B

Total revenue is a quadratic function. To maximize the total revenue, locate the vertex by using the formula $h = -b/2a$. This value of h represents the number of units that must be sold to maximize total revenue. Use the x-value to find the price.

$$\text{Total Revenue} = \text{Price } \times \text{Quantity}$$

$$\text{Total Revenue} = P \times (x)$$

$$\text{Total Revenue} = (52 - 2x)(x)$$

$$\text{Total Revenue} = 52x - 2x^2$$

$$h = -\frac{52}{2(-1)} \text{ or } 13$$

$$P = 52 - 2(13)$$

$$P = 26$$

7. A

The base of the exponent is 2 and the exponent itself reaches integer values when t is a multiple of 20. Since the base of 2 deviates from one by 1, it corresponds to a 100% increase with every next whole number exponent. Another way to view the base of 2 is that the population doubles, which is a 100% increase, not 200%. Therefore, answer choices B and D are incorrect. Answer choice C is incorrect because it provides that the 100% increase occurs every year, when in fact it would take twenty years to according to the model.

8. C

Find the zeros of $f(x)$, then apply them to $g(x)$ to find the possible values of the constant, b.

$$f(x) = 2x^2 - 3x - 9$$

$$0 = 2x^2 - 3x - 9$$

$$0 = (2x + 3)(x - 3)$$

$$x = -\frac{3}{2} \text{ or } 3$$

$$g(x) = x^2 + bx - 10$$

$$0 = (3)^2 + b(3) - 10$$

$$0 = 9 + 3b - 10$$

$$0 = 3b - 1$$

$$-3b = -1$$

$$b = 1/3$$

9. B

Expand out the equation and bring all terms with C to one side. Factor out C and solve.

$$\frac{H - 2C}{5} = kC + 2$$

$$H - 2C = 5kC + 10$$

$$H - 10 = 5kC + 2C$$

$$H - 10 = C(5k + 2)$$

$$C = \frac{H - 10}{5k + 2}$$

10. D

Set $f(a) = g(a)$ and solve the quadratic through factoring.

$$f(a) = g(a)$$

$$a + 17 = \frac{1}{2}(a - 4)^2 + 17$$

$$a = \frac{1}{2}(a - 4)^2$$

$$2a = (a - 4)^2$$

$$2a = a^2 - 8a + 16$$

$$0 = a^2 - 10a + 16$$

$$0 = (a - 8)(a - 2)$$

$$a = 2 \text{ or } 8$$

11. **2, 5**

Solve the system by setting the equations equal to each other. Find values of x, then plug in to get possible values of y.

$$y = y$$

$$x^2 - 4x + 5 = 5 - x$$

$$x^2 - 3x = 0$$

$$x(x - 3) = 0$$

$$x = 0 \text{ or } 3$$

$$y = 2 \text{ or } 5$$

12. **5**

Set $f(x) = g(x)$, then bring all terms to one side and use Root Product formula or factor then multiply the roots.

$$f(x) = g(x)$$

$$x^2 + 11 = -x^2 + 9x + 1$$

$$2x^2 - 9x + 10 = 0$$

Product of the Roots: $c/a = 10/2 = 5$

13. **9**

The lowest output to a square root is zero. Therefore, the lowest value of the function is 9.

$$f(x) = 2(0) + 9 = 9$$

14. **2**

Solve for b to evaluate x.

$$2b = 64$$

$$b = 32$$

$$b = 2^5$$

$$x = \sqrt[5]{b}$$

$$x = \sqrt[5]{2^5} = 2$$

15. **3**

Since the remainder is zero, $x - a$ must be a factor of $f(x)$. Factor by grouping, then report the rational zero.

$$f(x) = x^3 - 3x^2 - 5x + 15$$

$$f(x) = (x^3 - 3x^2) + (-5x + 15)$$

$$f(x) = x^2(x - 3) - 5(x - 3)$$

$$f(x) = (x - 3)(x^2 - 5)$$

$$x = 3 \text{ or } \pm\sqrt{5}$$

$$a = 3$$

INTERPRETING DATA

Rates and Probability

Rates and Ratios

Percent and Percent Change

Line of Best Fit

Conditional Probability

Data Trends and Statistics

Basic Measurements

Advanced Statistics

Experimental Design

Box Plots

Introduction

As it stands now, all questions under the Problem Solving and Data Analysis umbrella fall into section four of the test, the second math portion where students have access to a calculator. Students are expected to efficiently convert units, apply ratios, make predictions based on a line of best fit, calculate probability, find statistical measurements, and draw conclusions based on specific sampling techniques. Expect to see several tables, graphs, and charts that organize a considerable amount of information; however, despite the sheer volume of data, the individual questions often only target a small fraction of what is provided to reference. In total, Problem Solving and Data Analysis accounts for 29% of all math questions on the test. Since all of these questions allow for a calculator, they make up a hefty portion of section four.

3.1 RATES AND PROBABILITY

a. Rates and Ratios

Rates and ratios define the relationship between two measurements, and in doing so provide sufficient information to make unit conversions. Almost all of the unit conversion questions that you will see on the test come with equivalency statements and can be solved using proportions or dimensional analysis. Recall proportions can be used to ensure equivalent ratios. The benefit to using a **proportion** lies mostly in its simplicity of organization in that it is only a fraction equal to another fraction that is set up in a corresponding way. When proportions are used to make unit conversions, one fraction can represent the known relationship between the two measurements and the other fraction can represent the conditional relationship between the two measurements that reflect the unique information outlined by the question.

Another way to convert units or calculate unknown quantities based on ratios and rates is dimensional analysis. **Dimensional analysis** can be used to make a series of unit conversions all at once through multiplying by the known ratio between two measurements so as to cancel out units. Essentially, given information is multiplied by fraction after fraction until desired measurements are reached. Dimensional analysis is a technique that is worth practicing because it can streamline lengthy calculations while providing a built-in organizational framework that makes it easier to stay on track despite performing a good number of conversions.

The following example shows how proportions or dimensional analysis can be used to meet the same goals. Identify the similarities and differences of each method, then determine which one suits your strengths or works better for a given circumstance.

Q+A

Rates and Ratios

Diagnostic Exam | Calculator Q6

A car traveled at an average speed of 24 miles per hour for 1.9 hours and consumed fuel at a rate of 17 miles per gallon. Approximately how many gallons of fuel did the car use for the entire 1.9-hour trip?

A. 1.96

B. 2.54

C. 2.68

D. 3.12

Workspace

Solution Outline

Strategy 1

Step 1: Find the total distance traveled using a proportion.

$$\frac{24\text{mi}}{1\text{hr}} = \frac{x\text{ mi}}{1.9\text{hr}}$$

$$x = 45.6\text{ mi}$$

Step 2: Find the total gallons consumed using a proportion.

$$\frac{17\text{ mi}}{1\text{ gal}} = \frac{45.6\text{ mi}}{y\text{ gal}}$$

$$y = 2.68$$

Answer: C

Strategy 2

Step 1: Start with given information and multiply by the known ratios (provided in the question) in a way that will cancel out units.

$$\frac{1.9\cancel{\text{hrs}}}{1} * \frac{24\cancel{\text{mi}}}{1\cancel{\text{hr}}} * \frac{1\text{gal}}{17\cancel{\text{mi}}} = 2.68\text{gal}$$

Notes

Additional Notes: Ultimately, the question above asks for the number of gallons consumed. The rate at which the vehicle consumes fuel is 1 gallon for every 17 miles. Additionally, to find the number of gallons used for the entire trip, we need to know the total distance traveled. Since we know how fast and how long the car travels, we can solve for distance. Many students come in with the knowledge that distance equals rate times time, $d = rt$, but proportions and dimensional analysis will yield the same result. Another topic emphasized on the test is ratios, which are used to relate two more quantities. Remember that ratios can be rewritten as fractions.

Q+A

Rates and Ratios **Diagnostic Exam | Calculator Q29**

An animal feed company sells a brand of dog food that contains a ratio of 2 grams of fat to 9 grams of carbohydrate and 8 grams of carbohydrate to 5 grams of protein. What is the ratio of fat to protein in grams?

A. 2 : 5

B. 5 : 2

C. 16 : 45

D. 45 : 16

Workspace

Solution Outline

Step 1: Use dimensional analysis to cancel out units through multiplication and division.

$$\frac{2\text{g fat}}{9\text{g carb}} \cdot \frac{8\text{g carb}}{5\text{g protein}} = \frac{16\text{g fat}}{45\text{g protein}}$$

Answer: C

Notes

In some cases, ratios can be used to parse out the relationship of individual parts that together create a collective whole. The following example describes a situation where we can use this idea of ratios to quickly determine the number of individuals belonging to a particular group.

Q+A

Rates and Ratios **Diagnostic Exam | Calculator Q30**

The ratio of boys to girls in a high school physics class is 4 : 5. If there are 27 students total in the class, how many are girls?

A. 12

B. 15

C. 18

D. 21

Workspace

Solution Outline

Step 1: Relate the number of girls in the class to the total number of students. Using the reduced ratio of 4 boys to 5 girls shows us that 5 out of 9 students are girls. In other words, five-ninths of the total number of students are girls.

Step 2: Find the number of girls.

$$\frac{5 \text{ girls}}{9 \text{ total}} * \frac{27 \text{ total}}{1} = 15 \text{ girls}$$

Answer: B

Notes

Other questions may push us further in regards to problem solving and calculations. The next example shows how to use dimensional analysis to perform a number of unit conversions for us to then be able to find the difference between two quantities. It is important to note the phrase, "How many more…" signals us to subtract, which will help us find the overall gap between two numbers.

Q+A

Unit Conversions — Diagnostic Exam | Calculator Q30

A café sells two different brews of coffee: Wake-Up and Chill-Out. The Wake-Up coffee contains 5.83 milligrams (mg) of caffeine in every 10 milliliters (mL) whereas the Chill-Out coffee contains 3.96mg of caffeine in 12mL. How many more milligrams of caffeine to the nearest integer are in a 12-ounce (oz) cup of Wake-Up than there are in a 12-ounce cup of Chill-Out?

$$8\text{oz} = 240m\text{L}$$

Workspace

Solution Outline

Step 1: Use dimensional analysis to find the amount of caffeine for each brew. The phrase, "How many more/less…" is a common occurrence on the test – it means find the difference and should not be confused with another common phrase, "How many times more/less…" which means find the quotient.

Wake Up: $\frac{5.83mg}{10mL} * \frac{240mL}{8\text{oz}} * \frac{12\text{oz}}{1} = 209.88mg$

Chill Out: $\frac{3.96mg}{12mL} * \frac{240mL}{8\text{oz}} * \frac{12\text{oz}}{1} = 118.80mg$

Step 2: Take the difference.

$209.88 - 118.80 = 91.08$

Answer: 91

Notes

b. Percent and Percent Change

Many of the percent change questions on the test require students to find the original or new value after a given percent increase or decrease. The most common percent increase or decrease questions use everyday scenarios, like discounts, taxes, and tips. For example, when a sweater is marked 20% off, that is a percent decrease of 20% where the sweater now goes for 80% of its original value. In general, the equation: New value $= (1 \pm r)$(initial value) can be used with percent change questions, where r represents the percent increase or decrease written as a fraction or decimal (e.g. $20\% = 0.20$). In this case, we are applying the factor of change $1 \pm r$ to find a new value.

Application of Percent Change	New value $= (1 \pm r) \times$ Initial Value

The following example allows us to see how to apply multiple percent changes when both the initial and new values are unknown. Remember that we can use our own variables as placeholder even when they are not specified in the question.

Q+A

Percent Change **Diagnostic Exam | Calculator Q2**

Marcelo has to buy a textbook for his intro chemistry class. He finds out he can get a discount that takes 18% off the original price. He will still have to pay a 7% sales tax on the discounted price. If he pays h dollars at the register, what expression represents the original price of the textbook?

A. $0.89h$ B. $\frac{h}{0.89}$ C. $(0.82)(1.07)h$ D. $\frac{h}{(0.82)(1.07)}$

Workspace

Solution Outline

Step 1: Find the value of the discounted book. We don't know the original price, so use the variable x.

$$\text{New value} = (1 \pm r) * \text{initial value}$$

$$\text{New value} = (1 - 0.18)x = 0.82x$$

Step 2: Find the value of the discounted book, which we found to be $0.82x$ in the previous step, with the sales tax. This value is the final price, h.

$$\text{New value} = (1 + 0.07) * 0.82x$$

$$\text{New value} = (1.07)(0.82)x$$

$$h = (1.07)(0.82)x$$

Step 3: Solve for the initial value, x.

Answer: D

Notes

Overall, when multiple percent changes occur, we can actually apply them all at once by multiplying the initial value by the factor of change $1 \pm r$ for each percent increase or decrease. So when there are several percent increases and decreases applied to a single value, the new value can be calculated using a string of multiplication that looks like: $\text{New Value} = (1 \pm r_1)(1 \pm r_2)(1 \pm r_3) \ldots (1 \pm r_n) * \text{Initial Value}$. Keep in mind that the factor's deviation from 1 reflects the percent change. For example, the factor 0.96 is 0.04 below one, which indicates a 4% decrease.

Multiple Percent Changes	$\text{New value} = (1 \pm r_1)(1 \pm r_2)(1 \pm r_3) \ldots (1 \pm r_n) \times \text{Initial Value}$

The particular method off using the factor $1 \pm r$ to apply a percent change shows itself in various ways, but none more specific than compound interest. For example, when money is saved, the principal may accrue interest to grow the investment. The compound interest formula allows us to calculate the amount A that the principal P reaches after t years with an interest rate r compounded n number of times per year.

Compound Interest	$A = P\left(1 + \frac{r}{n}\right)^{nt}$

Key Terms: Annually $n = 1$, Semiannually $n = 2$, Quarterly $n = 4$, Monthly $n = 12$.

Keep in mind that the accumulated amount of money is reached because interest is being generated off the principal, meaning that at any point in time: Interest $= A - P$.

Q+A

Percent Change **Calculator**

Amy puts x dollars into a savings account that generates 4% interest compounded semiannually. Which equation can be used to find the amount of time it will take for her savings to double?

A. $1.02^t = 2$ B. $1.02^{2t} = 2$ C. $1.04^t = 2$ D. $1.04^{2t} = 2$

Workspace

Solution Outline

Step 1: Define all variables in the compound interest formula.

$A = 2x \quad P = x \quad r = 0.04 \quad n = 2$

Step 2: Substitute these values into the compound interest formula.

$$A = P\left(1 + \frac{r}{n}\right)^{nt}$$

$$2x = x\left(1 + \frac{0.04}{2}\right)^{2t}$$

$$2 = 1.02^{2t}$$

Answer: B

Notes

Now that the application of percent change has been used to directly calculate a new value from an old value after any given percent change, we can introduce the percent change formula. In general terms, percent change measures the percent deviation between two values relative to the starting value. Because of their similarities, we will also take a look at the percent error formula, which is used when a value deviates from an accepted valued. Table 3.11 compares the percent change formula with the percent error formula to show their similarities.

Table 3.11 | Percent Change and Percent Error Formulas

Percent Change Formula	Percent Error Formula
$\text{Percent Change} = \frac{\text{New} - \text{Original}}{\text{Original}} * 100$	$\text{Percent Error} = \frac{\text{Value} - \text{Accepted}}{\text{Accepted}} * 100$

Q+A

Percent Change in Science **Calculator**

As part of a chemistry experiment, Alissa collected b grams of benzoic acid after separating it from a heterogeneous mixture. Her obtained amount of benzoic acid was 2.46 grams less than the accepted value. Which of the following represents the magnitude of the percent error in her investigation?

A. $\frac{2.46}{b + 2.46}(100)$ B. $\frac{b}{b + 2.46}(100)$ C. $\frac{2.46}{b - 2.46}(100)$ D. $\frac{b}{b - 2.46}(100)$

Workspace

Solution Outline

Step 1: The numerator of the percent error formula represents a difference. Alissa obtained a value that was off the anticipated amount by 2.46 grams.

Step 2: The question states that Alissa's measurement of b was too low. Therefore, the accepted value must be more. The correct expression for the accepted value is $b + 2.46$ grams.

Step 3: Create the fraction and multiply by 100 to turn it into a percent.

$$\text{Percent Error} = \frac{2.46}{b + 2.46}(100)$$

Answer: A

Notes

c. Line of Best Fit

Scatter plots allow us to see trends in data based on two quantitative variables. A quantitative variable is one that takes on numerical values with some natural unit of measure. The most common trend we need to know for the test is the linear correlation where bivariate data follows a straight path on a coordinate plane. To better analyze and use scatter plots with linear correlations, we can accompany the data with a line of best fit.

A **line of best fit** is drawn through the data in such a way that minimizes the amount by which the data points vary from the line. As a result, the line of best fit may pass through all, some, or none of the data points. The benefit in having a line of best fit in conjunction with scatter plots showing a linear correlation is that it can be used to find **predicted values** of one variable based on the quantity of the other variable. In most cases, we predict the value of dependent variable along the y-axis based off a certain value of the independent variable along the x-axis. Here, it is important for us to note the line of best fit contains predicted values, whereas the data points themselves represent **actual values**. The difference between the actual value and predicted value at any point in time is called a residual.

Residual	Actual − Predicted

Since the residual represents a difference between two values, we may have to use it to find its percent error relative to the referenced quantity, usually the actual value. Before we delve more deeply into this idea, we should acknowledge that not all data contain linear correlations. Sometimes, data points follow paths that can be more appropriately modeled by polynomial or exponential functions. If this is the case, then we should look to the parent functions discussed in 2.3 Function Properties to ensure we use a function that suits the trajectory of the data. Even still, data may fit a function closely, which would indicate a strong correlation, or loosely, which would indicate a weak correlation. In particular, we must be able to identify strong or weak and positive or negative linear correlations, which are summarized in Table 3.11.

Table 3.11 | Linear Correlation Summary

	Positive	Negative
Strong	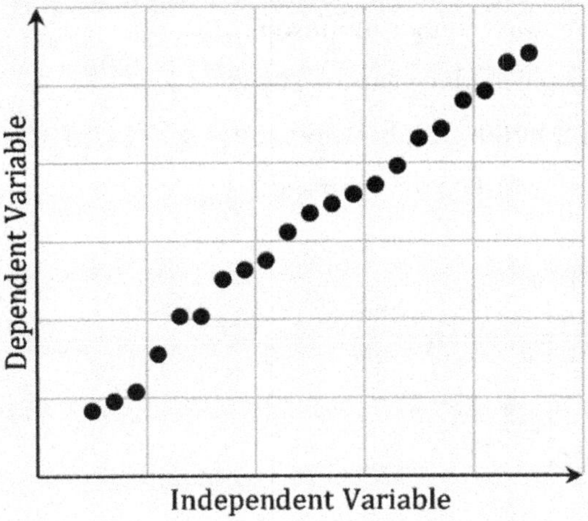	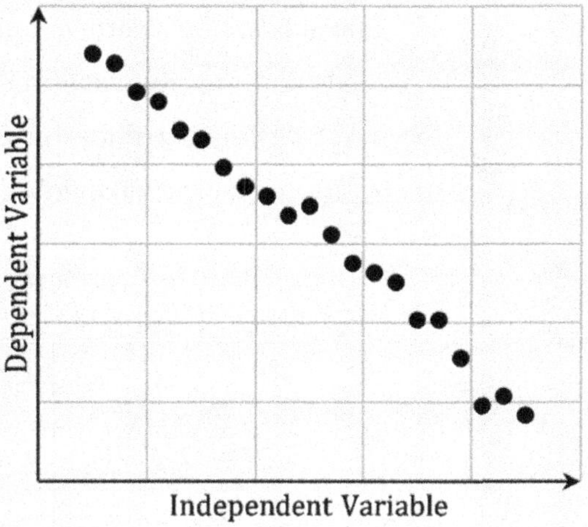
Weak	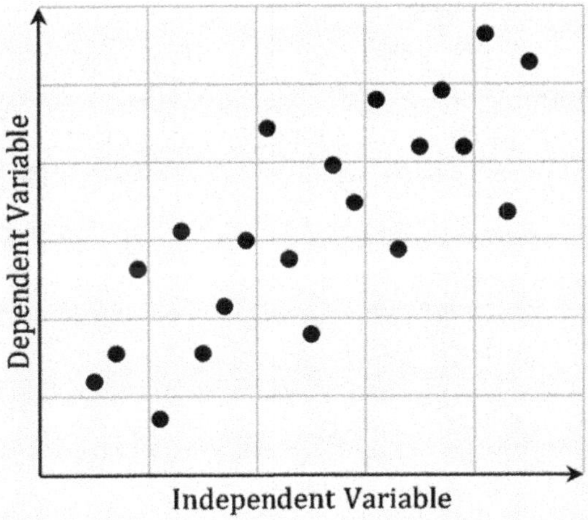	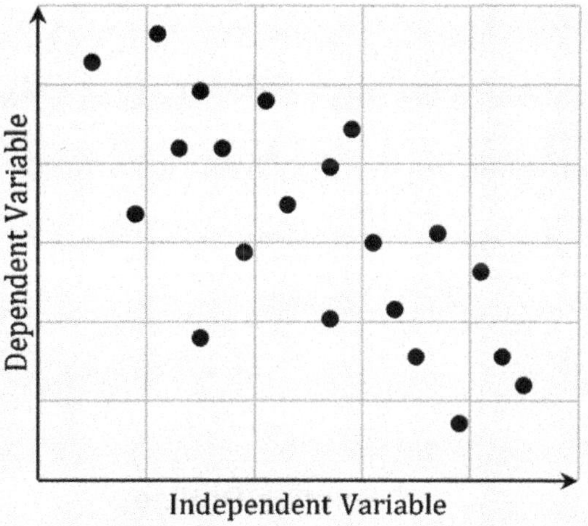

As we know, the independent variable is always placed on the x-axis and the dependent variable goes on the y-axis. Sometimes, the independent variable is referred to as the explanatory variable, and the dependent variable is called the response variable. These terms are often used when we design experiments to explore the relationship between two quantities. For example, an investigation might be designed to determine how the amount of fertilizer affects the growth of a particular plant species. The explanatory variable is the amount of fertilizer, and the response variable is the amount of plant growth. We will look more closely at experimental design in Section 3.2 Data Trends and Statistics. For now, let's use the following example from the Diagnostic Exam to see how the test will challenge us to use a scatter plot with a linear correlation and accompanying line of best fit to determine percent error. Note that estimating predicted and actual values will often allow us to get close enough to the exact answer to answer the question correctly, while saving us time since the numbers will be easier to work with.

Q+A

Rates and Ratios **Diagnostic Exam | Calculator Q14**

The number of annual museum visits in Town A in the years following 1980 was recorded and graphed in scatterplot below. A line of best fit was drawn to help estimate the number of museum visits that the city would receive for any particular year. In 2004, what was the percent error to the nearest hundredth between the estimated and actual number of museum visits Town A received?

A. 0.14

B. 2.52

C. 5.00

D. 14.29

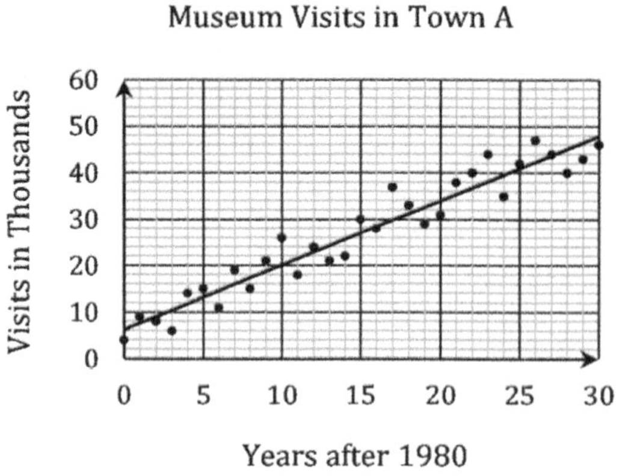

Workspace

Solution Outline

Step 1: Find actual and predicted data points. The scatterplot indicates the actual number of annual museum visits Town A received and the line of best fit indicates the predicted number of annual museum visits. Since 2004 is 24 years after 1980, we can use the actual point (24,35) on the scatter plot and the predicted point (24,40) on the line of best fit. Given the units of y-axis, this means that the actual number of annual museum visits in 2004 was 35 thousand, but 40 thousand was predicted. However, it is okay to ignore this fact when calculating the percent error.

Step 2: Calculate percent error.

$$\frac{(40-35)}{35} * 100 = 14.29\%$$

Answer: D

Notes

d. Conditional Probability

Almost all probability questions on the test can be answered with a few simple ideas. In most cases, these problems do not go far beyond "part over total" and some general algebra skills. However, students must know how to tackle "and", "or", and conditional probability questions. Basic probability notation and rules are summarized in Table 3.14 below.

Table 3.14 | Probability Notation and Rules Summary

Notation	Meaning	Formula
$\boldsymbol{P(A)}$	Probability that event A occurs	$P(A) = \frac{\text{Part}}{\text{Total}}$
$\boldsymbol{P(A \cap B)}$	Probability that events A **AND** B occur	$P(A \cap B) = P(A) \times P(B)$
$\boldsymbol{P(A \cup B)}$	Probability that event A **OR** event B occurs	$P(A \cup B) = P(A) + P(B) - P(A \cap B)$
$\boldsymbol{P(A \mid B)}$	Probability the evet A occurs, **GIVEN** event B occurs	$P(A \mid B) = \frac{P(A \cap B)}{P(B)}$

Notes: AND means MULTIPLY; OR means ADD

In most cases, data is presented in two-way tables, where rows and columns represent different categories. As stated above, finding the probability that an event occurs requires dividing the "part" by the "total". Beware that the "total" is often **NOT** the total of the entire table, but the total for the known information presented in the question. When we already know that one event occurs, this limits our parameters to the category that is given. This idea is called **Conditional Probability** and is a frequently tested topic. Look out for key phrases that signal us to find a conditional probability, such as "if...", "given that...", and "provided that..." statements, which indicate to us a specific group of interest. The next two examples come from the diagnostic exam and require the use of the chart below. The first example involves ratios and the second, conditional probability.

Examples 11 and 12 use the following information

A survey of 280 randomly selected 12th grade students attending high school in County A was conducted to gather data on their college plans. The results of the survey are categorized by student GPA and preferred college location as shown in the two-way table.

Number of 12th Grade Students

	In-State	Out-of-State	Total
≥ 3.77 GPA	81	45	126
< 3.77 GPA	x	y	154

Q+A

Rates and Ratios **Diagnostic Exam | Calculator Q11**

If the ratio of x to y is 5 to 2, the how many students in the survey plan to attend an In-State college?

A. 110 B. 154 C. 176 D. 191

Workspace

Solution Outline

Step 1: Understand that the total number of students who plan stay In-State for college is represented by $81 + x$.

Step 2: Manipulate the given ratio. Since x and y are the only two components contributing to 154 in the chart, we can quickly change the ratio to relate the variable x to the total, rather than to y.

Using the ratio: $x = 5$ and $y = 2$

Total: $x + y = 7$

Therefore, the ratio of x to the total is 5 to 7. Another way to say this is that x is five-sevenths of the total.

Step 3: Find x by taking five-sevenths of the total. Remember that "of" usually means multiply, so five-sevenths *of* the total means five-sevenths times the total.

$x = (5/7)154$ or 110

Step 4: Find the sum of 81 and x.

Answer: D

Notes

Q+A

Conditional Probability **Diagnostic Exam | Calculator Q12**

What is the probability that a student with a GPA of at least 3.77 plans to attend an Out-of-State college? (Table Required)

A. 5/14 B. 5/9 C. 9/14 D. 14/9

Workspace

Solution Outline

Step 1: Determine if any "given" information is provided. Given information is in fact provided since the question specifies that the student must have at least a 3.77 GPA. This is a conditional probability question.

Step 2: Isolate the row or column that represents the given probability. The first row of data displays information for students with a GPA at or above a 3.77. It is from this row then that the part and the total will be found.

Step 3: Divide the part by the total. The part the question asks for is the number of students with a GPA of 3.77 of more who plan to attend college out-of-state. The total is the number of students with at least a 3.77 GPA.

Part: 45 | Total: 126

$$\frac{45}{126} = \frac{5}{14}$$

Answer: A

Notes

Tip: Use the fraction operation in your calculator to quickly find a reduced fraction.

Another new testing favorite asks, "How many times more likely it is that event A will occur than event B," or some variation thereof. This requires taking a ratio of each event's probability. Finding the probabilities of the two events is imperative to determine this likelihood since, generally, categories will not be equally represented, meaning that the nominal values we see in the chart cannot be used on their own to compute chance favorability.

Q+A

Conditional Probability **Diagnostic Exam | Calculator Q36**

Track and Field Events by Category

	Sprints	Distance	Field	Total
Male	12	b	8	37
Female	a	23	11	d
Total	26	40	c	85

A high school track team has 85 students. The coach has organized the students using the chart above, where letters a-d represent constants. How many times more likely is it that a female track team member focuses on distance events than it is a male track team member focuses on distance events? Round final answer to the nearest hundredth.

Workspace

Solution Outline

Step 1: Determine each probability referred to in the question independently. Note that these probabilities are both conditional probabilities because of the known information presented in each statement. For example, another way of stating, "What is the probability that a female track team member focuses on distance events," is "Given that a track team member is female, what is the probability she focuses on distance events?"

There are two probabilities we need to find before we can answer the final question:

$P(\text{Female focuses on Distance}) = 23/48$

$P(\text{Male focuses on Distance}) = 17/37$

Recall to only use the row(s) or column(s) of the given information to find the part and total when calculating conditional probability. Note that each row and column must add to the total, which is why b must equal 17 and d must equal 48.

Step 2: Find the ratio of the two probabilities found above. Order is important here. Generally, follow the order as it appears in the questions. Think: *first over last.*

First: P(Female focuses on Distance)

Last: P(Male focuses on Distance)

Therefore, first over last is:

$$\frac{P(\text{Female focuses on Distance})}{P(\text{Male focuses on Distance})} = \frac{23/48}{17/37} = 1.04$$

Answer: 1.04

Notes

Problem Set: Rates and Probability

10 Questions

1 Calculator Allowed

Number of News Sources

	1	2-3	4-5	6
Group 1	25	37	12	1
Group 2	9	34	24	8

The table above shows the results of a survey where 150 people were asked how many different sources that they rely on to receive the news. Group 1 consisted of individuals ages 40 and older, while Group 2 consisted of individuals between the ages of 18 and 39. Given that an individual in the survey uses at least 2 news sources, what is the probability he or she is 40 or older?

A. 25/58

B. 25/150

C. 11/25

D. 33/58

2 Calculator Allowed

Quarter	Profit (in thousands)
Q1	24.2
Q2	22.6
Q3	x
Q4	26.5

A software company's quarterly profits are recorded in the chart above. If the company sees a 12% increase from Q2 to Q3, then which is closest to the value of x?

A. 21.0

B. 23.7

C. 25.3

D. 34.6

3 Calculator Allowed

$$Q = nq_e$$

$$q_e = -1.6 \times 10^{-19} C$$

$$1 \mu C = 10^{-6} C$$

The total charge Q in coulombs (C) can be found by multiplying the number of electrons n by the fundamental charge q_e as shown in the equations above. What is the magnitude of the charge in microcoulombs (μC) of 2.5×10^{14} electrons?

A. 4×10^{-5}

B. 4

C. 40

D. 4×10^{6}

4

The scientists of a leading athletic training association determine that the optimal recovery supplement following aerobic exercise has a mass ratio of carbohydrates to protein of 4 :1. If the recovery supplement contains 120 grams of just carbohydrates and protein, how many grams of carbohydrates are there?

A. 24

B. 30

C. 96

D. 100

5 Calculator Allowed

If the stock price of a company that produces digital design software goes down 20%, then by how much must the new stock price increase to get back to where it was before the decrease?

A. 20%

B. 25%

C. 30%

D. 35%

Questions 6-7 use the given information

Beer-Lambert's Law relates a solution's concentration to its absorbance, a property that measures the extent to which a solution impedes the transmission of light. After performing an experiment to investigate the absorbance that solutions of the Iron (II) cation, Fe^{2+}, produce, Tasha creates a scatter plot with x-axis measuring Fe^{2+} concentration in parts per million (ppm) and y-axis measuring absorbance. She then determines that the line of best fit has an equation in the form $A = bx$, where b is a constant that can be used to calculate the predicted absorbance A for any Fe^{2+} solution with concentration of x ppm. Her data is summarized in the scatter plot below.

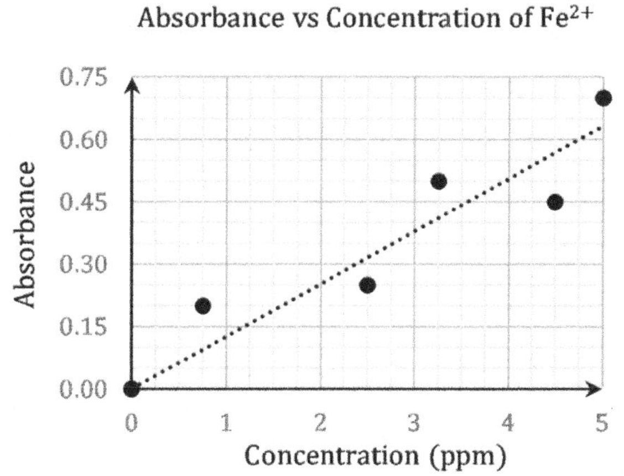

6 Calculator Allowed

Which of the following is closest to the percent error between the predicted and actual absorbance from Tasha's data for the 2.5ppm Fe^{2+} solution?

A. 10%

B. 25%

C. 40%

D. 55%

7 Calculator Allowed

An Fe^{2+} solution of unknown concentration has an absorbance measuring 0.45. According to the line of best fit, what is the expected concentration in ppm of this solution?

A. 3.5

B. 4.0

C. 4.5

D. 5.0

Questions 8-10 use the following information

The table below shows the number of students currently enrolled at a private university who come from different regions across the country and are pursuing degrees in science, technology, engineering, and mathematics (STEM).

Number of Students Enrolled in STEM Degree Areas

Region	Science	Technology	Engineering	Math	Total
Northeast	572	346	1,655	421	2,994
Midwest	129	355	292	59	835
South	488	953	620	236	2,297
West	1,644	246	409	829	3,128
Total	2,833	1,900	2,976	1,545	9,254

8 Calculator Allowed

What is the probability that a student at this private university who is seeking a degree in science is from either the West or Northeast?

A. 0.57

B. 0.61

C. 0.78

D. 0.83

9 Calculator Allowed

If the breakdown of students from the table is representative of other universities as well, then approximately how many students would be pursuing an engineering degree at an institution that has 24,880 students enrolled in STEM areas?

A. 3,200

B. 8,000

C. 12,500

D. 15,800

10 Calculator Allowed

How many times more likely is it that a student from the Northeast pursues an engineering degree than it is a student from the South pursues a math degree?

A. 3.64

B. 4.92

C. 5.38

D. 7.01

Explanations

Problem Set: Rates and Probability

1. A

This is a conditional probability question. The total number of individuals who use two or more news sources is 116. Of these, only 50 belong to Group 1.

$$\frac{\text{Part}}{\text{Total}} = \frac{50}{116} \quad \text{or} \quad \frac{25}{58}$$

2. C

Since profits in Q3 are 12% higher than in Q2, then they are all of Q2'sprofits plus an additional 12%. This idea is best summarized by the equation below, where r is the percent change represented as a decimal.

$$\text{New Value} = (1 \pm r) * \text{Old Value}$$

$$x = 22.6(1 + 0.12)$$

$$x = 22.6(1.12)$$

$$x = 25.3$$

3. C

Use the formulas and information provided to find the charge Q. Then use dimensional analysis or other conversion technique to convert coulombs to microcoulombs.

$$Q = nq_e$$

$$Q = 2.5 \times 10^{14} * 1.6 \times 10^{-19}$$

$$Q = 4 \times 10^{-5}\ C$$

$$4 \times 10^{-5}\ C * \frac{1\ \mu C}{10^{-6}\ C} = 40\mu C$$

4. C

According to the information in the question, there are 4 grams of carbs to 1 gram of protein. Since the supplement only contains carbs and protein, four out of every five grams of the supplement come from carbohydrates. In other words, four-fifths of the total number of grams are from carbs.

$$\frac{4}{5} * 120 = 96$$

5. B

The factors used to apply a percent change are deviations from 1. A 20% decrease leaves 80% leftover, or 0.8. Such applied factors would have to multiply to 1 to effectively cancel each other out.

$$0.80b = 1$$

$$b = 1.25$$

25% increase

6. B

Use the percent error formula in conjunction with the information outlined in the passage. In this context, the percent error formula takes this form:

$$\text{Percent Error} = \frac{\text{Predicted} - \text{Actual}}{\text{Actual}} * 100$$

Predicted absorbance is found using the line of best f, which shows an absorbance of about 0.31 at a concentration of 2.5ppm. The actual absorbance is the value described by the data point at 2.5ppm, which is

0.25. Even rough estimates are sufficient to come to conclusion regarding the percent error since the answer choices are so different.

$$\text{Percent Error} = \frac{0.31 - 0.25}{0.25} * 100$$

$$\text{Percent Error} \approx 25\%$$

7. A

Use the line of best fit to predict the concentration by first going to an absorbance of 0.45 on the y-axis, then moving horizontally to reach the line. The solution concentration reached is about 3.5ppm.

8. C

Determine the conditional probability by finding the part and total according to the "given" information. The question states that the student must be seeking a science degree, which refers to a specific column in the table. The part and total should be gleaned from this column.

Condition: Science Degree

Part: West and Northeast Science Students

Total: All Science Degree Students

$$\frac{\text{Part}}{\text{Total}} = \frac{2{,}216}{2{,}833} \quad \text{or} \quad 0.78$$

9. B

Set up a proportion to find the number of engineering students at another university.

At this university:

Part = 2,976 engineering students

Total = 9,254 STEM students

At other university:

Part = x engineering students

Total = 24,880 STEM students

$$\frac{2{,}976}{9{,}254} = \frac{x}{24{,}880}$$

$$x \approx 8{,}000$$

10. C

To find how many times more likely one event is than another, take the ratio between the two probabilities.
Probability Event 1 or $P(1)$

Condition: Northeast Student (ROW)

Part: 1,655 Northeast, engineering

Total: 2,994 Northeast students

$$P(1) \approx 0.553$$

Probability Event 2 or $P(2)$

Condition: South Student (ROW)

Part: 236 South, math

Total: 2,297 South students

$$P(2) \approx 0.1027$$

$$\frac{P(1)}{P(2)} = \frac{0.553}{0.1027} \quad \text{or} \quad 5.38$$

3.2 DATA TRENDS AND STATISTICS

a. Basic Measurements

To quickly summarize a collection of data, we often turn measures of central tendency and descriptors that provide insight into the distribution's spread and variability. In this section, we will first review the definitions of these measurements and see how to interpret them when given frequency tables and dot plots. Then we will move onto more advanced statistical measurements and sampling techniques. While measures like mean, median, mode, and range are easy enough to calculate when given a small data set, the usual procedures get somewhat obfuscated when the data is lumped into summary tables, like frequency histograms. As a consequence, this can make a basic question about a data's average much harder – but it does not need to be. In fact, frequency charts can make finding measures of central tendency and spread even easier because the charts have already organized and synthesized the data. Equally valuable, since these questions occur most often in section four of the exam, we can have our calculators do some heavy lifting.

The **arithmetic mean** is found by taking the sum of all data values, then dividing it by the number of data values added together. Remember that the mean is greatly affected by outliers, or values that differ substantially from the rest of the data. When outliers are present, it may be best to turn to the **median**, which is the middlemost data point in an ordered data set, because it is less influenced by extreme values. If two numbers are located in the middle, then take their average to find the median. Data may also have values that are more prevalent than others, in which case the **mode** is the value that occurs the most. Distributions with one peak are called unimodal, and distributions with two peaks are called bimodal. The spread of the data is measured by its **range**, which is the maximum value minus the minimum.

Q+A

Frequency Chart **Diagnostic Exam | Calculator 15**

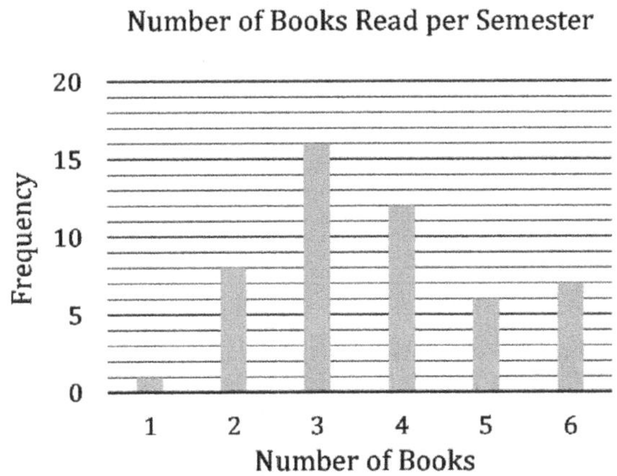

The frequency chart above summarizes the number of books read per semester by a random sample of fifty students attending a local college. What is the difference between the mean and median number of books read in this sample?

A. 0.2

B. 0.7

C. 1.2

D. 1.5

Workspace

Solution Outline

Strategy 1

Step 1: Most students are used to having a list of values to work with when answering questions about the mean and median. Frequency diagrams like this one are like compact lists where the frequency tells you about the number of times a certain value occurs. According to this histogram, the value of 3 occurs 16 times. It can be helpful to write the frequency above each column when doing these calculations.

Step 2: Find the mean. Consolidate most of the addition through multiplying the values by their frequencies. For example, instead of adding 16 three different times, we can simply multiply 16 by three.

$$\frac{1(1) + 2(8) + 3(16) + 4(12) + 5(6) + 6(7)}{50} = 3.7$$

Step 3: Find the median. To do this, we need to know what "place" between 1 and 50 represents the exact middle, then we can find the corresponding value from the histogram. Be careful here: many of us will say the middle student out of 50 is 25. It is not. To understand this, let's look at five students: 1, 2, 3, 4, 5

Can you identify the middle student? It's student number 3, right? Which is not half of 5. If we need to find the middle between two numbers, take a simple average. Therefore, the middle of 1 and 5 is 3. In the same way, the middle of 1 and 50 is 25.5, or right between the 25th and 26th student. Sometimes, there is not be a big difference between 25 and 25.5, but other times there is. The 25th student read 3 books and the 26th student read 4 books. So, the median occurs between 3 and 4:

$$\frac{3 + 4}{2} = \frac{7}{2} = 3.5$$

Step 4: Find the difference between the mean and median.

$$3.7 - 3.5 = 0.2$$

Strategy 2

Step 1: Use graphing calculator to input data values and their frequencies into two different lists. In a TI-84 calculator, for example, press STAT, then Edit to reach list editor. Input data values in List 1 and their frequencies into List 2.

Step 2: Calculate one variable statistics. In the TI-84 calculator, press STAT, then right arrow to Calc, and enter 1-Var Stats for List 1, with frequency of List 2. The first measure $\bar{x}$ is the mean, which is 3.7. Scroll down, and the median (labeled Med) is 3.5. The difference is just 0.2.

L1	L2
1	1
2	8
3	16
4	12
5	6
6	7

Answer: A

Notes

b. Advanced Statistics

The advanced statistics that students must know to perform well on the test are standard deviation, margin of error, and confidence intervals. Though these statistics do not show up on every test, they are worth reviewing as they strengthen a student's ability to understand and interpret data and their distributions. Fortunately, students do not need to calculate **standard deviation, margin of error,** or **confidence intervals**, but they do need to understand and compare data in these terms.

Standard deviation (σ) measures the spread of the data away from the mean. The further away data points are from each other, the larger the standard deviation; the closer together the data points are, the smaller the standard deviation. The test will not require students to calculate standard deviation, but rather it will challenge students to compare the standard deviations between two data sets. These data sets may vary in frequency distribution or spread, as displayed in the cases that follow in Table 3.21, which uses bell curves to analyze standard deviation. To gain a better sense for the value of the standard deviation, normal distributions like these span three standard deviations each way – six in total – from the mean that occurs in the middle until the area under the curve is negligible.

In some sense, the standard deviation measures a gap in data, and therefore does not change when the data points shift together with addition or subtraction. To start seeing this idea, let's follow a loosely related example: Molly has $15 and Zeke has $10. If both receive $20, Molly is still ahead $5. On the other hand, the standard deviation is affected when data points are changed by multiplication and division. For example, if we double Molly and Zeke's dollar amounts so that Molly has $30 and Zeke has $20, we see that Molly now has $10 more than does Zeke, up from the $5 she was ahead by at the start. Though, standard deviation is a measure of variability of data, this quirky example gives us a sense of how standard deviation changes with linear transformations.

Table 3.21 | Comparing Standard Deviations

Differences in Spread

Distribution 1

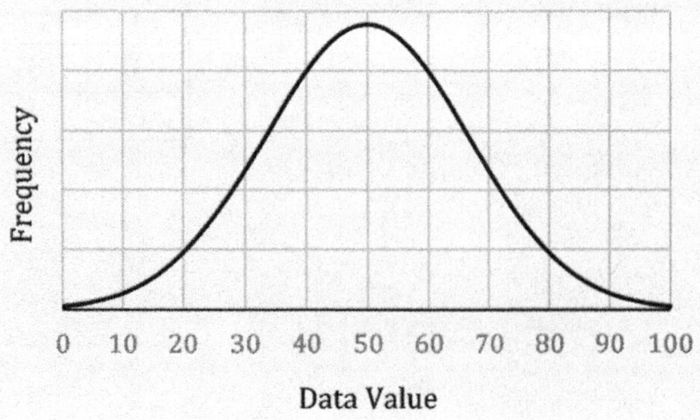

Standard Deviation (σ_1) Large

Distribution 2

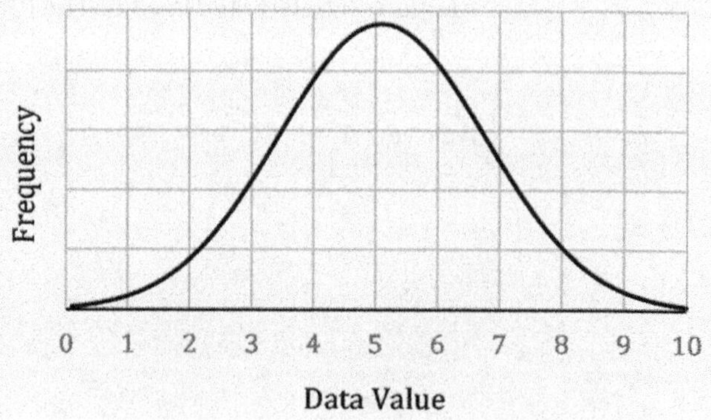

Standard Deviation (σ_2) Small

Summary: $\sigma_1 > \sigma_2$

Differences in Frequency

Distribution 1

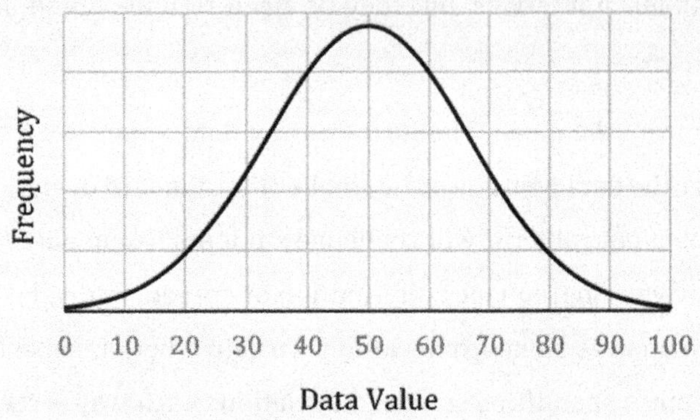

Standard Deviation (σ_1) Large

Distribution 2

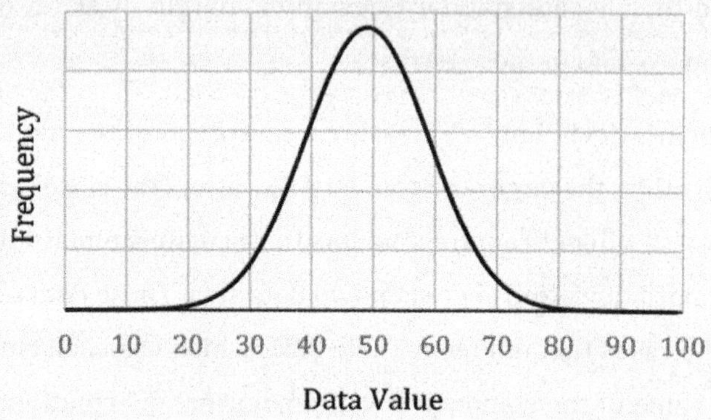

Standard Deviation (σ_2) Small

Summary: $\sigma_1 > \sigma_2$

When given distributions that look same, as in our first set of graphs in the table above, they are often graphed on a different scale, allowing us to see clear differences in spread. Since standard deviation in some ways measures the spread of data, we can infer that the standard deviation increases as the spread increases because the data points are farther apart. Conversely, when the data lies in a narrow range of values, the standard deviation will be smaller. When distributions are graphed on the same scale, like our second set of graphs in the table above, the distributions will often look quite different, allowing us determine if the data lies closer together based on shape alone. Essentially, a narrow curve has a smaller standard deviation than a wider curve because the majority of the data lies closer to the mean.

Q+A

Standard Deviation **Diagnostic Exam | Calculator Q18**

Entry-Level Salaries at Company A

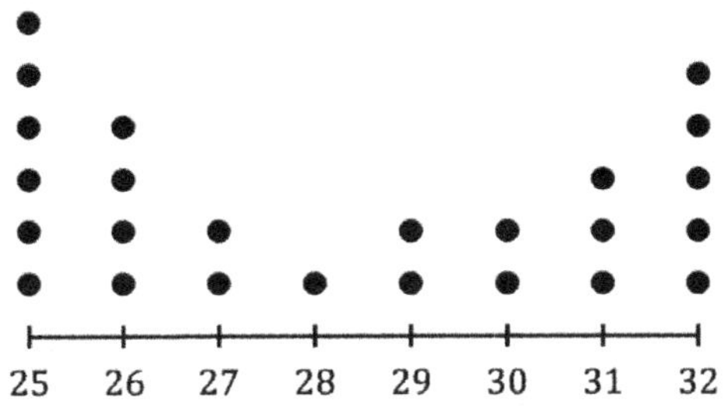

Annual Salary (Thousands of Dollars)

Entry-Level Salaries at Company B

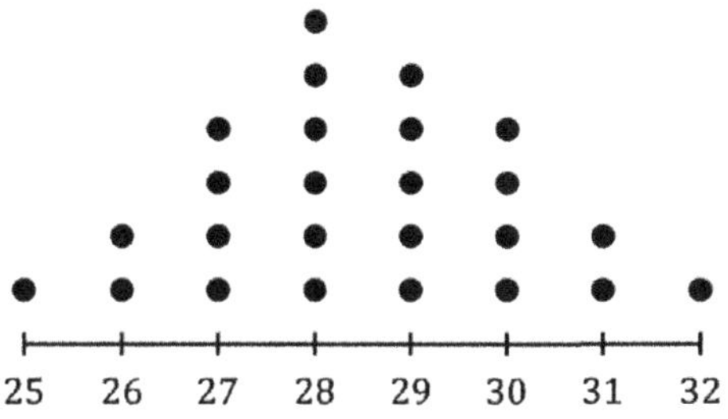

Annual Salary (Thousands of Dollars)

The dot plots above represent the salaries in US dollars of the 25 entry-level employees currently working at either Company A or Company B. Which of the following correctly compares the standard deviation of the salaries among the entry-level employees in each of the two companies?

A. The standard deviation of the entry-level salaries at Company A is greater.

B. The standard deviation of the entry-level salaries at Company B is greater.

C. The standard deviations of the entry-level salaries at Company A and Company B are equal.

D. No relationship can be determined from the provided information.

Workspace

Solution Outline

Step 1: Notice that the spread of the data is the same in both cases, where we can look at the shape of the distributions to compare the standard deviations. Most of the data points in Company B are in the same general area, whereas Company A has two clusters of data in very different regions, which creates more deviation from the average that lies closer to the middle. Consequently, the standard deviation of the data in Company A is greater.

Answer: A

Notes

Margin of Error is another important statistical measurement on the test. It expresses the maximum expected difference between the true population parameter and its estimate. In short, it tells us how "off" the data might be. Students should know how margin of error and standard deviation are affected by sample size. Table 3.21 summarizes the relationships that standard deviation and margin of error have with sample size.

Table 3.21 | Relationship between Sample Size and Standard Deviation or Margin of Error

Sample Size	Standard Deviation	Margin of Error
Decreases	Increases	Increases
Increases	Decreases	Decreases

Summary: Both standard deviation and margin of error are inversely related to the sample size.

This relationship should make sense. For example, think of the average student height in your math class. If you only took the heights of two students in your 30-student class, would these values likely reflect the true average height? What if these two students were the shortest or tallest in the whole class? Would their values *deviate* much from the typical height? Would there be some degree of *error* as a result? Since this sample size is small, we would expect larger deviations and more error. Conversely, larger samples are more likely to reflect the population – so long as participants are selected randomly from the population of interest – which will reduce both variability and error.

Q+A

Margin of Error **Diagnostic Exam | Calculator Q17**

A survey was conducted in a district to determine the percentage of registered voters who plan to participate in the next local election. The survey only sampled registered voters in the district, recording each respondent's level of education and whether or not the respondent would vote in the next local election. The results of the survey concluded that 63 ± 2.9 percent of those holding at most a Bachelor's degree and that 78 ± 6.1 percent of those holding at least a Master's degree planned to vote in the next local election. If the survey results were calculated using the same level of confidence for both groups, which of the following is true?

A. Between 60.1 and 65.9 percent of registered voters holding at most a Bachelor's Degree in the district will vote in the in the next local election.

B. No more than 84.1 percent of registered voters holding at least a Master's Degree in the district will vote in the next local election.

C. The sample size of registered voters holding at most a Bachelor's degree was likely larger than the sample size of registered voters holding at least a Master's degree.

D. The sample size of registered voters holding at most a Bachelor's degree was likely smaller than the sample size of registered voters holding at least a Master's degree.

Workspace

Solution Outline

Step 1: Essentially, what we know here is that we have two groups with different margins of error, indicated by the ± 2.9 and $\pm$ 6.1 portion of the results. The relationship we need to know is that margin of error has an inverse relationship with sample size. That means that as one goes up, the other goes down. So, the group with the larger margin of error has the smaller sample size and the group with the smaller margin of error has the larger sample size.

Answer: C

Notes

Confidence Intervals provide a range of values for a population parameter – like means and proportions – and come with a certain level of confidence. The most common of these are 90, 95, and 99 percent confidence levels. Confidence levels provide greater insight into the certainty that the true value of the population parameter falls within the confidence interval. For example, if the 95% confidence interval for the average annual rainfall (in inches) in a particular area is found to be (46, 51), then we are 95% confident that the true average annual rainfall in this area is between 46 and 51 inches. Another way to frame and interpret a confidence level as it relates to a confidence interval is that it is the percentage of intervals that would contain the true population parameter after repeating the same sampling process a many number of times. Certainty – or confidence – comes with knowing the long-term capture rate of the true population parameter by these intervals.

The width of a confidence interval that is constructed using a sample depends on several factors, namely the level of confidence and the sample size. In short, higher levels of confidence and smaller sample sizes lead to wider intervals, whereas lower levels of confidence and larger sample sizes lead to narrower confidence intervals. The reason being that it is more probable that wider confidence intervals would contain the true value; thus, higher levels of confidence are associated with wider intervals. An interval that is too narrow casts more doubt because the true value of the parameter can too easily fall outside the given range, leading to a lower level of confidence. As the width confidence intervals relates to sample size, refer back to the discussion on margin of error and standard deviation. The higher margin of error and standard deviation caused by smaller sample size would expand out the interval, with the converse also being true.

Table 3.22 | Confidence Intervals in Relation to Confidence Level and Sample Size

Interval Width	Confidence Level	Sample Size
Decreases	Decreases	Increases
Increases	Increases	Decreases

c. Experimental Design

Designing an experiment or study begins with identifying the **target population**, which is the entire group of subjects that is of interest. Such populations can range widely in specificity, but they must be clearly parameterized. For example, researchers may only want to investigate adults over the age of forty-five who experience chronic lower back pain to determine if a new treatment might alleviate their symptoms. After identifying the target population, we would want to obtain a **random sample** of participants from the target population. Random samples decrease sampling bias, leading to a group of subjects that is more representative of the target population.

Some sampling techniques often lead to bias that will skew the results of an experiment or study. For example, a *convenience sample* is a sample that comes from a group of people easy to reach, perhaps due to location or known contacts. So, if a study was done to determine the average number of kids per household in a certain city, then subjects should not solely be chosen from the area's parks, where large families tend to visit with greater frequency. Surveys or opinion polls that rely on volunteers should also come with increased scrutiny because of the built-in *self-selection bias* that occurs when people choose to participate in an investigation. One area we see heightened self-selection bias is in product reviews: consumers are more likely to review a product they hate rather than enjoy.

In experimental design, once there is a random sample from the target population, there should be **random assignment** of subjects into a treatment group and control group. The **treatment group** receives the agent under investigation, such as a therapeutic. The **control group** experiences the same set of variables that the treatment group does, except for the treatment. Instead, the control group may receive a *placebo* to mimic the idea of taking a treatment. By using a control group for comparison, researchers can determine whether a treatment is effective, ineffective, or inconclusive. When other precautions are taken in addition to the aforementioned ones, like creating a *double-blind study* where neither the patient nor physician knows who is in the treatment group, researchers can start to determine **causality.**

There are major differences between correlation and causation. The general rule here is that correlation does not imply causation. **Correlation** is an association between two variables. For example, there is the correlation that people in countries with more televisions per capita have better health outcomes. This might occur because countries with more TVs per capita may be wealthier countries where people have greater access to more advanced healthcare, not simply because there are more TVs around. **Causation** is when one variable is responsible for the outcome of another variable. When researchers use random selection and assignment of subjects into treatment and control groups in double-blind studies, they can start to determine a cause-and-effect relationship.

The outcomes of an experiment or study can only be **generalized** to the population that the sample originated from. That means that if a survey only reached high school juniors in a particular county, we can only make conclusions from the survey results about the high school juniors in that county. In the same way, if we want to get an idea about how students at a state university feel about a particular issue, we would only sample students from that state university, not just anyone from the town or state that the university is located in. In short, clearly identify who was selected as part of the sampling process to make appropriate conclusions and generalizations from the results of an experiment or study to a broader population.

d. Box Plots

Box plots are another chart that helps organize information. Students must know the general features of box plots as summarized in the diagram below. Note that box plots are percentage based – that is, it goes by quartiles. Therefore, every section of the box plot represents 25% of the data beginning with minimum and ending with the maximum. The box plot itself is not specific enough to indicate the mean or mode. However, there is enough information to infer whether the mean is less than or greater than the median. Box plots also do not depict sample size. To create a box plot on your own from raw data, arrange the values from least to greatest and find the median. This divides the data in half. Then find the median of the first half of the data to find the first quartile, or Q_1, and find the median of the second half of the data to find the third quartile, or Q_3.

Box plots give us a quick five-point summary of a data set. Those five points are the minimum, first quartile, median, third quartile, and maximum. With these five points we can determine the range (maximum minus minimum), interquartile range (Q_3 minus Q_1), and the presence of outliers. The chart below labels where we can find these five points on the box plot.

Interquartile Range	$\text{IQR} = Q_3 - Q_1$

Using the diagram below, the regions between the minimum and first quartile, first quartile and median, median and third quartile, and third quartile and max always represent twenty-five percent of the data, regardless of length.

Labeled Box Diagram

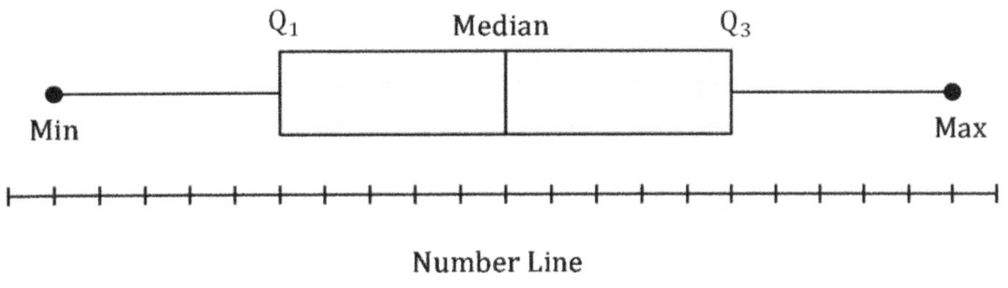

To test for outliers, we will take $1.5 \times \text{IQR}$ and subtract that value from first quartile and add it to third quartile, which will create our lower and upper fences, respectively. All values between the lower and upper fences are not outliers, whereas all values outside of the lower and upper fence are outliers.

Lower and Upper Fences	$\text{Lower Fence} = Q_1 - 1.5 \times \text{IQR}$ $\text{Upper Fence} = Q_3 + 1.5 \times \text{IQR}$

Q+A

Box Plots

Calculator

Average Number of Texts per Day Among Participants Ages 13-24

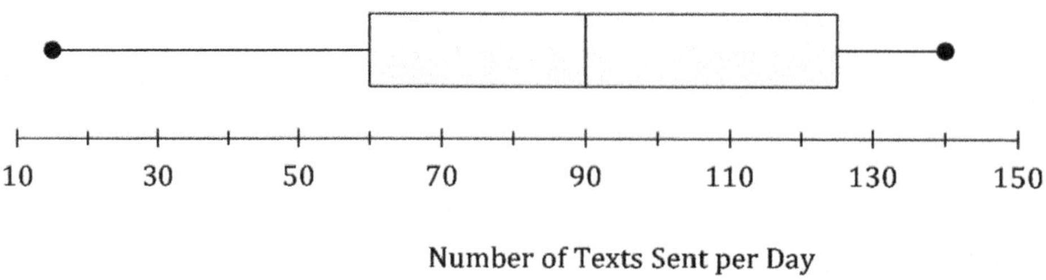

Researchers conducting a study on the number of texts per day sent by teenagers and young adults recorded their data in the box plot above. Which of the following must be true according to their results?

A. Approximately 140 participants were included in the study

B. Most participants in the study sent an average of about 15 to 60 texts per day

C. A quarter of the participants in the study sent an average of 125 to 140 texts per day

D. The mean of the data is 90 texts per day

Workspace

Solution Outline

Step 1: Box plots do not convey enough information to determine the number of participants, nor do they contain enough information to determine the mean or mode of the data. Each section of the box plots contains the same amount of data: 25%. Therefore, a quarter of the participants in the study sent an average of 125 to 140 texts per day.

Answer: C

Notes

Problem Set: Data Trends and Statistics

10 Questions

1

A proposal to expand a community's protected forest is on the ballot in a local election. Prior to the vote, Adam surveys 120 wildlife enthusiasts and bird watchers on the forest grounds. According to his survey results, 87% of the respondents are in favor the proposal. Which of the following is most likely true?

A. Exactly 87% of all voters will be in favor of the proposal.

B. There is an inadequate sample size to make reasonable predictions about the proposal's favorability among bird watchers and wildlife enthusiasts.

C. The sampling method likely creates a bias that misrepresents the proposal's actual favorability.

D. The sampling method was designed in a way that mitigates bias and accurately portrays the proposal's actual favorability with all registered voters.

2

Track Club	
Minutes Run	Number of Runners
40-44	2
45-49	8
50-54	15
55-59	5

Running Group	
Minutes Run	Number of Runners
15-24	2
25-34	8
35-44	15
45-54	5

A competitive track club and local running group each have 30 distance runners. The charts above record how many minutes on average each of their runners jog per day. Which of the following statements best relates the data?

A. The standard deviation for the minutes run among the members of the track club is larger.

B. The standard deviation for the minutes run among the members of the track club is smaller.

C. The standard deviation for the minutes run in each group is the same.

D. There is not enough information to compare the standard deviations between the two groups.

3 Calculator Allowed

The 25 students in a chemistry class perform a titration between an unknown weak acid and the strong base, sodium hydroxide (NaOH). Each student records the volume in milliliters (mL) of sodium hydroxide needed to reach the endpoint of the titration. According to the results summarized in the frequency table shown, which is closest to the average volume in mL of sodium hydroxide used in each titration?

Titration Results	
Volume NaOH (mL)	Frequency
12.5	2
13.7	16
14.0	6
16.2	1

A. 6.3

B. 13.8

C. 14.1

D. 15.7

4

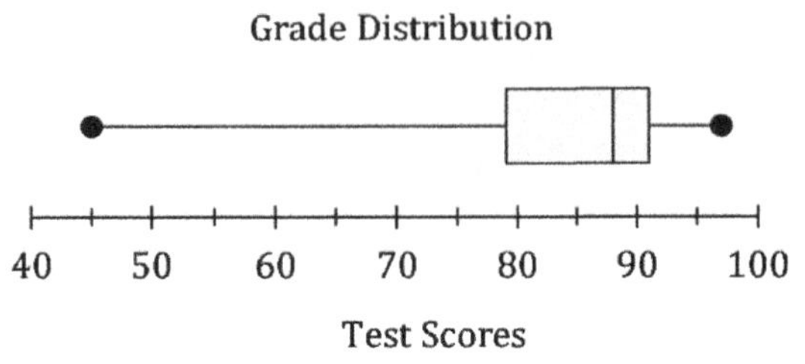

Mrs. Garcia uses the box plot above to represent her students' scores on their most recent history test. According to the data, which of the following statements is NOT correct?

A. Most scores are between 45 and 79

B. The median score is 88

C. The interquartile range is 12

D. The highest score is 97

5

A team of researchers wants to test the ability of a new enzyme inhibitor to effectively treat individuals suffering from the flu. Over 200 randomly selected participants with the flu were randomly assigned to one of two groups in a double-blind study. One group was given the new enzyme inhibitor and the other group was given the standard course of treatment. All other variables were controlled for. The researchers observed that the new enzyme inhibitor was just as effective in treating the flu as the standard course of treatment. Which of the following is a reliable conclusion from the study as it is described?

A. The enzyme inhibitor might be suitable as an alternative therapy to treat the flu.

B. The enzyme inhibitor can be used to prevent the transmission of the flu from one person to the other.

C. The enzyme inhibitor is a better therapy for the flu than the standard course of treatment.

D. No conclusion can be drawn from the data due to the insufficient sample size and restricted scope of the study.

6

A group of marine biologists samples a population of fish in a protected coastal area to determine the percentage of fish in the population that is within a defined length. Their data suggests a normal distribution with a standard deviation of 4.5cm. They also find that the average fish length for this population is 28cm with margin of error of 2.8cm with 95% confidence. What will the marine biologists likely see as a result of increasing the sample size?

I. An increase in the standard deviation

II. A decrease in the margin of error

III. A decrease in the average length

A. I only

B. II only

C. I and II

D. I, II, and III

7 Calculator Allowed

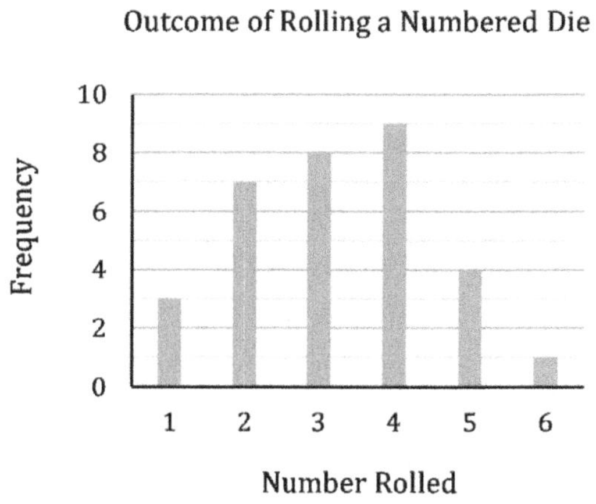

Elaine rolls a numbered die 32 times. The bar graph above shows the frequency of each number she rolled. Which is closest to the mean of this list of numbers?

A. 3.2

B. 3.5

C. 3.8

D. 4.1

8 Calculator Allowed

$$81, 85, 88, x, y$$

The list above shows Daniel's five test scores. If Daniel wants a test average equal to 90, what must be the sum of x and y?

A. 90

B. 98

C. 196

D. 198

9

A study shows that the average yearly household income in a town of 2,484 residents is \$88,000, whereas the median yearly household income is \$49,000. What is the best explanation to these findings?

A. The town's population is too small to use measures of central tendency.

B. The vast majority of residents have a yearly household income under \$49,000.

C. Nearly all residents have yearly household incomes of exactly \$88,000.

D. Some households in the town have much higher yearly incomes that inflate the average.

10 Calculator Allowed

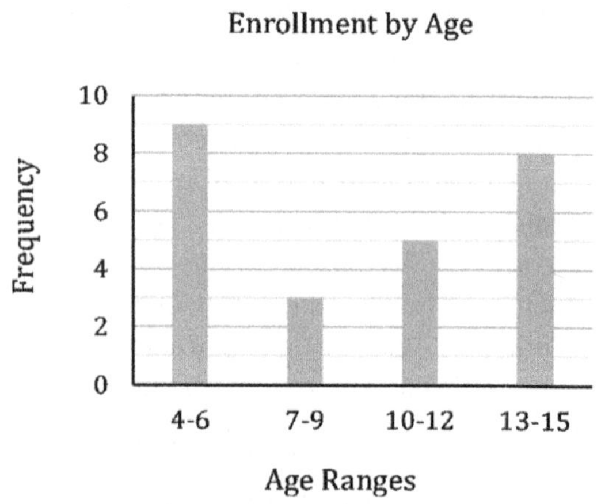

A summer camp enrolls a total of 25 boys and girls between the ages of 4 and 15. What could be the median age given the age distribution in the chart above?

A. 6

B. 8

C. 11

D. 13

Explanations

Problem Set: Data Trends and Statistics

1. C

The sample is likely to contain bias because the respondents are a select group of people that may not be representative of the larger community.

2. B

The frequency distributions are the same, but the spread is not. The track club have a smaller spread of data (40 to 59) and therefore has a smaller standard deviation.

3. B

The average from a frequency list can be found by adding together the product of each value and its respective frequency, then dividing the sum by the total frequency.

$$\frac{12.5(2) + 13.7(16) + 14.0(6) + 16.2(1)}{25}$$

$$\text{Average} = 13.776 \quad \text{or} \quad 13.8$$

Alternatively, input values into List 1 and frequencies into List 2 to calculate 1-Variable Stats, which will display values like the average and median.

4. A

Every section of a box plot represents a quarter of all the data. Longer sections do NOT represent more values.

5. A

The study is well designed and can be used to make correlations. It seems likely that the treatment is suitable as an alternative therapy. The experiment does not disclose that the treatment is effective against transmission; instead it shows infected patients responded to the inhibitor with results similar to the standard course of treatment.

6. B

An increase in sample size will lower both the standard deviation and margin of error.

7. A

The average from a frequency list can be found by adding together the product of each value and its respective frequency, then dividing the sum by the total frequency.

$$\frac{1(3) + 2(7) + 3(8) + 4(9) + 5(4) + 6(1)}{32}$$

$$\text{Mean} = 3.2$$

Alternatively, input values into List 1 and frequencies into List 2 to calculate 1-Variable Stats, which will display values like the average and median.

8. C

Set up a simple average calculation by adding all the values together, then dividing by the total number of test scores. The average should equal 90.

$$\frac{81 + 85 + 88 + x + y}{5} = 90$$

$$\frac{254 + x + y}{5} = 90$$

$$254 + x + y = 450$$

$$x + y = 196$$

9. D

When the mean is much smaller or larger than the median, there are often outliers in that same direction. Since the mean is greater than the median in this question, there are likely a few households with much higher incomes than most.

10. C

The 13th (middle) student occurs in the age group 10-12, so age 11 could be the median.

INTERPRETING DATA EXAM

DIRECTIONS

For questions 1-20, solve every problem and either circle or record the best answer from the answer choices provided. **For questions 21-25**, solve every problem and record your answer in the appropriate format. On the real test, you must grid in your answers according to the directions. You may use any available space in your workbook for scratch work.

NOTES

1. Calculator usage is allowed.
2. Assume real numbers unless otherwise indicated.
3. Figures may not be drawn to scale.
4. All figures are planar, unless otherwise indicated.
5. Domains represent all real numbers of for which f(x) is also a real number.

REFERENCE

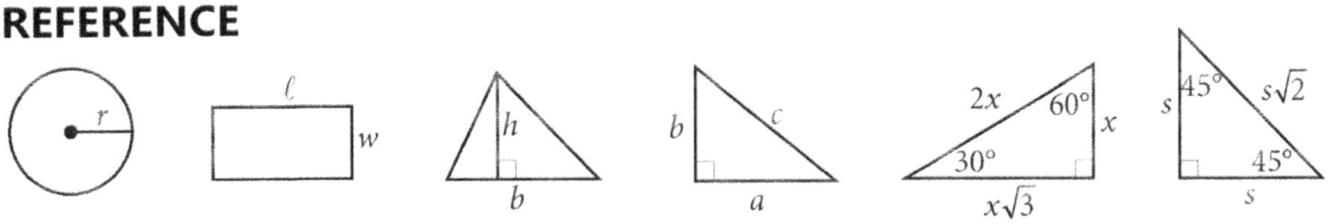

$A = \pi r^2$ $C = 2\pi r$ | $A = \ell w$ | $A = \frac{1}{2}bh$ | $c^2 = a^2 + b^2$ | Special Right Triangles

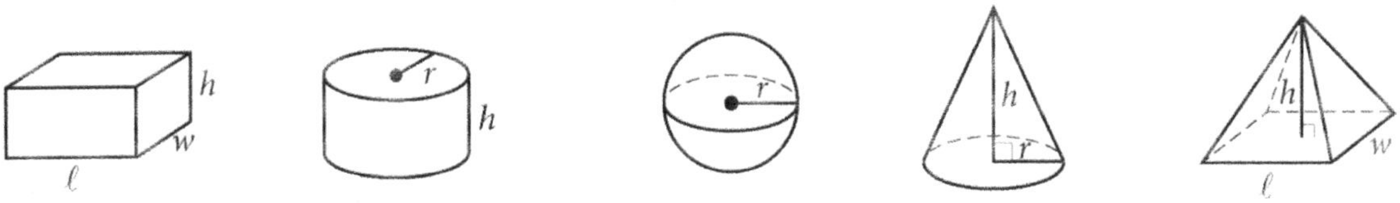

$V = \ell wh$ | $V = \pi r^2 h$ | $V = \frac{4}{3}\pi r^3$ | $V = \frac{1}{3}\pi r^2 h$ | $V = \frac{1}{3}\ell wh$

The arc of a circle contains 360 degrees.

The arc of a circle contains 2π radians.

The interior angles of a triangle add up to 180 degrees.

Calculator

36 Minutes, 25 Questions

1

Number of Arrivals for Commuter Rails A and B in August

	On Time	Delayed	Total
Rail A	1,066	203	1,269
Rail B	899	452	1,351
Total	1,965	655	2,620

Based on the table above, what fraction of the arrivals for Commuter Rail B were made on time?

A. $\frac{899}{1{,}066}$

B. $\frac{899}{1{,}351}$

C. $\frac{899}{1{,}965}$

D. $\frac{899}{2{,}620}$

2

A local university wants to determine the amount of public support for a new ordinance among residents in Town A, where the ordinance was passed. Which of the following details the best method for obtaining reliable results for this university?

A. Survey 200 randomly selected residents of Town A

B. Poll 500 randomly selected students from the university

C. Interview 1,500 randomly selected adults throughout the state

D. Survey 5,200 randomly selected citizens of the United States

3

A group of researchers conducts a five-year study to measure the amount of screen time teenagers spend daily across all of their devices. The study finds that in 2015, the teenagers aged 14-15 averaged x hours of screen time per day, up 11% from 2014. Which of the following expressions represents the average hours of screen time spent per day by teenagers in this age group in 2014?

A. $0.89x$

B. $1.11x$

C. $\frac{x}{0.89}$

D. $\frac{x}{1.11}$

4

A new housing development costs \$130 per square foot to build. Approximately how much does it cost per square meter?

1 meter = 3.28 feet

A. \$425

B. \$850

C. \$1,175

D. \$1,400

5

The school record at Eleanor Prep in the 100-meter sprint is 10.89 seconds. How fast is this in miles per hour?

1609 meters = 1 mile

A. 16.4

B. 18.8

C. 20.5

D. 24.1

Questions 6 -8 use the following information.

The figure below shows the average total cost spent per commuter using a ridesharing app in relation to the distance of the trip. A line of best fit was drawn in to help determine the expected cost for a tri taken by a commuter using this particular ridesharing app.

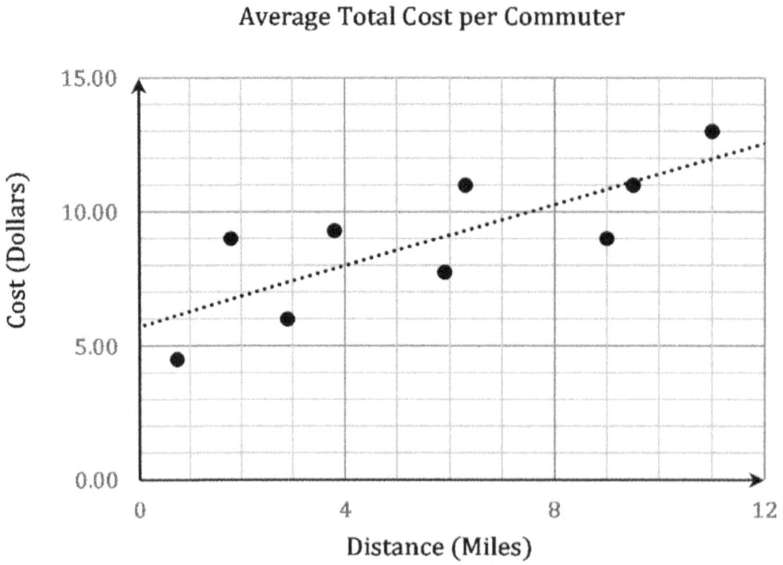

6

What is the expected cost per mile in dollars that a commuter should expect to pay in using the ridesharing app?

A. 0.57

B. 1.83

C. 3.05

D. 6.00

7

About how many more dollars should a user of the ridesharing app expect to pay for an 11-mile trip than for a 2-mile trip?

A. $5.00

B. $7.00

C. $10.00

D. $12.00

8

If the data point with the greatest residual is removed from the scatter plot and a new line of best fit is created, how will the expected cost per mile change for a user of the ridesharing app?

A. The cost per mile will not change

B. The cost per mile will decrease

C. The cost per mile will increase

D. No conclusion can be drawn

Questions 9, 10, and 11 use the following information

The box plots shown below summarize data collected on the monthly rent prices of one-bedroom apartments across multiple districts in both City A and City B. The data was collected in an effort to compare regional differences in rent prices for apartments of similar size and standards.

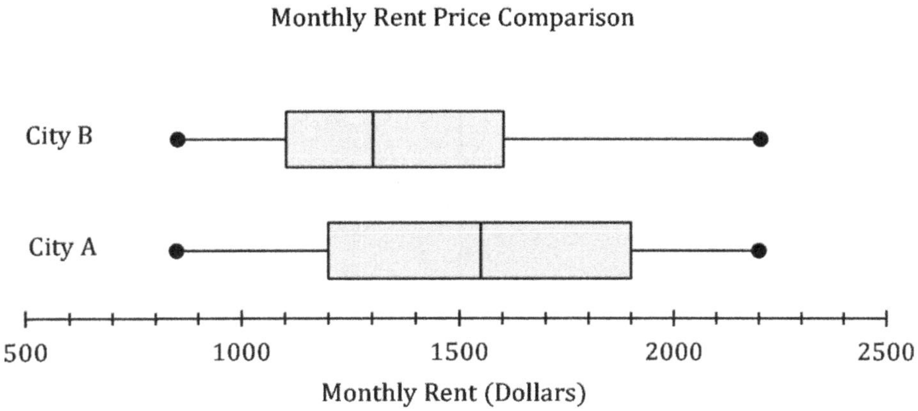

9

Which statement correctly describes the relationship found in the data of rent prices in City A and City B?

A. The interquartile range is smaller for City A

B. The interquartile range is smaller for City B

C. The interquartile range is the same for both

D. The interquartile range cannot be compared

10

Which statement correctly describes the relationship found in the data of rent prices in City A and City B?

A. The sample size is smaller for City A

B. The sample size is smaller for City B

C. The sample size is the same for both

D. The sample sizes cannot be compared

11

Which single statistical measure best summarizes the data of City B

A. mean

B. median

C. mode

D. range

12

If the ratio $1.25 : b$ is equivalent to $a : 10.2$, then what is the product of a and b?

A. 12.75

B. 13.25

C. 14.50

D. 15.75

13

Two savings accounts with the same initial deposit of x dollars grow every year. The first savings account increases by a factor of a annually, and the second increases by a factor of b annually. Which expression shows the difference between the second and first savings accounts after the second year?

A. $x(b + a)$

B. $x(b + a)(b - a)$

C. $x(b - a)$

D. $x(b - a)(b - a)$

14

City A Zones	Total Water Usage
Zone 1	8%
Zone 2	49%
Zone 3	17%
Zone 4	26%

The chart above details the amount of water usage per day in different zones City A. If 18 thousand gallons of water are used in Zone 4, approximately how many thousands of gallons are used Zones 1 and 2 combined?

A. 16.8

B. 39.5

C. 52.1

D. 69.2

15

Researchers studying the gray wolf population in the Midwest find that the male to female ratio is $3 : 7$. From their collected data, they calculate that 452 males are needed to make the male and female populations equal. How many female gray wolves are there according to the researcher's data?

A. 339

B. 462

C. 791

D. 1,130

16

	15-34 min	35-54 min
Company 1	1,562	897
Company 2	1,139	1,294

The table above records the number of deliveries made within certain time intervals for two companies. What fraction of the deliveries made between 35 and 54 minutes were from Company 2?

A. $\frac{1,294}{4,892}$

B. $\frac{1,294}{2,433}$

C. $\frac{1,294}{2,191}$

D. $\frac{1,294}{897}$

17

A sweater is marked 15% off, but is still subject to a 10% sales tax. If the final purchase price is P dollars, then which expression represents the original price of the sweater?

A. $0.95P$

B. $(0.85)(1.10)P$

C. $\frac{P}{0.95}$

D. $\frac{P}{(0.85)(1.10)}$

18

$$1\text{N} = 1\text{kg} \cdot \text{m/s}^2$$

$$1\text{kg} = 1{,}000\text{g}$$

$$1\text{cm} = 10^{-2}\text{m}$$

The force of an object is measured in newtons (N) which is equivalent to kilograms (kg) times meters (m) per second squared (s^2). How many newtons are there for a force of $258\text{g} \cdot \text{cm/s}^2$?

A. 0.00258

B. 0.0258

C. 0.258

D. 2.58

19

Terrestrial Biome	Annual Precipitation
Boreal Forest	180.1
Savanna	146.3
Temperate Forest	155.8
Tropical Rain Forest	483.9
Tundra	13.2
Woodland	126.3

An ecologist measures the annual precipitation in centimeters for several different terrestrial biomes. She records her observations in the chart above. Which two biomes should be excluded to decrease the standard deviation of her data set?

A. Boreal Forest and Woodland

B. Tundra and Woodland

C. Tropical Rain Forest and Tundra

D. Boreal Forest and Tropical Rain Forest

20

Data Set A consists of 50 positive integers and has a minimum value 34. Data Set B contains all of Data Set A and only one additional value of 15. Which of the following must be smaller for Data Set A than for Data Set B?

I. Mean
II. Median
III. Range
IV. Standard Deviation

A. I only

B. I and III only

C. III and IV only

D. II, III, and IV

21

A dog show accommodates a variety of breed categories. Of the participating dogs, 25% are sporting, 40% are hound, 20% are herding, and the remaining 9 dogs are in the terroir group. How many dogs are in the herding group?

22

Test Score (%)	Frequency
84	3
x	5
92	4
98	1

The chart above represents all 13 equally weighted test scores that Nico took in the first semester of his biology class. If his test score average was an 88%, what is the value of x to the nearest integer?

23

Tons of Timber Sold

	Company A	Company B
Paper	2.2	5.3
Furniture	5.6	1.7
Construction	3.1	2.1

Two logging companies A and B ship several tons of timber to businesses that specialize in paper products, furniture, and construction, as summarized in the table above. What percentage by weight of all timber produced by both Company A and Company B goes to construction?

24

Type II diabetes is a chronic disease that affects millions of Americans. An epidemiological study finds that 42% of American adults currently suffer from some type of chronic disease. Of those suffering from chronic disease, 28% have Type II Diabetes. If the US has 265 million adults, how many times more likely is it that an American adult does NOT suffer from Type II Diabetes compared to an American adult who does? Round to the nearest tenth.

25

The student to teacher ratio at a large public university is 244 : 5. How many teachers are there at the university, which has 29,768 students?

Explanations

Interpreting Data Exam

1. B

This is a conditional probability question because it specifies a known constraint: Commuter Rail B. Therefore, only use values for the row that represents Rail B.

Part: On Time Rail B Arrivals

Total: All Rail B Arrivals

$$\frac{\text{Part}}{\text{Total}} = \frac{899}{1{,}351}$$

2. A

Though Choice A has the lowest sample size, it is the best answer because it targets the right demographic: Residents of Town A.

3. D

Hours of screen time in 2014: y

Hours of screen time in 2015: x

$$\text{New Value} = (1 \pm \text{r}) \times \text{Old Value}$$

$$x = (1 + 0.11)y$$

$$x = 1.11y$$

$$y = \frac{x}{1.11}$$

4. D

Use the given equivalency statement to determine how square meters relate to square feet. Then use dimensional analysis to convert units.

$$1 \text{ meter} = 3.28 \text{ feet}$$

$$(1 \text{ meter})^2 = (3.28 \text{ feet})^2$$

$$1 \text{ m}^2 = 10.7584 \text{ ft}^2$$

$$\frac{\$130}{1 \cancel{\text{ft}^2}} * \frac{10.7584 \cancel{\text{ft}^2}}{1 \text{ m}^2} \approx \frac{\$1{,}400}{1 \text{ m}^2}$$

5. C

Use dimensional analysis to convert meters into miles and seconds into hours.

$$\frac{100 \text{ meters}}{10.89 \text{ seconds}} * \frac{1 \text{ mile}}{1609 \text{ meters}} * \frac{3600 \text{ seconds}}{1 \text{ hour}}$$

$$\approx 20.5\text{mph}$$

6. A

The expected cost per mile when using the ridesharing app is found by analyzing the slope of the line of best of fit. This line has a slope of approximately ½, which makes choice A the best answer.

7. A

At 11 miles, the cost obtained from the line of best fit is about $12.00 and, at 2 miles, the cost obtained from the line of best fit is about \$7.00. The difference between the two costs is closest to \$5.00.

8. C

The data point with the greatest residual occurs for the trip of almost two miles. The data point is above the line of best fit, which brings up the line of best towards the left. Removing the data point would cause the left side of the line of best fits to swivel (or rotate) down closer to the remaining data points on that side, creating a steeper line that would indicate an increase in the expected cost per mile.

9. B

The interquartile range is represented by the width of the outer box, which is the gap between the first and third quartiles. Since the box portion of the box plot for rent prices in City B is smaller, it has a smaller interquartile range.

10. D

Box plots work on a percentile basis and do not provide enough information on their own to determine sample size.

11. B

Since the majority of the data falls to the left, the data is somewhat skewed. Medians provide a better indicator of the center of the data in skewed distributions than do other statistical measurements, like the mean.

12. A

Use a proportion to find the product of a and b.

$$\frac{1.25}{b} = \frac{a}{10.2}$$

$$ab = 12.75$$

13. B

Determine expressions for the first and second savings accounts by applying the stated factor of change twice to the initial deposit. Then take their difference and factor the expression.

Second Account: xb^2

First Account: xa^2

Difference

$$xb^2 - xa^2 \quad \text{or} \quad x(b^2 - a^2) \quad \text{or} \quad x(b + a)(b - a)$$

14. B

Use a proportion to solve for the number of gallons used in Zones 1 and 2.

Zone 4 number of gallons: 18 thousand

Zone 4 percent water usage: 26

Zones 1 and 2 number of gallons: x thousand

Zones 1 and 2 percent water usage: 57

$$\frac{18}{26} = \frac{x}{57}$$

$$x \approx 39.5 \text{ thousand}$$

15. C

Use the given ratio to create expressions that represent the number of males and females, then add 452 to the number of males so that it equals the number of females.

$$3x \text{ male} : 7x \text{ female}$$

$$3x + 452 = 7x$$

$$4x = 452$$

$$x = 113$$

$$7x = 791 \text{ female}$$

16. C

This is a conditional probability question because it specifies a known constraint: 35-54 minutes. Therefore, only use values for the column that represents 35-54 minutes

Part: Company 2 Deliveries (35-54 min)

Total: All Deliveries (35-54 min)

$$\frac{\text{Part}}{\text{Total}} = \frac{1{,}294}{2{,}191}$$

17. D

Use formula to calculate the result after any number of percent changes.

Sweater Original Price: x

Sweater Final Price: P

$$\text{New Value} = (1 \pm \text{r}_1)(1 \pm \text{r}_2) \times \text{Old Value}$$

$$P = (1 - 0.15)(1 + 0.10)x$$

$$P = (0.85)(1.10)x$$

$$x = \frac{P}{(0.85)(1.10)}$$

18. A

Use dimensional analysis to convert the given units into the derived units for of Newtons, N. This requires changing grams to kilograms and centimeters to meters.

$$\frac{258 \text{ g} \cdot \text{cm}}{\text{s}^2} * \frac{10^{-2}\text{m}}{1 \text{ cm}} * \frac{1 \text{ kg}}{1{,}000\text{g}}$$

$$0.00258 \frac{\text{kg} \cdot \text{m}}{\text{s}^2} \quad \text{or} \quad 0.00258 \text{ N}$$

19. C

To decrease the standard deviation, takeaway the most extreme (in this case the highest and lowest) data points: Tropical Rain Forest and Tundra

20. C

Since Data Set B is the same as Data Set A, but has one additional point that is lower than the minimum of Data Set A, it has a greater range and standard deviation, but a smaller mean. We do not have sufficient information to make a conclusion about the the median. This means that Data Set A has a smaller range and standard deviation, but a larger mean.

21. 12

Use a proportion to solve for the number of dogs in the Herding group.

Number of Terroirs: 9

Percentage of Terroirs: 15%

Number of Herding: x

Percentage of Herding: 20%

$$\frac{9}{15} = \frac{x}{20}$$

$$x = 12$$

22. 85

Construct a simple average calculation with the given information.

$$\frac{3(84) + 5x + 4(92) + 1(98)}{13} = 88$$

$$\frac{5x + 718}{13} = 88$$

$$5x + 718 = 1144$$

$$5x + 426$$

$$x = 85.2$$

23. 26

Find percentage by finding the part and total, then multiplying by 100 to convert the decimal into a percent.

Part: Tons of timber designated for construction

Total: Tons of timber for all products

$$\frac{3.1 + 2.1}{20}$$

$$\frac{5.2}{20} = 0.26 \quad \text{or} \quad 26\%$$

24. 7.5

Find the probabilities that an American adult has Type II Diabetes and the probability that one does not. Then take a ratio of the two probabilities to determine how many times more likely it is for that an American adult belongs to the first group compared to the second. Of the 42% who have a chronic disease, 28% have Type II Diabetes. Since 28% of 42% is 11.76%, that is the percentage that any American adult suffers from Type II Diabetes. All others do not have the condition, or 88.24% of Americans.

$$P(\text{Diabetes}) = 0.42(0.28) = 0.1176$$

$$P(\text{No Diabetes}) = 1 - 0.1176 = 0.8824$$

$$\frac{0.8824}{0.1176} \approx 7.5$$

25. 610

Use the given ratio to create expressions that represent the number of students and teachers, then solve for the number of teachers.

$$244x \text{ students} : 5x \text{ teachers}$$

$$244x = 29{,}768$$

$$x = 122$$

$$5x = 610$$

IV

REMAINING TOPICS

Complex Numbers

Geometry and Trigonometry

Circle Properties

Introduction

The topics under Additional Topics include imaginary numbers, geometry, trigonometry, and circles. Expect to find these questions in both the no calculator and calculator sections of the test. Altogether, the Additional Topics questions comprise 10% of all math related questions. Though this may not seem significant, it can make or break your test. Often, despite reaching into higher-level math, the questions rooted in these additional topics do not require a greater amount of time or work to complete. For the most part, students can increase their scores through simply understanding and applying a few basic concepts. Pay close attention to special right triangles and circles on the coordinate plane. These questions frequently occur on the test.

4.1 COMPLEX NUMBERS

a. Standard Form

A complex number in standard form can be represented in the form $a + bi$, where a and b are real numbers, generally integers, and $i = \sqrt{-1}$. Though the complex, or imaginary, number i is a constant, it can be treated like a variable. It is conventional to use the variable z to represent a complex number in standard form as seen below.

Standard Form	$z = a + bi$

When complex numbers are in $a + bi$ form makes applying general operations – addition, subtraction, multiplication, and division – a much more efficient process. However, in some cases, we will be given terms where i has an exponent other than one. When faced with such terms, we should first evaluate them so that we reach either an integer value or term where i has an exponent of one. To do this, we should realize that the values of i with higher exponents follow a cyclical pattern identified in Table 4.11 below.

Table 4.1) | Values of the Imaginary Number, i

$i^0 = 1$	$i^1 = i$	$i^2 = -1$	$i^3 = -i$

For exponents on i beyond these, all we need to do is divide the exponent by 4 and find the remainder. The remainder will correspond with the correct exponent on i so that you can simplify accordingly. Suppose we need to simplify i^{39}. Just divide 39 by 4 to get 36 with a remainder of 3. Therefore, $i^{39} = i^3 = -i$. The most important substitution to know here is that $i^2 = -1$. This replacement occurs more frequently than the others.

Q+A

Standard Form **Diagnostic Exam | No Calculator Q15**

$$(2 - 3i) - 4(5 - 2i)$$

Which of the following complex numbers in $a + bi$ form is equivalent to the expression above given that $i = \sqrt{-1}$?

A. $-18 + 5i$

B. $-18 - 5i$

C. $-18 - 11i$

D. $-22 + 11i$

Workspace

Solution Outline

Step 1: Use Distributive Property to expand the given expression. This allows you to "get rid of" the parentheses so that you can combine like-terms: $(2 - 3i) - 4(5 - 2i) = 2 - 3i - 20 + 8i$

Step 2: Combine like-terms. Add and subtract the whole numbers to each other and then the coefficients to i. The result is $-18 + 5i$

Answer: A

Notes

Note that the example above did not require making any substitutions for any of the i terms; we could treat it almost exactly like any other polynomial expression. However, in the next example we will have to remember that imaginary terms with integer exponents can take on the values of $1, i, -1,$ and $-i$ as summarized in Table 4.11.

Q+A

Standard Form **No Calculator**

$$2i(5i - 4)$$

Which of the following complex numbers in $a + bi$ form is equivalent to the expression above given that $i = \sqrt{-1}$?

A. $-10 + 8i$

B. $10 - 8i$

C. $10 + 8i$

D. $-10 - 8i$

Workspace

Solution Outline

Step 1: Use the distributive property to expand out the expression.

$2i(5i - 4)$

$10i^2 - 8i$

Step 2: Substitute $i^2 = -1$, then simplify.

$10(-1) - 8i$

$-10 - 8i$

Answer: D

Notes

b. Rationalize Denominator

Another testing focal point is rationalizing the denominator of a complex number in standard form. Remember that rational numbers are real numbers that can be written as the fraction of integers or numbers with decimals that either end or repeat. When we talk about rationalizing the denominator, generally that also means we want the denominator to be a whole number. In this section, the imaginary number i should not occur in the denominator. Since i is actually a square root, the ideas discussed earlier in this book about rationalizing denominators that contain square roots still hold. However, we will recap these rules as they relate the complex number, i.

Since standard form takes the shape $a + bi$ and a can equal zero, the denominator may be either a monomial or binomial. When there is a singular imaginary term in the denominator, all we need to do is create an equivalent fraction by multiplying both the numerator and denominator by i.

Q+A

Rationalize the Denominator **No Calculator**

$$\frac{-i + 4}{2i}$$

If the expression above is written in the form $a + bi$ and $i = \sqrt{-1}$, then what is the value of a?

A. -2

B. $-1/2$

C. 1

D. 2

Workspace

Solution Outline

Step 1: Multiply by numerator and denominator by i.

$$\frac{(-i + 4)}{2i} * \frac{i}{i} = \frac{-i^2 + 4i}{2i^2}$$

Answer: B

Step 2: Substitute $i^2 = -1$, then simplify.

$$\frac{-i^2 + 4i}{2i^2} = \frac{-(-1) + 4i}{2(-1)} = \frac{1 + 4i}{-2}$$

$$a + bi = -\frac{1}{2} - 2i$$

Notes

When the denominator is instead a binomial in the standard form $a + bi$, multiply the numerator and the denominator by its conjugate, $a - bi$.

Complex Conjugates	$a + bi$ and $a - bi$

Remember to distribute when multiplying by expressions with multiple terms. To save time, realize that the denominator will end up as a difference of squares, which means we only need to focus on multiplying out the first and last terms.

Q+A

Rationalize the Denominator **No Calculator**

$$\frac{2 - 7i}{1 + 2i}$$

If the expression above is written in the form $a + bi$ and $i = \sqrt{-1}$, then what is the value of b?

A. $-12/5$

B. -12

C. $-11/5$

D. -11

Workspace

Solution Outline

Step 1: Multiply the numerator and the dominator by the conjugate of $1 + 2i$. The conjugate of $1 + 2i$ is $1 - 2i$.

$$\frac{(2 - 7i)}{(1 + 2i)} * \frac{(1 - 2i)}{(1 - 2i)}$$

$$\frac{2 - 4i - 7i + 14i^2}{1 - 4i^2}$$

Answer: C

Step 2: Use the substitutions $i^2 = -1$ and combine like-terms.

$$\frac{2 - 4i - 7i - 14}{1 + 4} = \frac{-12 - 11i}{5}$$

$$a + bi = -\frac{12}{5} - \frac{11i}{5}$$

Notes

Problem Set: Complex Numbers

10 Questions

1

What is the difference between the complex numbers $4 - 3i$ and $3 - 4i$, given that $i = \sqrt{-1}$?

A. $7 - 7i$

B. $7 + 7i$

C. $1 - i$

D. $1 + i$

2

$$i^6 + i^8 + i^{10} + i^{12}$$

Which of the following is equivalent to the expression above, given that $i = \sqrt{-1}$?

A. 0

B. $2i$

C. 4

D. $4i$

3

$$\frac{1 + 2i}{3 - 2i}$$

The expression above is equal to the complex number $a + bi$, where a and b are constants and $i = \sqrt{-1}$. What is the value of b?

A. $-1/13$

B. $7/13$

C. $-8/13$

D. $8/13$

4

$$z_1 = -12 + i$$

$$z_2 = -12 - i$$

What is the product of the complex numbers z_1 and z_2 provided above, given that $i = \sqrt{-1}$?

A. 143

B. 144

C. 145

D. 146

5

$$\frac{-15 + 10i}{5i}$$

What is an equivalent expression to the one above, given $i = \sqrt{-1}$?

A. $2 + 3i$

B. $2 - 3i$

C. $3 + 2i$

D. $3 - 2i$

6

$$\frac{5+3i}{1+7i}$$

The expression above is equal to the complex number $a + bi$, where a and b are constants and $i = \sqrt{-1}$. What is the value of a?

A. -32

B. $-16/25$

C. $13/25$

D. 26

7

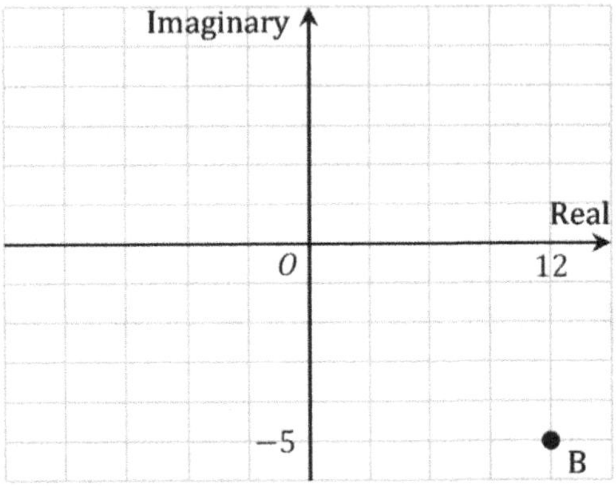

The complex number $12 - 5i$ is plotted as point B on the real-imaginary axis above. What is the magnitude of the complex number? (Note that the magnitude is equivalent to the distance between the point and the origin.)

A. 7

B. 13

C. 17

D. 23

8

$$i^{49}$$

Given that i is the imaginary number $\sqrt{-1}$, which of the following expressions is equivalent to the one above?

A. -1

B. 1

C. $-i$

D. i

9

$$2i(8i + 8i^3)$$

Given that i is the imaginary number $\sqrt{-1}$, which of the following expressions is equivalent to the one above?

A. 0

B. 16

C. 32

D. 48

10

$$z_1 = -3 + i$$
$$z_2 = 8 - 9i$$

What is the sum of the complex numbers $z_1$and z_2 provide above, given that $i = \sqrt{-1}$?

A. $5 - 8i$

B. $5 + 8i$

C. $-24 - 9i$

D. $-24 + 9i$

Explanations

Problem Set: Complex Numbers

1. D

Subtract the given quantities by distributing, then combining like-terms.

$$4 - 3i - (3 - 4i)$$

$$4 - 3i - 3 + 4i$$

$$1 + i$$

2. A

Reduce each term of i to find the final value of the expression. One way to do this is by replacing each exponent with its remainder after dividing it by four. Then substitute: $i^0 = 1$ and $i^2 = -1$.

$$i^6 + i^8 + i^{10} + i^{12}$$

$$i^2 + i^0 + i^2 + i^0$$

$$(-1) + 1 + (-1) + 1 = 0$$

3. D

Rationalize the denominator by creating an equivalent fraction using the conjugate. The conjugate to the complex number $a + bi$ is $a - bi$. Recall that $i^2 = -1$.

$$\frac{1 + 2i}{3 - 2i} * \frac{3 + 2i}{3 + 2i}$$

$$-\frac{1}{13} + \frac{8}{13}i$$

$$b = 8/13$$

4. C

Multiply the given complex numbers and substitute $i^2 = -1$.

$$(-12 + i)(-12 - i)$$

$$144 - i^2$$

$$144 + 1$$

$$145$$

5. A

Rationalize the denominator by creating an equivalent fraction using i, then substitute $i^2 = -1$. Separate the fraction and simplify.

$$\frac{-15 + 10i}{5i}$$

$$\frac{-15 + 10i}{5i} * \frac{i}{i}$$

$$\frac{-15i + 10i^2}{5i^2}$$

$$\frac{-15i + 10(-1)}{5(-1)}$$

$$\frac{-15i - 10}{-5}$$

$$3i + 2$$

6. C

Rationalize the denominator by creating an equivalent fraction using the conjugate. The conjugate to the complex number $a + bi$ is $a - bi$. Recall that $i^2 = -1$.

$$\frac{5+3i}{1+7i} * \frac{1-7i}{1-7i}$$

$$\frac{13}{25} - \frac{16}{25}i$$

7. B

Use the Pythagorean Theorem, where 12 and 5 are a and b, to find the magnitude.

$$a^2 + b^2 = c^2$$

$$12^2 + 5^2 = c^2$$

$$c = 13$$

8. D

Replace the exponent with its remainder after dividing it by four. Simplify accordingly.

$$\frac{49}{4} = 12, R = 1$$

$$i^1 = i$$

9. A

Make the substitution $i^3 = -i$, then simplify accordingly.

$$2i(8i + 8i^3)$$

$$2i(8i - 8i)$$

$$2i(0)$$

$$0$$

10. A

Add the given complex numbers and simplify by combining like-terms.

$$-3 + i + (8 - 9i)$$

$$-3 + i + 8 - 9i$$

$$5 - 8i$$

4.2 GEOMETRY AND TRIGONOMETRY

a. Lines and Angles

This section begins the geometry portion of the test, where we will review properties of lines, angles, and 2D & 3D figures. The most important line and angle properties are summarized in the pages that follow, which include the relationships of parallel lines and transversals as well as adjacent angles.

Before we delve into line and angle theorems, we should first recap some basic vocabulary terms. Figures that are **congruent** are identical in both size and shape, such that congruent angles have the same degree measure. Similarity, on the other hand, implies same shape but not same size. Angle pairs that are **complementary** must add up to 90 degrees and those that are **supplementary** must sum to 180 degrees. As we learned in the first chapter on lines, **parallel** lines lie on the same plane and never cross, whereas **perpendicular** lines cross at a right angle. Line **segments** are a portion of a line with distinct endpoints and **bisectors** include any line or segment that divide something – such as a segment or angle – into two equal parts. This means that an **angle bisector** divides an angle perfectly in half and a **perpendicular bisector** divides a line segment in half while crossing that segment to form a right angle.

Parallel Lines and Transversals

A **transversal** is a line that crosses two or more other lines. A transversal intersects parallel lines in the exact same way, creating both congruent and supplementary angle pairs. The word alternate means on different sides of the transversal. The word interior means inside the parallel lines. The word exterior means outside of the parallel lines. Importantly, if $m \parallel n$ and l is a transversal, then alternate exterior, alternate interior, and corresponding angles are congruent. Vertical angles are always congruent. Table 4.21 summaraizes these angle pairs and displays a diagram.

Table 4.21 | Parallel Lines and Transversal Summary

Angle Pairs		
Alternate Exterior	$\angle 1 \cong \angle 8$	$\angle 2 \cong \angle 7$
Alternate Interior	$\angle 3 \cong \angle 6$	$\angle 4 \cong \angle 5$
Corresponding	$\angle 1 \cong \angle 5$	$\angle 3 \cong \angle 7$
	$\angle 2 \cong \angle 6$	$\angle 4 \cong \angle 8$
Vertical	$\angle 1 \cong \angle 4$	$\angle 5 \cong \angle 8$
	$\angle 2 \cong \angle 3$	$\angle 2 \cong \angle 7$

Diagram

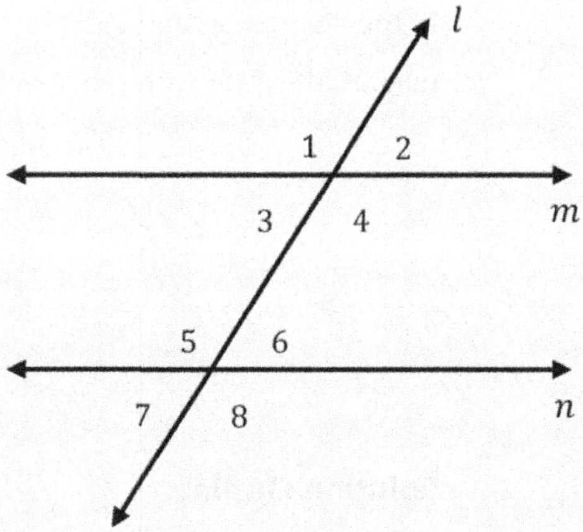

Adjacent Angles

Angles that share a vertex and a side such that the lie next to each other are called **adjacent angles.** The sets of adjacent angles that are frequently tested include those that either form one side of a line or radiate out in all directions. For example, a **linear pair** includes the two adjacent angles that form one side of a line and, consequently, have a sum of 180 degrees. Still, adjacent angles always follow the **Angle Sum Postulate,** which states that angle measurements are additive, such that the outermost angle is equal to the sum of the interior adjacent angles that comprise it. Table 4.22 summarizes the most important angle sum properties.

Table 4.22 | Angle Sum Summary

Angle Sum Postulate	One Side of Line	All Directions
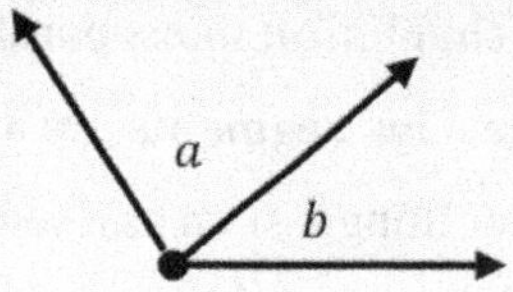	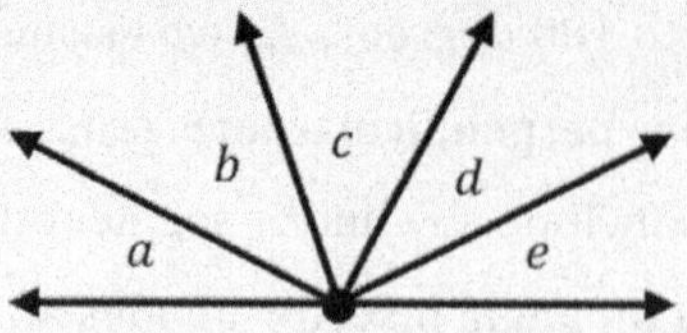	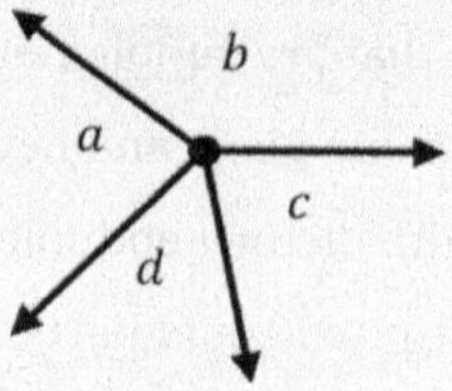
Full angle $= a + b$	$a + b + c + d + e = 180$	$a + b + c + d = 360$

Q+A

Angle Sum Postulate

No Calculator

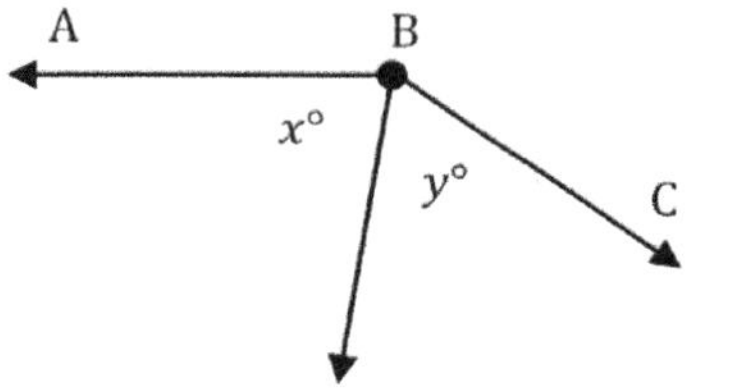

A. 28

B. 30

C. 82

D. 112

In the diagram above, $\angle ABC = 142°, x = 4z,$ and $y = z + 2$. What is magnitude of the difference between x and y, $|x - y|$?

Workspace

Solution Outline

Step 1: Write the expression for the magnitude of the difference between x and y in terms of z.

$$|x - y| = |4z - (z + 2)| = |3z - 2|$$

Step 2: Find z by adding x and y to equal 142.

$$x + y = 142$$
$$4z + z + 2 = 142$$
$$5z = 140$$
$$z = 28$$

Step3: Plug z into the expression generated in Step 1, $|3z - 2|$.

$$|3(28) - 2)| = 82$$

Answer: C

Notes

b. Polygons and 3D Figures

Two-dimensional figures with straight sides that make enclosed shapes are called **polygons**. Students must know general interior and exterior angle properties of polygons, as well as some other basic nomenclature used to describe their general features. To begin the review, an **interior angle** is an angle inside a polygon that is formed by its consecutive sides, whereas an **exterior angle** is an angle formed outside a polygon when one of its sides is extended. Together the interior angle and exterior angle create a linear pair. A polygon with all side of equal length are called **equilateral** and those with all angles of equal measure are called **equiangular.** A polygon that is both equilateral and equiangular is called **regular.** All regular polygons, therefore, are **convex,** meaning all interior angles are less than 180°. A **concave** polygon has at least one interior angle that is greater than 180°. Table 4.23 below includes the names of several polygons and the sum of their interior angles.

Table 4.23 | Polygon Name and Interior Angle Sum

Number of Sides	Name of Polygon	Interior Angle Sum
3	Triangle	180°
4	Quadrilateral	360°
5	Pentagon	540°
6	Hexagon	720°
7	Heptagon	900°
8	Octagon	1080°
9	Nonagon	1260°
10	Decagon	1440°
11	Undecagon	1620°
12	Dodecagon	1800°
n	n-gon	$(n - 2)180°$

The sum of the interior angles and exterior angles are summarized in the Table 4.24 below, which uses a pentagon as an example. Though the sum of the interior angles is dependent on the number of sides it contains, the sum of the exterior angles is not as it is always 360 degrees. This is a unique feature of polygons that frequently finds its way onto the test.

Table 4.24 | Polygon Names and Interior Angle Sums

Interior Angle Sum		Exterior Angle Sum	
Formula	Sum $= (n-2)180°$	Formula	Sum $= 360°$
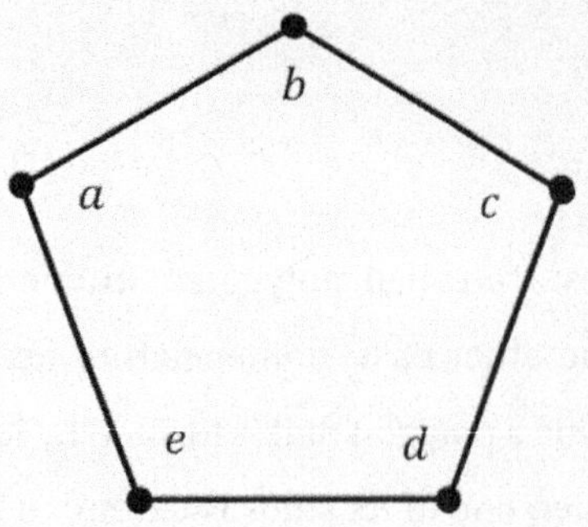		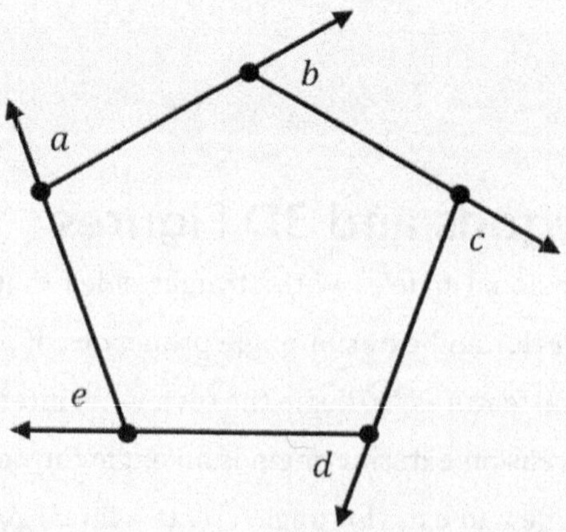	
Sum $= (5-2)180°$	$a+b+c+d+e = 540°$	Sum $= 360°$	$a+b+c+d+e = 360°$

Q+A

Interior and Exterior Angle Properties

No Calculator

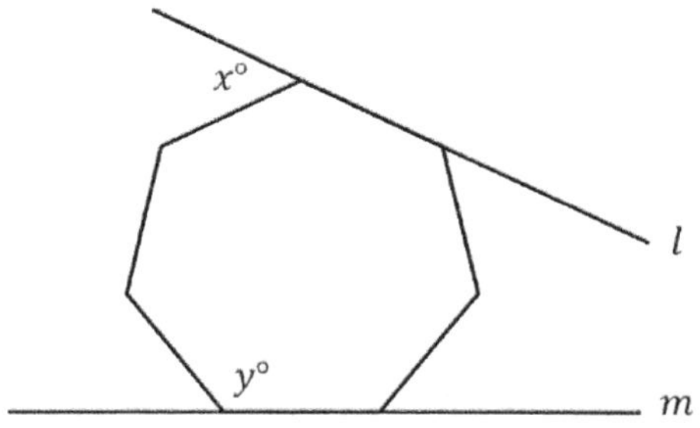

The lines l and m form two sides of a regular heptagon as shown in the diagram above. Line l creates the exterior angle of measure $x°$ and line m creates the interior angle $y°$. What is the sum of x and y?

A. 102

B. 129

C. 165

D. 180

Workspace

Solution Outline

Step 1: Since the heptagon is regular, all interior angles are congruent to each other and all exterior angles are congruent to each other.

Step 2: Any unique interior angle with its corresponding exterior angle form a linear pair and are, thus, supplementary. Therefore, the sum of x and y must be 180.

Answer: D

Notes

In addition to finding angle measures, we will also need to evaluate other quantities, such as perimeter, area, and volume. Recall that **perimeter** refers to the length around an object and that both **area** and **volume** quantify the amount of space the object occupies in two and three dimensions, respectively. For a 3D figure, we may also wish to find its **surface area**, which is the sum total area of each face. Tables 4.25a through 4.25c summarize the formulas of various 2D and 3D figures. Though it is not polygon, circles are two dimensional shapes that are included in this section, but will be discussed in greater detail later on. Note that the length around a circle is called the **circumference**, rather than perimeter.

Table 4.25a | Perimeter and Circumference Formulas

Figure	Polygon	Circle
Formula	$P =$ Sum of all the sides	$C = 2\pi r$

Table 4.25b | Area Formulas

Figure	Triangle	Parallelogram	Trapezoid	Circle
Formula	$A = \frac{1}{2}bh$	$A = bh$	$A = \frac{1}{2}(b_1 + b_2)h$	$A = \pi r^2$

Table 4.25c | Volume Formulas

Figure	Prism	Pyramid	Sphere
Formula	$V = BH$	$V = \frac{1}{3}BH$	$V = \frac{4}{3}\pi r^3$

In the Tables 4.25a through 4.25c, r is the length of a radius, b is the length of a base, h is the height of a 2D shape, B is the area of the base to a 3D figure, and H is the height of 3D figure. It is also useful to point out that the volume formula of a cylinder matches that of a prism and the volume formula of a cylinder matches that of a pyramid, where the area of the base is that of a circle.

Q+A

Change in Volume **Diagnostic Exam | No Calculator Q12**

The volume of a right square pyramid is 42 cubic meters. If the side lengths to the base double, what is the volume, in cubic meters, of the pyramid?

A. 21 B. 42 C. 84 D. 168

Workspace

Solution Outline

Step 1: Use the volume formula of a square pyramid and identify the dimensions that change stated in the question. Remove all constants.

$$V = \frac{1}{3}s^2h$$

$$V \propto s^2$$

Step 2: Find the factor of change that results from doubling the side lengths to the base from s to $2s$.

$$V \propto (2s)^2$$

$$V \propto 4s^2$$

Step 3: The factor of change is four. That means the new volume is four times the original.

$$4(42) = 168$$

Answer: D

Notes

We can also determine the factor of change for the volume when there is a percent change that affects one or more dimension to the 3D figure. Recall that factor $(1 \pm r)$ can be used to implement a percent change, which we will see in the next example from the Diagnostic Exam.

Q+A

Change in Volume **Diagnostic Exam | Calculator Q32**

The radius of a right cylinder increases by 15%. If the height remains constant, by what factor, rounded to the nearest tenth, does the volume change?

Workspace

Solution Outline

Step 1: Use the volume equation of a right cylinder

$$V = \pi r^2 H$$

Step 2: Ignore elements of the equation that do not change.

$$V \propto r^2$$

Step3: Apply the percent increase so that r goes to $1.15r$. The resulting coefficient represents the factor of change

$$V \propto (1.15r)^2$$

$$V \propto 1.3225r^2$$

Answer: 1.3

Notes

c. Triangles and Trigonometry

Arguably, the most important shape to know on the test is the triangle. All triangles share a few general properties, like the sum of the interior angles being 180° and the sum of the lengths of the smallest two sides being greater than the length of the largest side. However, there are several types of triangles, and each comes with its own unique properties. By and large, the properties of equilateral, isosceles, and right triangles predominate the test. Their properties are summarized in Tables 4.26a and 4.26b that follow. Test-takers must also demonstrate proficiency in using trigonometric properties in both right and similar triangles.

In brief, recall that **equilateral** triangles have three congruent sides, leading to three congruent angles of sixty degrees, while **isosceles** triangles only have two congruent sides, leading to two congruent base angles. Triangles that have no congruent sides or angles are called **scalene.** **Right** triangles have one angle of ninety degrees, and follow the **Pythagorean Theorem** ($a^2 + b^2 = c^2$) and the **Trigonometric Ratios**, like sine, cosine, and tangent. The two special right triangles we will review in this section have access to on the formula sheet provided when taking the test are the 30-60-90 and the 45-45-90 triangles.

Table 4.26a | Triangle Properties

All Triangles	
Interior Angle Sum	180°
Relationship Among Sides	$a + b > c$ $a + c > b$ $b + c > a$
Sides and Angles	If $A < B < C$, then $a < b < c$
Area Formula	$A = \frac{1}{2}bh$

Diagram

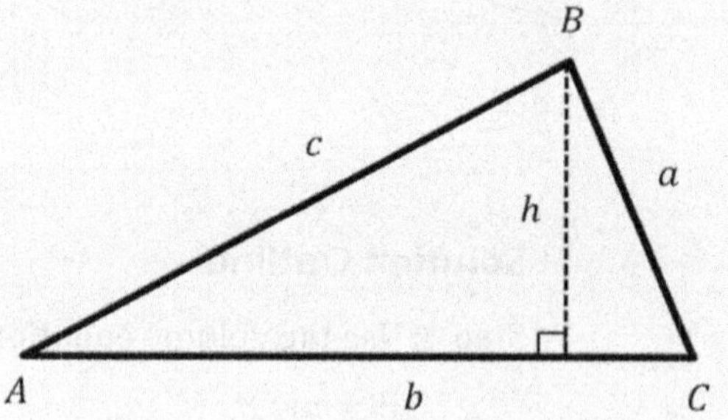

Isosceles Triangles	
Base Angles	Congruent
Legs	Congruent
Altitude from Vertex	Bisects vertex and base

Diagram

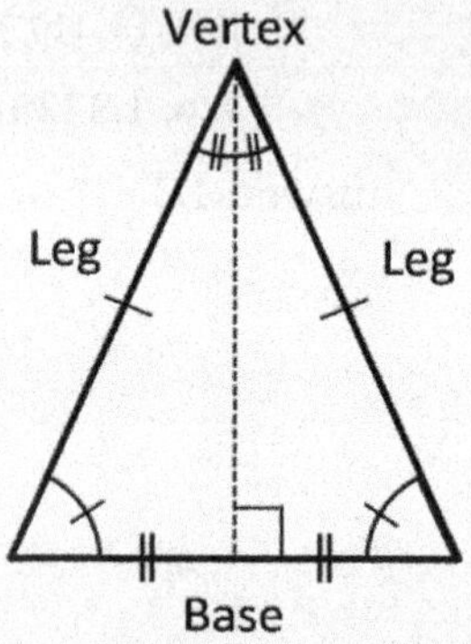

Equilateral Triangles	
Angles	60-60-60
Sides	Congruent
Altitude from Vertex	Bisects vertex and base

Diagram

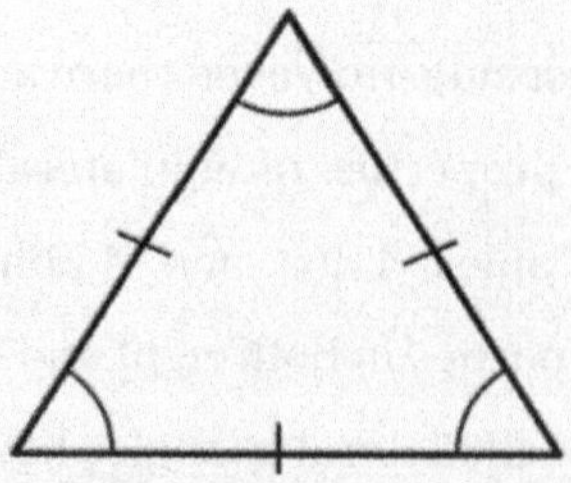

Table 4.26b | Triangle and Trigonometry Summary

Right Triangles		Diagram
Complementary Angles	$A + B = 90°$	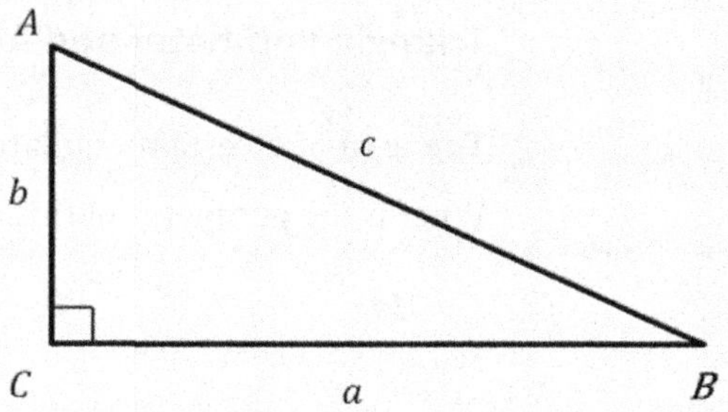
Pythagorean Theorem	$a^2 + b^2 = c^2$	
Trigonometric Ratios SOH – CAH – TOA	$sinA = \frac{opp}{hyp} = \frac{a}{c}$ $cosA = \frac{adj}{hyp} = \frac{b}{c}$ $tanA = \frac{opp}{adj} = \frac{a}{b}$	

Pythagorean Triples and Multiples	
3 – 4 – 5	6 – 8 – 10
5 – 12 – 13	10 – 24 – 26
7 – 24 – 25	14 – 48 – 50

Special Right Triangles	
$30° - 60° - 90°$	$45° - 45° - 90°$
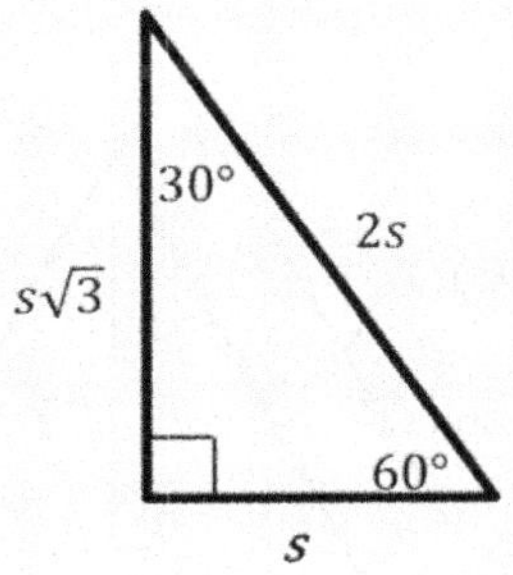	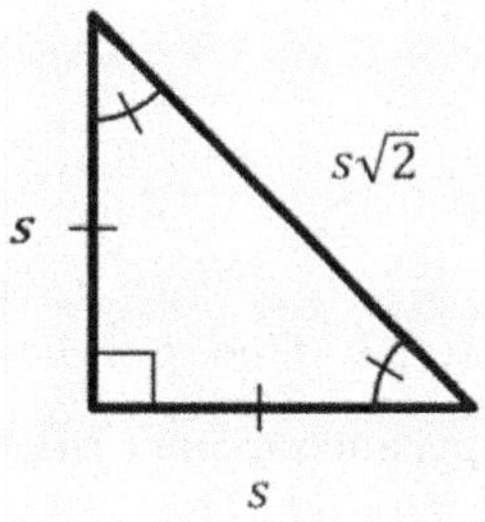

In addition to knowing the general properties of right triangles summarized in the table above, we should also quickly discuss a couple related trigonometric ones as well. For example, when two triangles are similar, corresponding angles are congruent, which would, thus, have equivalent ratios for any given trigonometric function. Just as importantly, sine and cosine are equal when their angles are complementary. Since right triangles have a built-in pair of complementary angles – the two acute angles – we see this property commonly applied to them, but it is true for any two angles that add up to ninety degrees.

Co-Function Property	$\sin(A) = \cos(B)$ if A and B are complementary

The examples that follow come from the Diagnostic Exam and have us relate many properties of triangles at once. In Calculator Question 9, we will have to dissect an equilateral triangle into two 30-60-90 special right triangles to determine the relationship between its perimeter and area.

Q+A

Triangle Perimeter and Area

Diagnostic Exam | Calculator Q9

The area of a certain equilateral triangle with side lengths of 8 centimeters (cm) is $s^2\sqrt{3}$ cm^2. What is the perimeter of this triangle in terms of the constant s?

A. $3s$ B. $6s$ C. $9s$ D. $12s$

Workspace

Solution Outline

Step 1: Break down an equilateral triangle into a 30°, 60°, 90° triangle to see its base and height.

Step 2: Use area formula to find out how s relates to the side length, 8cm.

$$A = \frac{1}{2}bh$$

$$s^2\sqrt{3} = \frac{1}{2}(8)(4\sqrt{3})$$

$$s^2\sqrt{3} = 16\sqrt{3}$$

$$s = 4$$

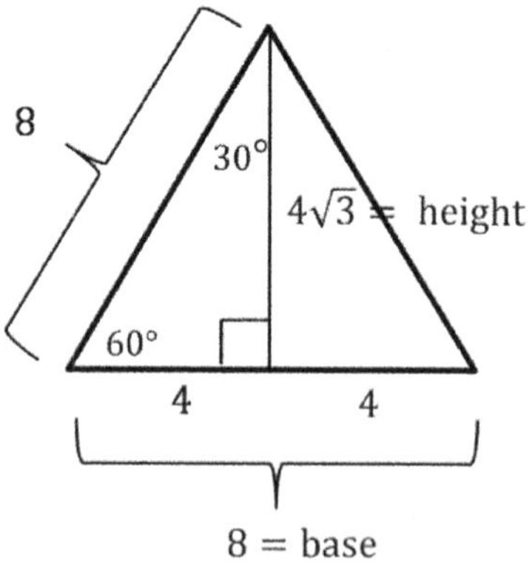

Step3: Find out how many times s goes into the perimeter. Since the triangle is equilateral, all sides are 8cm and the perimeter is 24.

$$4n = 24$$

$$n = 6$$

$$P = 6s$$

Answer: B

Notes

The next two examples will require the trigonometric properties (discussed following Table 4.26b) regarding similar right triangles as well as complementary angles.

Q+A

Trigonometry and Similarity

Diagnostic Exam | Calculator Q23

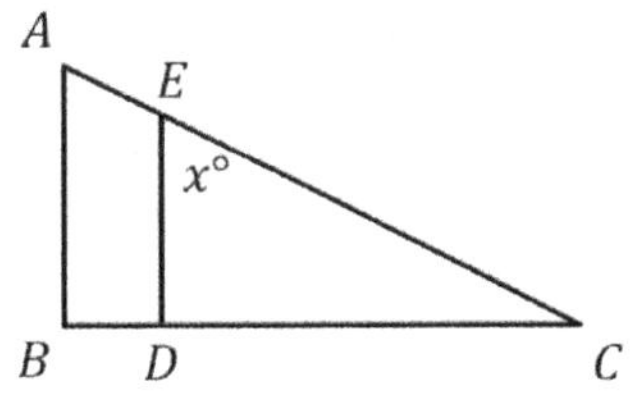

A. 5/12

B. 5/13

C. 12/5

D. 12/13

In triangle ABC, above, $\angle B$ is 90°, $AB = 10$, and $BC = 24$. Provided $\overline{AB} \parallel \overline{DE}$, $CD = 18$, and $\angle DEC = x°$, what is the value of $\cos x°$?

Workspace

Solution Outline

Step 1: Identify that $\Delta ABC \sim \Delta EDC$ due to them having all congruent angles. Both triangles have a right angle and shared angle at C, so their third angle must also be congruent to each other. Similar figures always have congruent angles.

Step 2: The corresponding angles of similar triangles have the same trigonometric ratios because those angles are congruent. In this question, that means $\cos(x°) = \cos(A)$. Seeing as triangle ABC is a right triangle with side lengths that are multiples of the 5, 12, 13 Pythagorean Triple, $\cos(x°) = 5/13$

Answer: B

Notes

As we will see in the next example, we can use trigonometric functions and their properties for angles that are not part of right triangles. In Diagnostic Exam No Calculator Question 18, for instance, a right angle is broken up into two adjacent angles, which are, therefore, complementary. Here we will use the Co-Function Property of Sine and Cosine.

Q+A

Co-Function Property **Diagnostic Exam | No Calculator Q18**

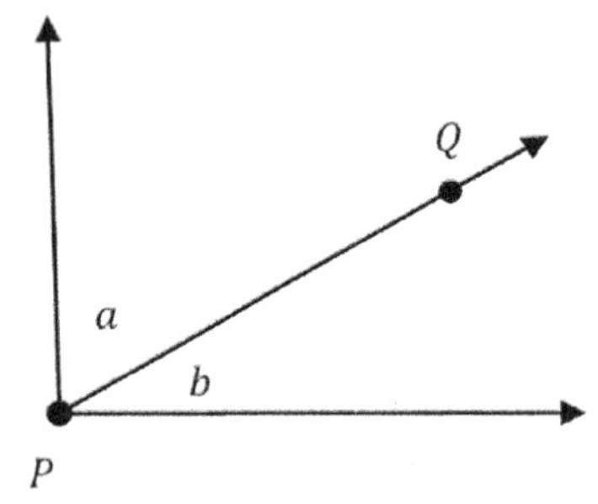

The ray $\overrightarrow{PQ}$ divides the right angle at P into two different angles, a and b, shown in the diagram to the left. What is the value of the expression $\sin a - \cos b$?

Workspace

Solution Outline

Step 1: Since angles a and b are complementarye, $\sin a = \cos b$. The expression is equal to zero as a result.

Answer: 0

Notes

Knowing the value of a trigonometric ratio gives us enough information to create our own right triangle, which is helpful when we are required to solve for other trigonometric ratios, especially in the No Calculator portion of the exam.

Q+A

Trigonometric Ratios **Diagnostic Exam | Calculator Q24**

Given that $\sin\theta = 2T$ and $\cos\theta = T$, what could be the value of T?

A. $^1/_2$ B. $^{\sqrt{2}}/_2$ C. $^{\sqrt{3}}/_2$ D. $^{\sqrt{5}}/_5$

Workspace

Solution Outline

Step 1: Label a right triangle using the given values of sine and cosine. It is helpful to view each with a denominator of one to "see" the hypotenuse.

$$sin\theta = \frac{2T}{1} \quad \text{and} \quad cos\theta = \frac{T}{1}$$

Step 2: Follow Pythagorean Theorem

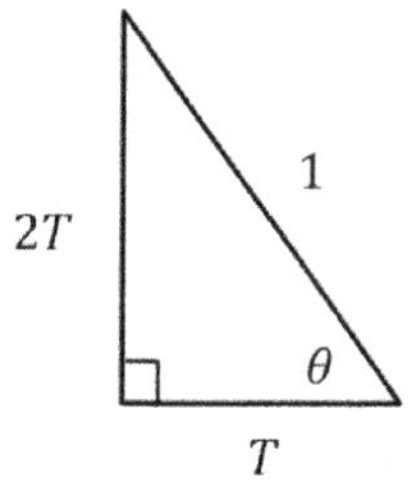

$$a^2 + b^2 = c^2$$

$$(2T)^2 + (T)^2 = (1)^2$$

$$4T^2 + T^2 = 1$$

$$5T^2 = 1$$

$$T = \frac{1}{\sqrt{5}} \quad \text{or} \quad \frac{\sqrt{5}}{5}$$

Answer: D

Notes

Degrees and Radians

There are two different units in which to express angle and arc measurements: Degrees and Radians. The relationship provided on the reference sheet provided when taking the exam is that 2π rad $= 360°$, but this can be reduced further to π rad $= 180°$. Use dimensional analysis or proportions to convert between degrees and radians.

Q+A

Degrees and Radians **Diagnostic Exam | No Calculator Q17**

If an angle measure of 330° is written in radians as $n\pi$, what is the value of n?

Workspace

Solution Outline

Strategy 1

$$\frac{n\pi \text{ rad}}{330°} = \frac{\pi \text{ rad}}{180°}$$

$$n = 11/6$$

Strategy 2

$$330° * \frac{\pi \text{ rad}}{180°} = \frac{11}{6}\pi \text{ rad}$$

Answer: 11/6

Notes

Trigonometric Values on the xy-Plane

Right triangles can be drawn on the coordinate axis using the origin and a known point by sketching in one line between the two and another from the known point to the x-axis to form a right angle. In this manner, the line that connects the origin to the known point acts as the hypotenuse to the right triangle, and the angle in question is the one formed between the hypotenuse and x-axis. This angle is called a **reference angle**, which is the smallest possible angle between a terminal side and the x-axis.

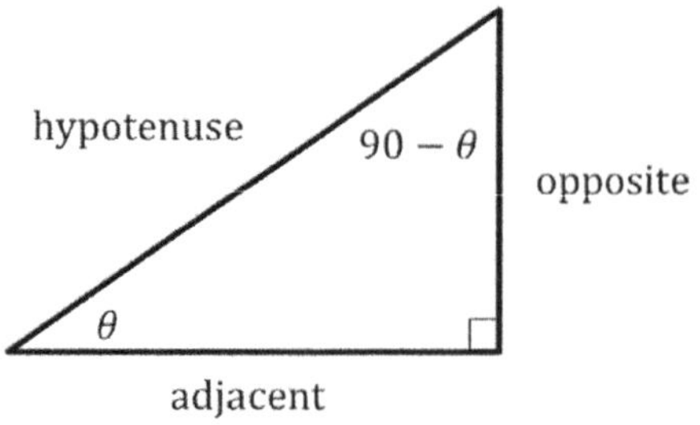

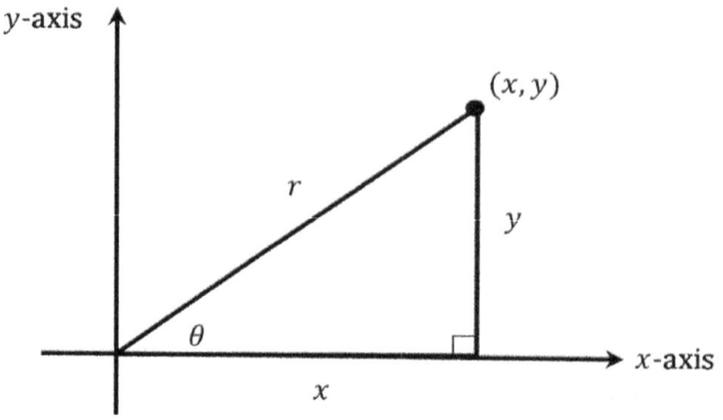

Additionally, angles on the xy-plane are expressed relative to the positive x-axis, which begins at 0° or 0 radians and moves counterclockwise. Every time an axis is reached is another 90° or $\pi/2$ radians. Use the diagram below to map out the axis values in degrees and radians as well as see how we can use a given reference angle to find new angles in any quadrant.

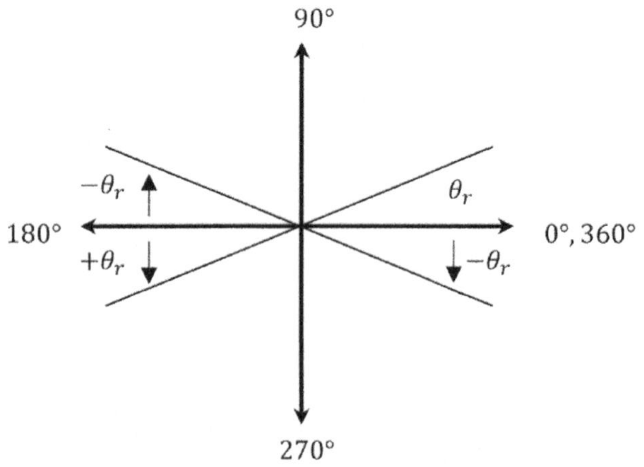

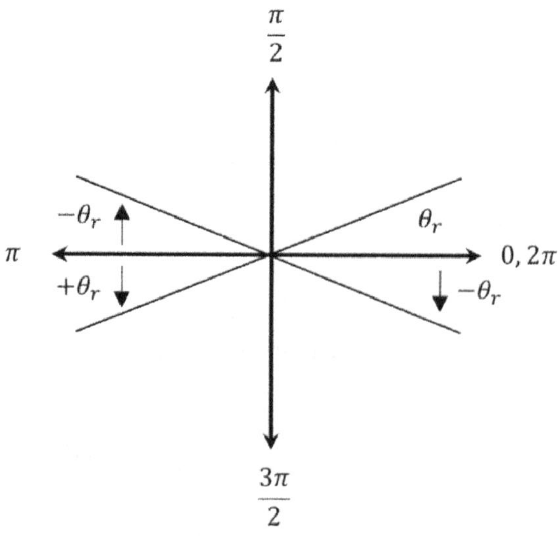

If a large angle measurement is given, then we can use a much smaller value by subtracting 360° or 2π rad because they will have coterminal sides and a shared reference angle. For example, the coterminal angle of 405° is 45° because 405° −

$360° = 45°$. Using the properties of the 45-45-90 special right triangle, we can find its trigonometric ratios. In fact, we can use special right triangles to find the values of sine, cosine, and tangent for angles thar have reference angles equal to $30°, 45°$, and $60°$ (or $\pi/6$, $\pi/4$, and $\pi/3$ respectively). Use the Table 4.28 to review the trigonometric ratios at these values and others.

Table 4.28 | Unit Circle Trigonometric Ratios

Radian	Degree	Cosine	Sine	Tangent
0	$0°$	1	0	0
$\pi/6$	$30°$	$\sqrt{3}/2$	$1/2$	$\sqrt{3}/3$
$\pi/4$	$45°$	$\sqrt{2}/2$	$\sqrt{2}/2$	1
$\pi/3$	$60°$	$1/2$	$\sqrt{3}/2$	$\sqrt{3}$
$\pi/2$	$90°$	0	1	undefined
π	$180°$	-1	0	0
$3\pi/2$	$270°$	0	-1	undefined
2π	$360°$	1	0	0

Q+A

Degrees and Radians

No Calculator

What is the value of $\cos 300°$?

Workspace

Solution Outline

Step 1: Determine the reference angle. Since $300°$ is only $60°$ away from x-axis, the reference angle is $60°$ and the 30-60-90 special right triangle can be used.

Step 2: Use the known 30-60-90 triangle to determine the value of $\cos 60°$, is $1/2$.

Step 3: Since $300°$ lies in IV quadrant, the x-value, and therefore cosine value, is positive.

Answer: $1/2$

Notes

d. Similarity

The corresponding angles of similar figures are congruent and their corresponding side lengths are proportional. Since the side lengths are proportional, there is a number you can multiply by – called the scale factor – to go from a side on one figure to the corresponding side on the other. If the scale factor is not readily identifiable, a proportion between corresponding sides can be used to solve for an unknown side length. Still, the scale factor will help in relating the perimeter, area, and volume of similar figures as summarized in Table 4.29a.

Table 4.29a | Similar Figures and Scale Factors

Perimeter	Area	Volume
$P_{new} = P_{old} * (\text{scale factor})$	$A_{new} = A_{old} * (\text{scale factor})^2$	$V_{new} = V_{old} * (\text{scale factor})^3$

Q+A

Similar Figures

Diagnostic Exam | No Calculator Q20

In the diagram to the right, quadrilateral ABCD is similar to quadrilateral EFGH. If quadrilateral ABCD has an area equal to 40 square meters, then what is the area, also in square meters, of the quadrilateral EFGH?

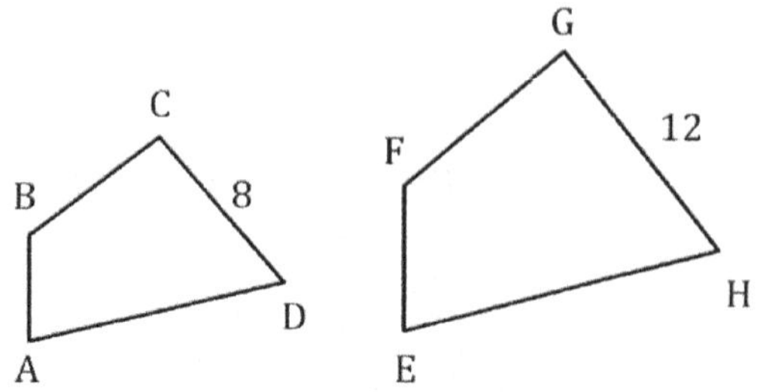

Workspace

Solution Outline

Step 1: Map out the corresponding sides, CD and GH.

Step 2: Find the scale factor, x.

$$8x = 12 \quad \text{or} \quad x = 1.5$$

Step 3: Use the scale factor to find the area of quadrilateral EFGH.

$$\text{Area} = 40(1.5)^2$$

Answer: 90

Notes

Similar triangles, like all similar figures, are proportional: All pairs of corresponding angles have the same measure, but the side lengths change according to the scale factor. In contrast, congruent triangles are identical, meaning that all pairs of corresponding sides and angles have the same measure. Several theorems can be used to determine triangle congruence: SSS, SAS, AAS, and ASA. However, as long as all the angles are the same, the triangles are at least similar. Without all angle measurements, similar triangles can still be identified through ensuring that the ratios of corresponding sides are equivalent. The test contains more questions on triangle similarity, so the statements in Table 4.29b are worth reviewing. Notice that the table does not include AAS and ASA since any two triangles that already have two angles in common necessarily have all congruent angles and would be similar as a result.

Table 4.29b | Triangle Similarity Statements

Angle-Angle	Side-Angle-Side	Side-Side-Side
All angles must be congruent. When triangles have two angles already in common, the third must be as well.	Two corresponding consecutive sides must be proportional and their contained angle must be congruent.	All corresponding sides must be proportional to each other, which may be observed by the scale factor.

Q+A

Triangle Similarity and Ratios

No Calculator

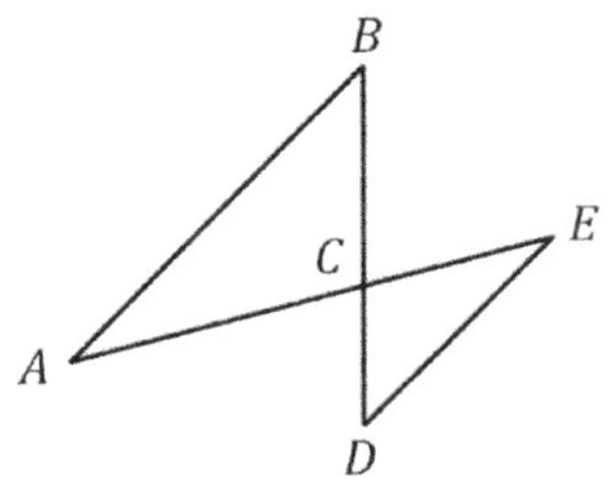

A. 150

B. 200

C. 250

D. 300

In the diagram above, the ratio of $\overline{AC}$ to $\overline{CE}$ is 5 to 3. If $\overline{BD}$ is 400 and $\overline{AB} \parallel \overline{DE}$, what is the length of $\overline{CD}$?

Workspace

Solution Outline

Step 1: Determine AA similarity.

Step 2: Find corresponding sides.

$\overline{AB} \sim \overline{DE}$; $\overline{AC} \sim \overline{CE}$; $\overline{BC} \sim \overline{CD}$

Step 3: Use ratio of $\overline{AC}$ to $\overline{CE}$ to find length of $\overline{CD}$. All corresponding sides of similar figure occur in the same ratio. Therefore, the ratio of $\overline{AC}$ to $\overline{CE}$ is the ratio of $\overline{BC}$ to $\overline{CD}$.

$5x + 3x = 400$

$8x = 400$

$x = 50$

$CD = 3x = 150$

Answer: C

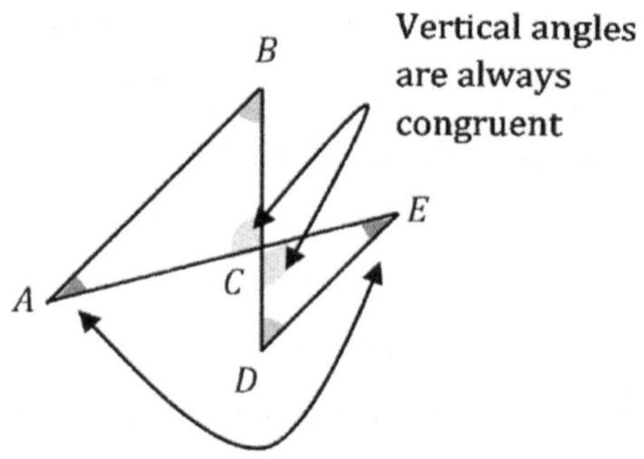

$\angle A \cong \angle E$ & $\angle B \cong \angle D$

Alternate Interior Angles

Notes

Problem Set: Geometry and Trigonometry

13 Questions

1 Calculator Allowed

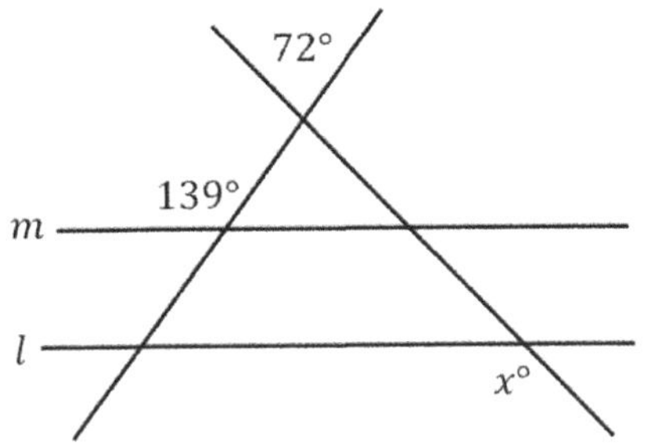

In the diagram above, lines l and m are parallel. What is the value of x?

A. 113

B. 126

C. 139

D. 152

2

The sides of a triangle with perimeter 54 meters have the ratio 5: 6: 7. What is the longest side in meters of this triangle?

A. 15

B. 18

C. 21

D. 24

3

$$\sin(a°) = \cos(b°)$$

$$a = 5x$$

$$b = 4x$$

Given the equations above and the fact that angles a and b are between 0° and 90°, what is the value of a?

A. 40

B. 50

C. 60

D. 70

4

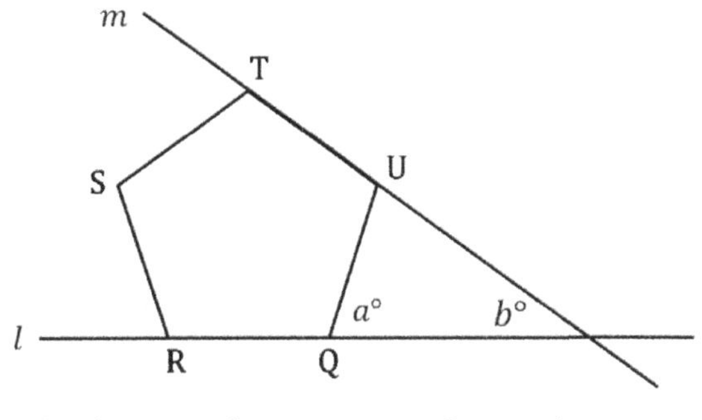

The diagram above contains the regular pentagon QRSTU with points R and Q on line l and points T and U on line m. What is the sum of a and b?

A. 98

B. 102

C. 106

D. 108

5

The volume of a cone is 48 cubic centimeters. If its radius doubles and its height is halved, what is the new volume of the cone in cm^3 ?

A. 48

B. 64

C. 80

D. 96

6

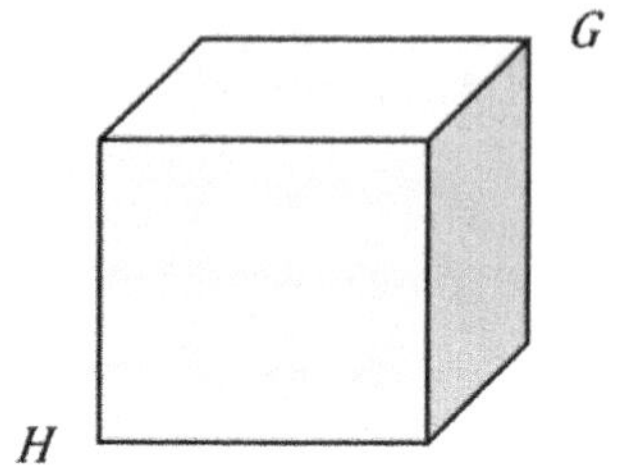

The cube above has side lengths of a units. In terms of a, what is the distance between the two vertices labeled, G and H?

A. $a\sqrt{2}$

B. $2a$

C. $a\sqrt{3}$

D. $3a$

7

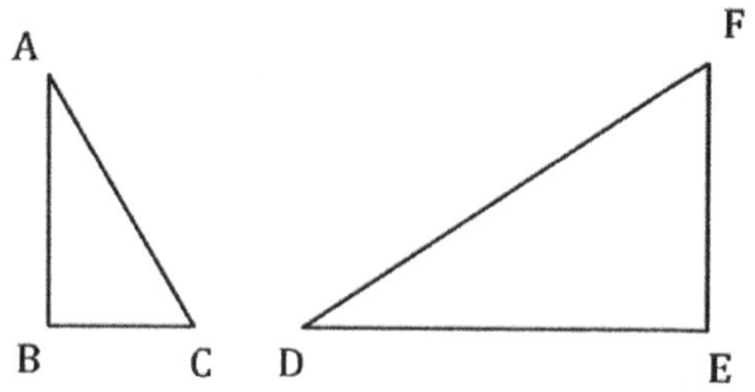

Triangles ABC and DEF are shown above, where angles B and E are both right angles and $\cos(A) = \sin(F)$. If $AB = 3$, $BC = 1$, and $DE = 6$, then what is DF?

A. 2

B. $\sqrt{10}$

C. $2\sqrt{10}$

D. 20

8 Calculator Allowed

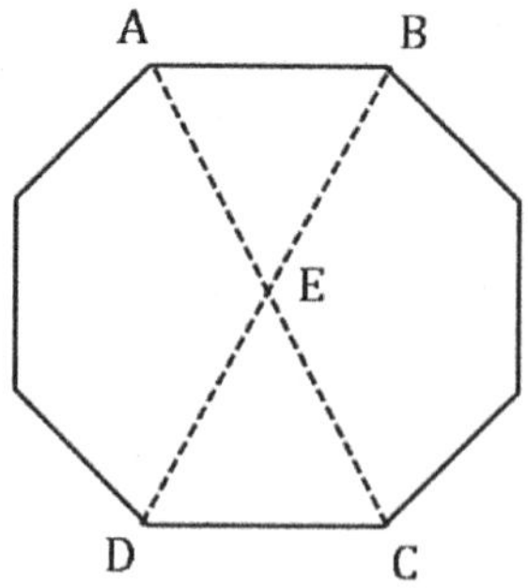

Line segments AC and BD are diagonals of the regular octagon above and intersect at point E. What is the measure of ∠AED in degrees?

A. 135

B. 140

C. 145

D. 150

9

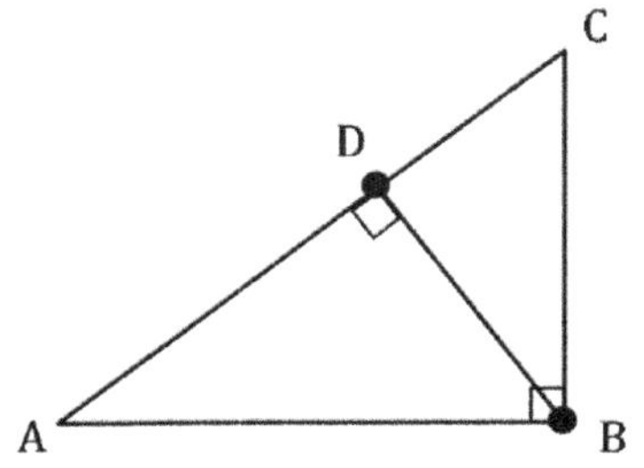

In the diagram above, AB=13 and BC=10. If $AC = a$ and $BD = b$, then which of the following represents the product of a and b?

A. 65

B. 80

C. 115

D. 130

10

Rectangle ABCD is located on the xy-plane where A is located at (0,1) and B is located at $(3,-2)$. If the rectangle has an area of 24, then what could be the coordinate pair of C?

A. (7,2)

B. $(4,-5)$

C. (1,8)

D. $(-2,-5)$

11

If the length of a rectangle increases by 25% while its area stays the same, then by how much must the width decrease?

A. 10%

B. 15%

C. 20%

D. 25%

12

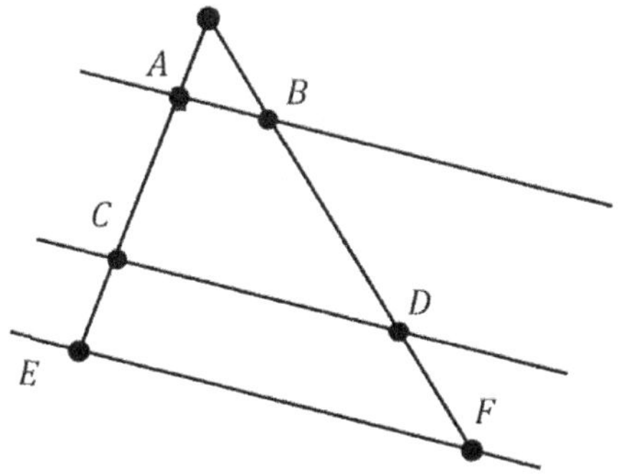

In the diagram above $\overline{AB} \parallel \overline{CD} \parallel \overline{EF}$. If $AC = 2$ and $BD = 3$, what is the ratio $DF: CE$?

A. 2/3

B. 3/2

C. 3

D. 6

13

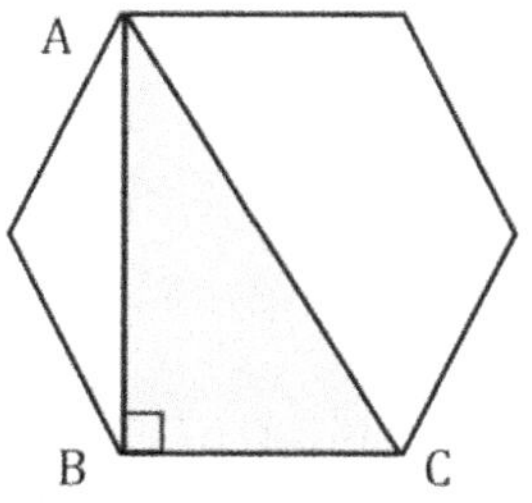

The regular hexagon above has side lengths of b units. Points A, B, and C are vertices of the hexagon and shaded triangle. What is the area of the shaded right triangle in terms of b?

A. $\frac{1}{2}b^2$

B. $\frac{\sqrt{3}}{2}b^2$

C. b^2

D. $\frac{3}{2}b^2$

Explanations

Problem Set: Geometry and Trigonometry

1. A

Use the small interior triangle to find the alternate exterior angle that shares the same measure as $x°$. Vertical angles are always congruent. Linear pairs and the interior angles of triangles always add up to 180 degrees.

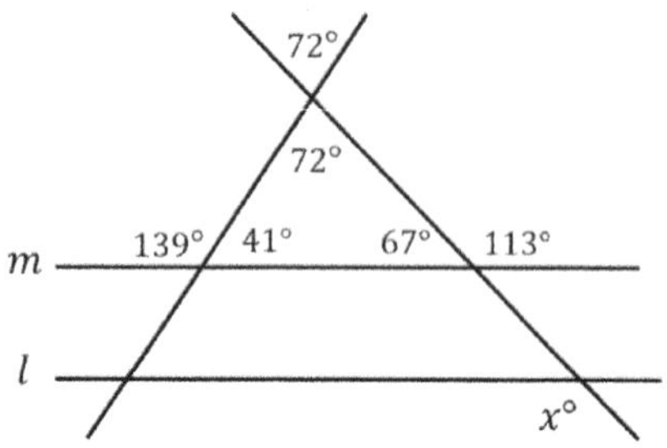

2. C

Find the fraction of the perimeter for the longest side. The given ratio relates the three sides of the triangle together, which provide the three parts to the total. Together the three "parts" add up to 18, with the longest side representing 7 out of the 18. That is, the longest side is seven-eighteenths of the perimeter.

$$\frac{7}{18} * 54 = 21$$

3. B

The sine and cosine of complementary angles are always equal. Therefore, a and b must add up to 90 degrees.

$$4x + 5x = 90$$

$$9x = 90$$

$$x = 10$$

$$a = 5(10) \text{ or } 50$$

4. D

The exterior angles of polygons always add up to 360 degrees. The angles of the triangle that are adjacent to the pentagon are two of the pentagon's exterior angles.

$$a = \frac{360}{5} \text{ or } 72$$

The interior angles of triangles add up to 180 degrees.

$$72 + 72 + b = 180$$

$$144 + b = 180$$

$$b = 36$$

$$a + b = 72 + 36 \quad \text{or} \quad 108$$

5. D

Find the factor of change by focusing on the dimensions that change as they relate to the volume equation. By doubling the radius and halving the height, the volume changes by a factor of 2.

$$V = \frac{1}{3}\pi r^2 h$$

$$V \propto r^2 h$$

$$V' \propto (2r)^2 \left(\frac{1}{2}h\right)$$

$$V' \propto 4r^2 * \frac{1}{2}h$$

$$V' \propto 2r^2 h$$

$$V' = 2(48)$$

$$V' = 96$$

6. C

Two right triangles must be used to find the length of $\overline{HG}$. These right triangles are formed by drawing in a diagonal across the base and a diagonal from H to G. Because the base is a square, the triangle formed by the diagonal is the special 45-45-90 right triangle. Therefore, if the legs are a and a, the hypotenuse is $a\sqrt{2}$. This is one of the legs of the second right triangle formed – the second being a.

Complete the Pythagorean theorem:

$$(a)^2 + \left(a\sqrt{2}\right)^2 = c^2$$

$$a^2 + 2a^2 = c^2$$

$$3a^2 = c^2$$

$$c = a\sqrt{3}$$

7. C

Triangles ABC and DEF are similar since angles A and F are complimentary, leaving both triangles with the same angle measurements.

The factor of change from triangle ABC to DEF is 2 because DE is double the length of AB. This means EF is 2 because that is double BC.

Complete the Pythagorean Theorem to find DF.

$$a^2 + b^2 = c^2$$

$$6^2 + 2^2 = c^2$$

$$36 + 4 = c^2$$

$$40 = c^2$$

$$c = 2\sqrt{10}$$

8. A

The angles AEB and DEC are each $360°/8 = 45°$. The angle AED is the supplement of 45°, or 135°.

9. D

The line segments $\overline{AB}$ & $\overline{BC}$ and $\overline{AC}$ & $\overline{BD}$ are both dimensional pairs that can be used to represent the base and height of the triangle. Since the triangle has a unique area that is only affected by the product of the base and the height, this product must be the same for every base and height pair.

Therefore, $ab = 13(10)$ or 130

10. A

The distance from A to B is $3\sqrt{2}$ because moving from point A to point B requires moving right 3 units and down 3 units, which forms a 45-45-90 right triangle where the hypotenuse is one side of the rectangle. Mapping the changes to the points is also important in finding the slope, which is $-3/3$ or -1. The opposite reciprocal of -1 is 1. This is important because rectangles have consecutive sides that are perpendicular to each other. This new slope must be used to get point C, which is found by moving away from point B.

$$A = bh$$

$$24 = 3\sqrt{2} * x$$

$$x = 4\sqrt{2}$$

The movement from point B has to be up 4 right 4 or down 4 left 4 so that the length of BC is $4\sqrt{2}$ and is perpendicular to AB. Possible points: (7,2) and $(-1, -6)$

11. C

The new area must be equal to the original area. That is, the factor of change for each dimension must together multiply to one so that the area remains unaffected. The factor of change for any percent increase or decrease is $(1 \pm r)$ where r is the percent expressed as a decimal.

$$1.25a = 1$$

$$a = 0.8$$

Since 0.8 is 0.2 below 1, the width must decrease by 20% so that the area stays the same.

12. B

The ratios of the corresponding segments formed between the parallel lines are equal.

$$BD{:}\,AC = DF{:}\,CE$$

$$3 : 2 = DF : CE$$

13. B

Hexagons have interior angles of 120 degrees. The diagonal AC is a line of symmetry that bisect the interior angle at vertex C. This means that the angle in the shaded triangle at point C is 60 degrees. The shaded triangle is therefore a 30-60-90 special right triangle. If the small leg b then the larger one is $b\sqrt{3}$.

$$A = \frac{1}{2}bh$$

$$A = \frac{1}{2} * b * b\sqrt{3}$$

$$A = \frac{\sqrt{3}}{2}b^2$$

4.3 CIRCLE PROPERTIES

a. Angle and Chord Properties

Several important circle theorems and equations have earned their spot on the test. The most important ones are about angle, chord, and arc measurements. Often, the angle and arc measurements used are in radians, rather than degrees. To review how to convert between these two measurements, visit 4.2 Geometry and Trigonometry. As it relates to circles, the important takeaway is that there are 2π radians – just as there are 360° – in a full circle. Keep in mind that many questions regarding circles on the test do not involve a full circle, but instead a portion of one.

Additionally, many questions on the test require knowledge across other geometry topics, especially triangle properties. The most common triangles produced in tandem with the following angle and chord properties are right triangles. However, quadrilaterals also make a strong showing as well. Knowing the sum of the interior angles in both triangles (180 degrees) and quadrilaterals (360 degrees) will make several common question types much easier.

Central and Inscribed Angles

The central angle of a circle is, as the term suggests, formed at the center of a circle. This angle corresponds to the measure of its intercepted arc and can be used to determine the fraction of the circle under consideration. For example, if the central angle is 120 degrees, then it encompasses one third of the total circle because that is equivalent to 120 out of the total 360 degrees. Inscribed angles are related to the central angle and the measure of its intercepted arc, though the intercepted arc is always double the inscribed angle. Use Table 4.31 below to see the relationship between these types of angles and their intercepted arcs.

Table 4.31 | Central and Inscribed Angles Summary

Central Angle	Inscribed Angle
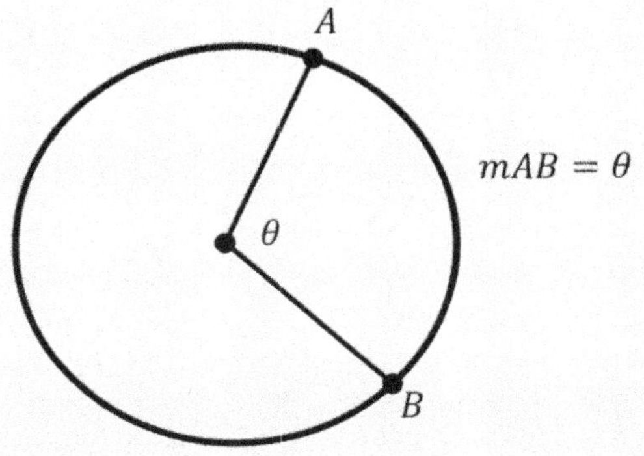	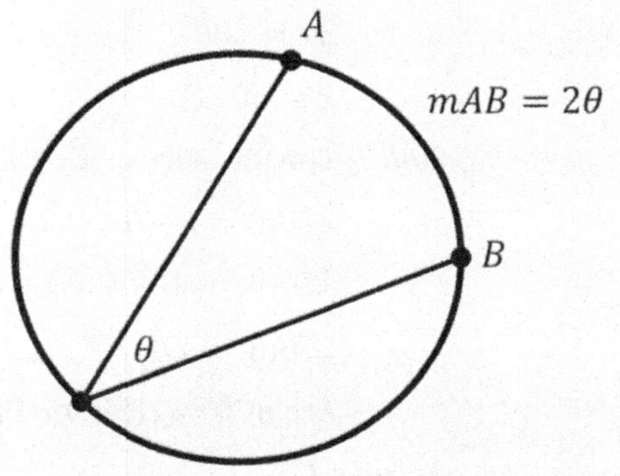
Remember: Angles are measured in either degrees or radians. A full circle has 360° or 2π radians.	Application: This theorem justifies why an angle inscribed in a semicircle is always 90°

Q+A

Inscribed Angle Property

No Calculator

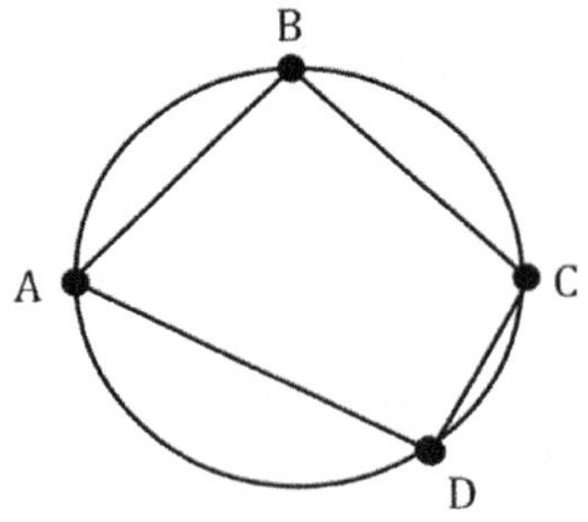

A. 59

B. 60

C. 118

D. 121

Points A, B, C, D are located on the circle above to create an inscribed quadrilateral. Given $\angle DAB = x - 1$ and $\angle BCD = 2x + 1$, what is the measure of arc BCD?

Workspace

Solution Outline

Step 1: Using the inscribed angle property, understand that $\angle DAB$ and $\angle BCD$ are supplementary. The reason being, both angles intercept arcs on opposite sides of the circle. Together, these arcs encompass the full circle, or all 360 degrees. Since inscribed angles are half the measure of their intercepted arcs, the two angles then must add up to 180.

$$(2x + 1) + (x - 1) = 180$$

$$3x = 180$$

$$x = 60$$

Step 2: The measure of arc BCD is double $\angle DAB$.

$$\angle DAB = x - 1$$

$$\angle DAB = 60 - 1$$

$$\angle DAB = 59$$

$$\text{Arc BCD} = 2(59) \text{ or } 118$$

Answer: C

Notes

Chords and Tangent Lines

Line segments inside circles are called chords. Their endpoints fall on the circle itself and have unique properties that that are summarized in Table 4.32. One key idea is that chords of the same length intercept arcs of the same measure. Moreover, a radius that bisects a chord is perpendicular to it. Similarly, a radius is perpendicular to the tangent line at the point of tangency. Though tangent lines are not chords, they are grouped in this section because of this unique property. Recall that a tangent line is a line that touches a curve at a single point.

Table 4.32 | Chord and Tangent Line Properties

Congruent Chords	Bisected Chord	Tangent Line
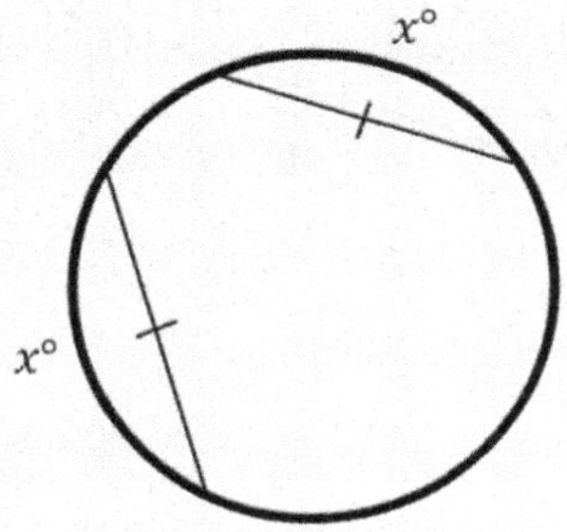	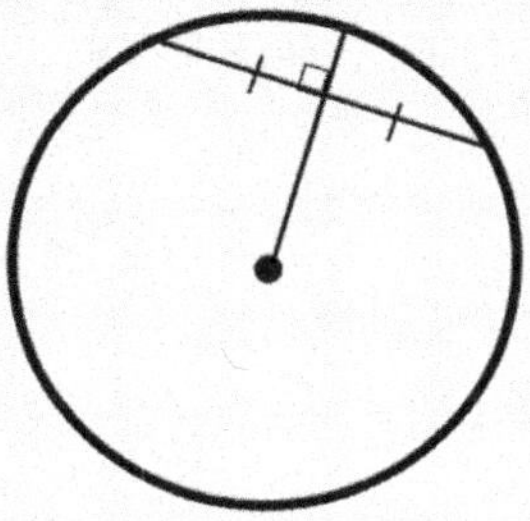	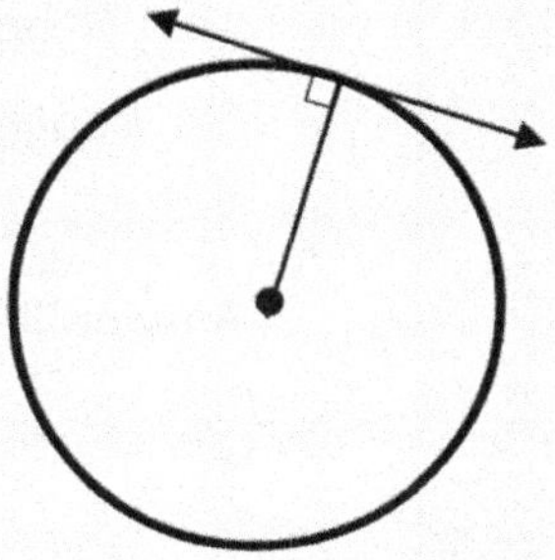

Q+A

Bisected Chord Theorem **Calculator**

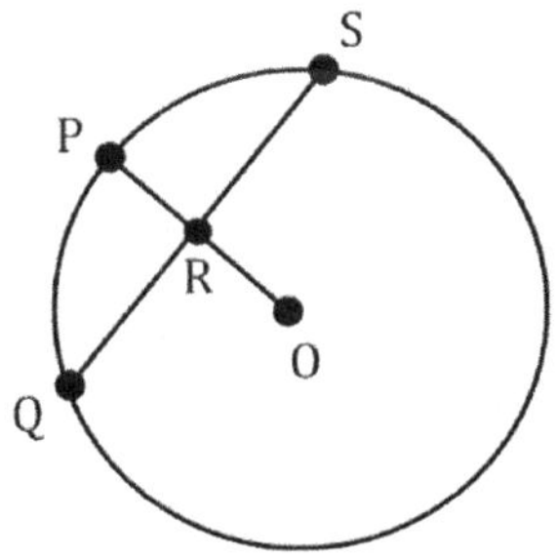

A. 25

B. 50

C. 75

D. 100

The measurements given to describe the image above are all in centimeters. Chord QS has a length of 48 and line segment PR has a length of 18. If point R is the midpoint of chord QS, then what is the diameter of Circle O?

Workspace

Solution Outline

Step 1: Label given information onto Circle O.

Step 2: Use bisected chord theorem and draw in a new radius connecting point O to point Q, creating an interior right triangle.

Step 3: Relate $\overline{\text{OR}}$ in terms of the radius, r, and use Pythagorean Theorem for triangle QRO to solve for r. Then find diameter. Since $\text{OP} = \text{PR} + \text{OR}$, we know that $\text{OR} = r - 18$.

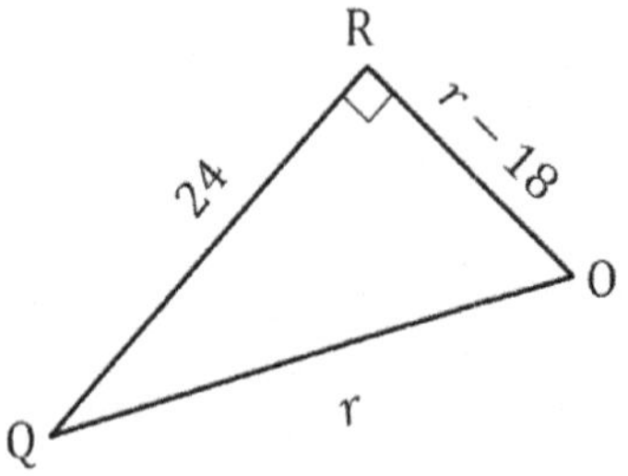

$a^2 + b^2 = c^2$

$24^2 + (r - 18)^2 = r^2$

$576 + r^2 - 36r + 324 = r^2$

$900 = 36r$

$r = 25\,; d = 50$

Answer: B

Notes

b. Arc Length and Sector Area

We can continue using the central angle to find both arc length and sector area. In short, the **arc length** is only a fraction of the circumference and **sector area** is only a fraction of the total area. This fraction can be found using the central angle over the total degrees or radians in a circle. In other cases, the fraction of the circle under consideration is easily identifiable without needing to know the central angle. For example, if the circle is divided in into five equal parts, each portion would contain one-fifth the total circumference and area. Table 4.33 summarizes the arc length and sector area formulas when the central angle is measured in either degrees or radians. Though students are often more comfortable working in degrees, the formulas in radians are sometimes easier to use.

In addition to the formulas outlined in Table 4.33, we can also use proportions to answer questions involving arc length and sector area. Understand that each fraction represents the fraction of the full circle under consideration and can be found using the fractions below.

Fractions of Entire Circle	$\dfrac{\text{Degrees}}{360} = \dfrac{\text{Radians}}{2\pi} = \dfrac{\text{Arc Length}}{\text{Circumference}} = \dfrac{\text{Sector Area}}{\text{Total Area}}$

The degrees and radians described above come from the central angle or its intercepted arc. The example that follows comes from the Diagnostic Exam, and shows how can we use the ideas and formulas described in Table 4.33 effectively to answer questions that relate arc length and sector area.

Table 4.33 | Arc Length and Sector Area Formulas

Arc Length		Sector Area	
Distance from A to B along the circle 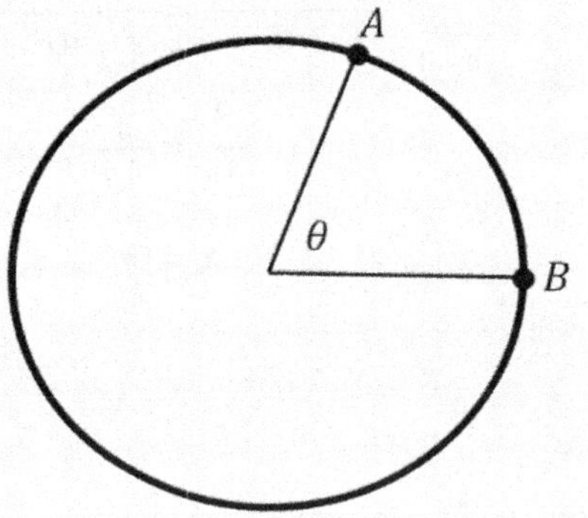		Area of shaded region 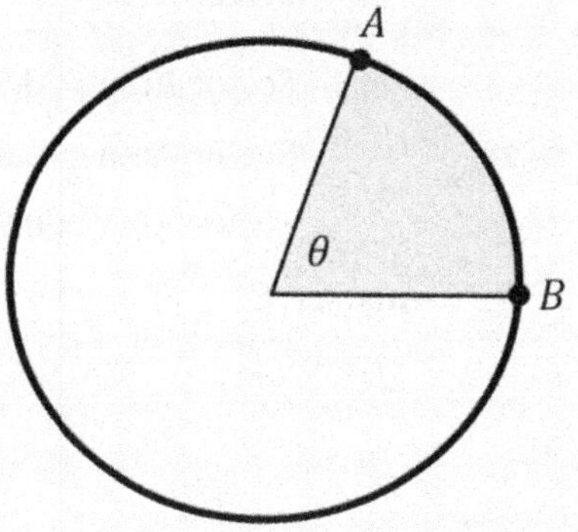	
Degrees	$\text{Arc Length} = \frac{\theta}{360} * 2\pi r$	Degrees	$\text{sector Area} = \frac{\theta}{360} * \pi r^2$
Radians	$\text{Arc Length} = \frac{\theta}{\cancel{2\pi}} * \cancel{2\pi} r$ $\text{Arc Length} = r\theta$	Radians	$\text{Sector Area} = \frac{\theta}{2\cancel{\pi}} * \cancel{\pi} r^2$ $\text{sector Area} = \frac{1}{2} r^2\theta$

Q+A

Arc Length and Sector Area

Diagnostic Exam | No Calculator Q8

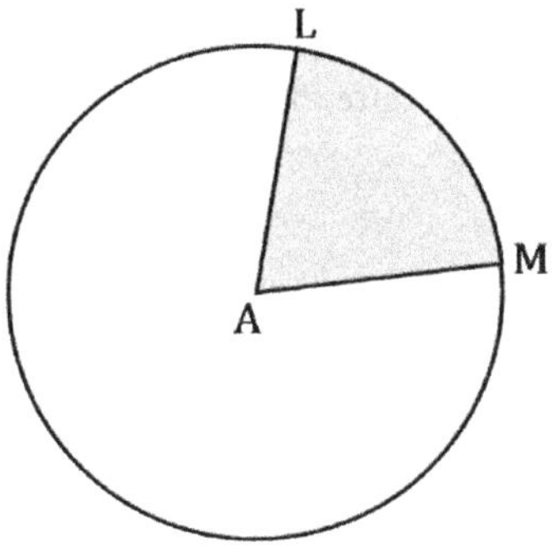

A. 1

B. 2

C. 3

D. 6

Circle A has a radius of 2 meters and arc LM has a length of 3 meters.

What is the area in square meters of the shaded sector?

Workspace

Solution Outline

Strategy 1

Dissect sector area equation and plug in.

$$\text{Sector Area} = \frac{1}{2}r^2\theta$$

$$\text{Sector Area} = 0.5 * r * r\theta$$

$$\text{Sector Area} = 0.5(\text{radius})(\text{arc length})$$

$$\text{Sector Area} = 0.5(2)(3) \text{ or } 3$$

Answer: C

Strategy 2

Use a proportion to solve for the Sector Area.

$$\frac{\text{Arc Length}}{2\pi r} = \frac{\text{Sector Area}}{\pi r^2}$$

$$\frac{3}{4\pi} = \frac{x}{4\pi}$$

$$x = 3$$

Notes

c. Standard Equation

The standard equation of a circle comes from the distance formula – which in turn has its roots in the Pythagorean Theorem – and it ensures that every point on the circle is equidistant from the circle's center. Completing the square is a common technique to bring the expanded equation of a circle into standard form. Use Table 4.34 to review the key elements of a circle graphed on the coordinate plane and its representation in standard form. The center of the circle is located at the (x, y) point (h, k). To reach points on the circle, move out the length of the radius in all directions from the center. In addition to the standard equation, the slope and midpoint formulas also have relevance in this section. For example, the midpoint between the endpoints that form a circle's diameter locates the circle's center.

Table 4.34 | Standard Equation of a Circle on the Coordinate Plane

Standard Equation		Diagram
$(x-h)^2 + (y-k)^2 = r^2$		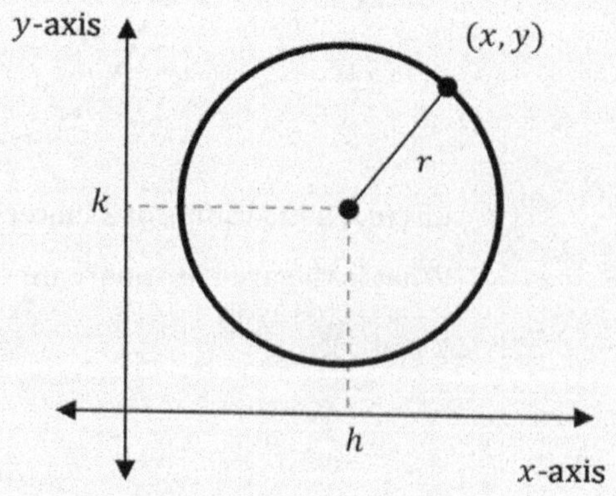
Center	(h, k)	
Radius	r	

Q+A

Standard Equation **Diagnostic Exam | Calculator Q27**

$$(x+2)^2 + (y-6)^2 = 20$$

In the xy-plane, a circle with the equation above contains the point $A(0,10)$. If $\overline{AB}$ forms a diameter to the circle, what coordinate pair represents B?

A. $(-4, 2)$

B. $(4, -2)$

C. $(0, 2)$

D. $(2, 0)$

Workspace

Solution Outline

Strategy 1

Use midpoint formula, where A and B are endpoints and the center is $(-2,6)$.

$$x_m = \frac{x_1 + x_2}{2}$$

$$-2 = \frac{0 + x}{2}$$

$$x = -4$$

Since only one answer choice has an x-coordinate of -4, choice A must be correct.

Answer: A

Strategy 2

Use the slope between point A and the center to reach point B. The movement from point A to the center of the circle is left 2 units, down 4 units. Apply that set of movements again, starting at the center takes us from $(-2,6)$ to $(-4,2)$.

Notes

The next example that we will discuss also requires the use of standard equation of a circle; however, we first need to complete the square to bring the equation into standard form. We first introduced this method in our discussion of quadratic functions. To review the process in more detail, return to Table 2.43b Complete the Square. As we will see in the following example, we will need to complete the square for both the x- and y-variables.

Q+A

Complete the Square **No Calculator**

$$x^2 + 6x + y^2 - 10y - 13 = 0$$

The equation of circle P in the xy-plane is given above. What is the distance between the center of circle P and the origin of the coordinate plane?

A. 4

B. $\sqrt{34}$

C. 6

D. $\sqrt{47}$

Workspace

Solution Outline

Step 1: Isolate and group variable terms to one side.

Step 2: Create perfect square trinomials by adding 9 and 25 to both sides of the equation.

Step 3: Factor each trinomial.

Step 4: Use standard equation to find center.

Step 5: Find distance between center and the origin using distance formula.

Answer: D

$$x^2 + 6x + y^2 - 10y = 13$$
$$(x^2 + 6x + 9) + (y^2 - 10y + 25) = 47$$
$$(x + 3)^2 + (y - 5)^2 = 47$$

Center: $(-3,5)$

Origin: $(0,0)$

$$\text{Distance} = \sqrt{9 + 25} = \sqrt{34}$$

Notes

And to wrap up this section, we will revisit a question from the Diagnostic Exam, Calculator section, which involves a circle and line graphed on the xy-plane. We can solve this problem using both algebraic and graphing methods. With the algebraic method, we will rely on substitution to create a univariate, quadratic equation that we do not actually need to solve. The reason being that the question asks for the number of solutions, which, for a quadratic equation, means we can use the discriminant. To review the use and implementation of the discriminant, return to Table 2.44 Discriminant Summary. Although the graphing method may be more relevant than the algebraic method to answer this particular question due to the proximity of the circle to the y-axis, there are some instances where the use of a rough sketch does not lead to an obvious answer.

Q+A

Standard Equation **Diagnostic Exam | Calculator Q5**

$$y = x + 4$$

$$(x - 2)^2 + (y - 4)^2 = 9$$

How many solutions are there to the system of equations provided above?

A. 0

B. 1

C. 2

D. More than 2

Workspace

Solution Outline

Step 1: Substitute the value for y found in the top equation into the bottom equation.

$(x - 2)^2 + (x + 4 - 4)^2 = 9$

$(x - 2)^2 + x^2 = 9$

Step 2: Expand and simplify the left-hand side of the equation. Then set equation equal to zero.

$x^2 - 4x + 4 + x^2 = 9$

$2x^2 - 4x + 4 = 9$

$2x^2 - 4x - 5 = 0$

Step 3: Use the discriminant to determine the number of solutions. Since the discriminant is positive, there are two solutions.

$b^2 - 4ac$

$(-4)^2 - 4(2)(-5)$

$16 + 40$

56

Answer: C

Notes

Problem Set: Circle Properties

12 Questions

1

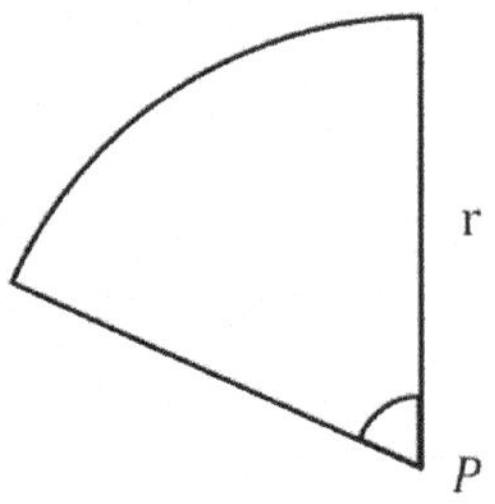

Shown above is a sector of circle P, which has a radius 4 meters. If the sector is formed by a central angle of 1.25 radians, then what is the perimeter in meters of the sector?

A. 11

B. 12

C. 13

D. 14

2

The circumference of circle A is 8π inches. If arcMN is 2π inches, then what is the measure of the central angle b in degrees that intercepts arcMN ?

A. 80°

B. 90°

C. 100°

D. 110°

3

$$(x-3)^2 + (y-1)^2 = 10$$

On the xy-plane, circle P takes on the standard equation shown above. If the point $(4, c)$ falls on circle P, then which of the following can be the value of the constant c?

A. -2

B. -1

C. 5

D. 6

4

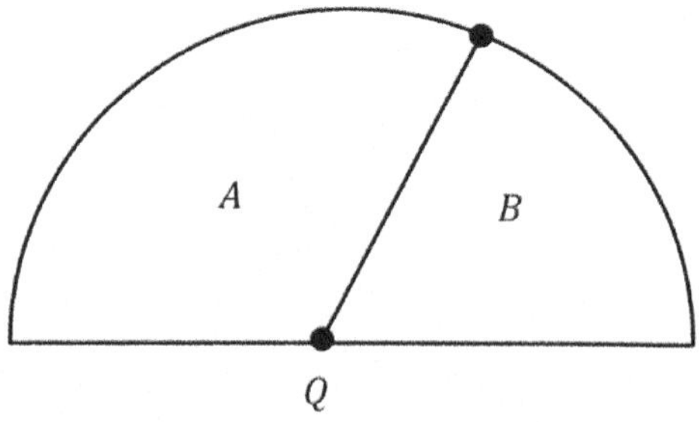

The semicircle with center point Q is shown above and contains two sectors with areas A and B. The radius of the semicircle is 8 inches. If $A + B = x\pi$ in^2, then what is the value of x?

A. 24

B. 32

C. 40

D. 48

5

What is the equation of the circle on the xy-plane that contains a diameter with endpoints $(3, -8)$ and $(9, -2)$?

A. $(x+6)^2 + (y-5)^2 = 9$

B. $(x-6)^2 + (y+5)^2 = 9$

C. $(x+6)^2 + (y-5)^2 = 18$

D. $(x-6)^2 + (y+5)^2 = 18$

6

The line with equation $y = 0.75x + 4$ is tangent to circle P at $x = 4$. If circle P has a radius of 5, then what could be the coordinates of center P?

A. (1,11)

B. (4,7)

C. (7,4)

D. (11,1)

7

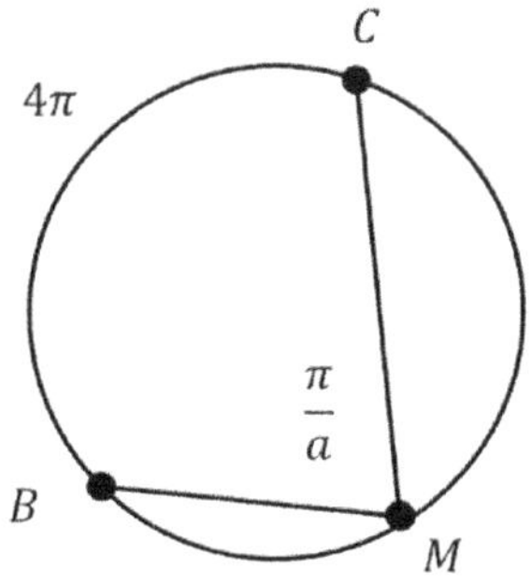

The circle above has a radius of 5 meters and its minor arc BC is 4π meters. What is the measure a if $\angle BMC$ is π/a radians?

A. 2/5

B. 4/5

C. 5/4

D. 5/2

8

$$(x+2)^2 + (y-9)^2 = 3$$
$$y = x + 8$$

How many solutions are there to the system of equations above?

A. 0

B. 1

C. 2

D. Cannot be determined

9 Calculator Allowed

What is the minimum number of complete revolutions a wheel with radius 26 inches would have to make to roll at least 2.5 miles?

$$5280 \text{ feet} = 1 \text{ mile}$$

A. 969

B. 970

C. 971

D. 972

10

A circle on the xy-plane is centered on the origin and contains the point $(3, 3\sqrt{3})$. What could be a y-intercept of this circle?

A. 3

B. 4

C. 5

D. 6

11

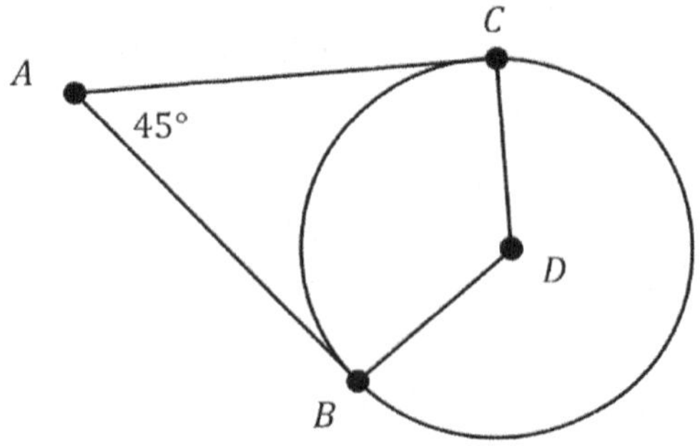

The line segments AB and AC are tangent to the circle above with center D and circumference of 40. If $\angle BAC$ is 45°, then what is the length of arc BC?

A. 5

B. 10

C. 15

D. 20

12

$$5x^2 + 5y^2 - 20x + 10y - 15 = 0$$

What are the coordinates of the center of a circle on the xy-plane with equation provided above?

A. $(-2, 1)$

B. $(-2, -1)$

C. $(2, 1)$

D. $(2, -1)$

Explanations

Problem Set: Circle Properties

1. C

Find the perimeter by adding the lengths of the two sides and provided arc. Since the central angle is measured in radians, the abbreviated arc length equation $l = r\theta$ may be used.

$$\text{Perimeter} = \text{r} + \text{r} + \text{r}\theta$$

$$\text{Perimeter} = 4 + 4 + 4(1.25)$$

$$\text{Perimeter} = 8 + 5$$

$$\text{Perimeter} = 13$$

2. B

Use the arc length formula or a proportion to find the central angle in degrees.

$$\frac{\text{arc length}}{\text{circumference}} = \frac{\theta}{360}$$

$$\theta = \frac{\text{arc length}}{\text{circumference}} * 360$$

$$\theta = \frac{2\pi}{8\pi} * 360$$

$$\theta = \frac{1}{4} * 360$$

$$\theta = 90$$

3. A

Substitute the given point into the provided circle equation. Since $(4, c)$ is an (x, y) pair, $x = 4$ and $y = c$.

$$(x-3)^2 + (y-1)^2 = 10$$

$$(4-3)^2 + (c-1)^2 = 10$$

$$1^2 + (c-1)^2 = 10$$

$$(c-1)^2 = 9$$

$$c - 1 = \pm 3$$

$$c = 1 \pm 3$$

$$c = -2 \text{ or } 4$$

4. B

Add A and B to reach the total area of the semicircle.

$$\text{Area} = \frac{1}{2}\pi r^2$$

$$\text{Area} = \frac{1}{2}\pi(8)^2$$

$$\text{Area} = 32\pi$$

$$A + B = 32\pi$$

$$x\pi = 32\pi$$

$$x = 32$$

5. D

The center (h, k) occurs at the midpoint to any diameter of the circle.

$$h = \frac{3+9}{2} \text{ or } 6$$

$$k = \frac{-8+(-2)}{2} \text{ or } -5$$

Standard Form:

$$(x-h)^2 + (y-k)^2 = r^2$$

$$(x-6)^2 + (y+5)^2 = r^2$$

Plug in the (x, y) point $(9, -2)$ to find r^2:

$$(9-6)^2 + (-2+5)^2 = r^2$$

$$3^2 + 3^2 = r^2$$

$$9 + 9 = r^2$$

$$18 = r^2$$

Final Equation:

$$(x-6)^2 + (y+5)^2 = 18$$

6. A

The slope of the line is $3/4$. Since the radius of a circle is always perpendicular to a tangent line at its point of tangency, we can use the opposite reciprocal $(-4/3)$ to move from the point of tangency to the center. But first, we need the (x, y) of the point of tangency. Since it occurs at $x = 4$ and is on the line $y = 0.75x + 4$, this point can be solved for.

$$y = 0.75x + 4$$

$$y = 0.75(4) + 4$$

$$y = 3 + 4$$

$$y = 7$$

The point of tangency occurs at (4,7). The movements down 4 and right 3, as well as up 4 left 3, can be used to map the center because they maintain a slope that is perpendicular with an overall shift of five units (think 3-4-5 right triangle). This leads to the possible center points at (7,3) and (1,11) .

7. D

Find the central angle that forms the arc of length 4π meters. Realize that the central angle (θ_c) is double the inscribed angle (θ), so $\theta_c = 2\theta$.

$$\text{arc length} = r\theta_c$$

$$\text{arc length} = r(2\theta)$$

$$4\pi = 5(2\theta)$$

$$4\pi = 10\theta$$

$$\theta = \frac{2\pi}{5} \quad \text{or} \quad \frac{\pi}{(5/2)}$$

$$a = \frac{5}{2}$$

8. A

Solve the system through using substitution and the discriminant of the resulting quadratic.

$$(x+2)^2+(y-9)^2=3$$

$$y=x+8$$

Substitute $y=x+8$ into the top equation:

$$(x+2)^2+(x+8-9)^2=3$$

$$(x+2)^2+(x-1)^2=3$$

$$x^2+4x+4+x^2-2x+1=3$$

$$2x^2+2x+5=3$$

$$2x^2+2x+2=0$$

$$x^2+x+1=0$$

Discriminant: b^2-4ac

$$(1)^2-4(1)(1)=-1$$

There are no real solutions when the discriminant is negative.

9. B

Find the number of revolutions by relating the circumference to the total distance travelled.

$$\text{Distance} = \text{revolutions} \times \text{circumference}$$

$$2.5(5280)=n*2\pi\left(\frac{26}{12}\right)$$

$$n=969.62$$

$$n=970$$

Have to round up - regardless of decimal value - to reach the full distance.

*All lengths were converted to feet.

*Any common unit will work.

10. D

The (x,y) point of $(3,3\sqrt{3})$ create the special 30-60-90 triangle with respect to the origin. This means the hypotenuse is 6. The circle crosses the y-axis at both positive and negative six.

11. C

The angles at A and D must be supplementary because B and C are right angles and the quadrilateral that is formed must have an interior angle sum of 360 degrees. Therefore, angle D is 135°.

$$\text{Arc length} = \frac{\theta}{360}*\text{circumference}$$

$$\text{Arc length} = \frac{135}{360}*40$$

$$\text{Arc length} = \frac{135}{9} \text{ or } 15$$

12. D

Complete the square to write the equation of the circle in standard form, where (h,k) represents the center.

$$5x^2+5y^2-20x+10y-15=0$$

$$5x^2-20x+y^2+10y=3$$

$$x^2-4x+4+y^2+2y+1=8$$

$$(x-2)^2+(y+1)^2=8$$

$$(2,-1)$$

REMAINING TOPICS EXAM

DIRECTIONS

For questions 1-10, solve every problem and either circle or record the best answer from the answer choices provided. **For questions 11-15**, solve every problem and record your answer in the appropriate format. On the real test, you must grid in your answers according to the directions. You may use any available space in your workbook for scratch work.

NOTES

1. Calculator usage is NOT allowed in Part 1.
2. Assume real numbers unless otherwise indicated.
3. Figures may not be drawn to scale.
4. All figures are planar, unless otherwise indicated.
5. Domains represent all real numbers of for which f(x) is also a real number.

REFERENCE

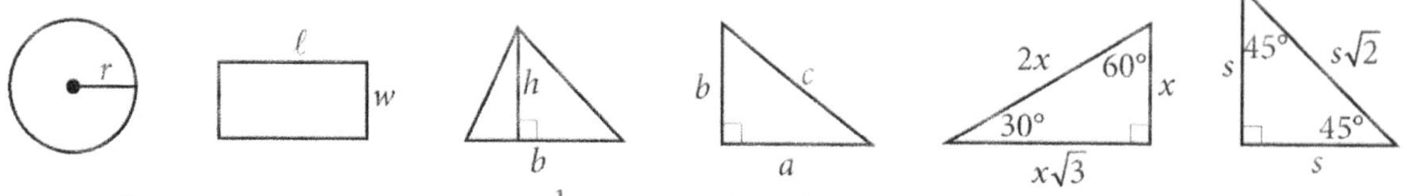

$A = \pi r^2$ $C = 2\pi r$ | $A = \ell w$ | $A = \frac{1}{2}bh$ | $c^2 = a^2 + b^2$ | Special Right Triangles

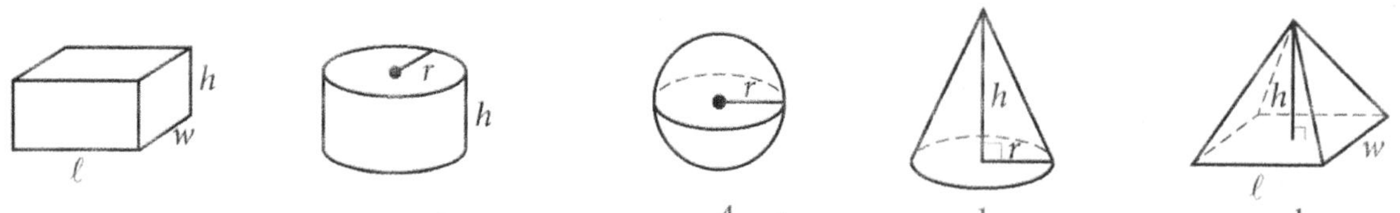

$V = \ell wh$ | $V = \pi r^2 h$ | $V = \frac{4}{3}\pi r^3$ | $V = \frac{1}{3}\pi r^2 h$ | $V = \frac{1}{3}\ell wh$

The arc of a circle contains 360 degrees.

The arc of a circle contains 2π radians.

The interior angles of a triangle add up to 180 degrees.

Part 1: No Calculator

19 Minutes, 15 Questions

1

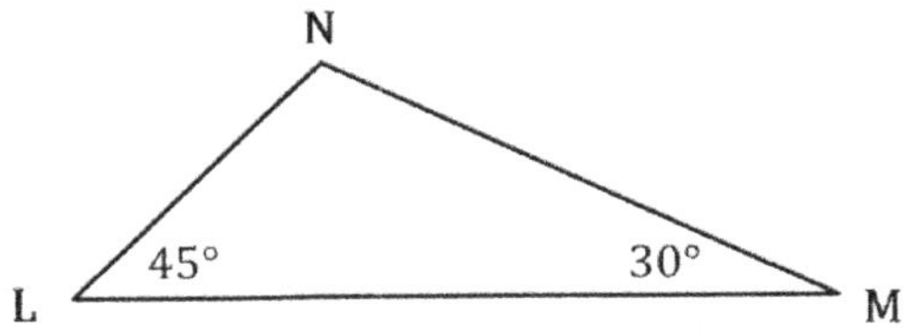

In the triangle above, $\angle M = 30°$, $\angle L = 45°$, and $LN = 10$. What is the length of MN?

A. $5\sqrt{2}$

B. $5\sqrt{3}$

C. $10\sqrt{2}$

D. $10\sqrt{3}$

2

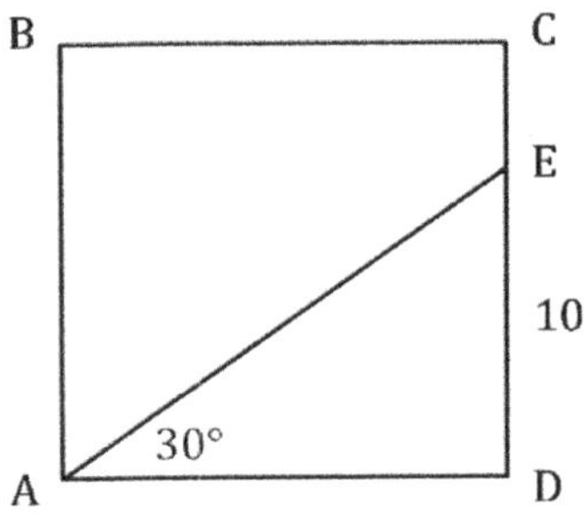

In the figure above, square ABCD contains the line segment $\overline{AE}$. If $\angle DAE = 30°$ and DE=10, what is the area of the square ABCD?

A. 100

B. $100\sqrt{3}$

C. 300

D. $300\sqrt{3}$

3

$$3(1-2i)-8(1-2i)$$

Which of the following is equivalent to the expression above given that $i = \sqrt{-1}$?

A. $-5+10i$

B. $5-10i$

C. $-11+22i$

D. $11-22i$

4

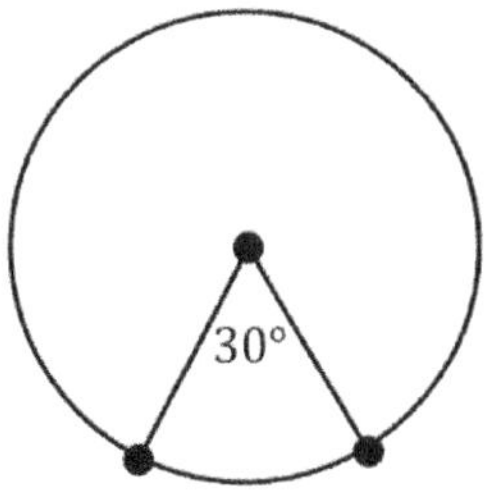

(Top view of cake)

A bakery sells a cylindrical cake by the slice. Each slice is a wedge that forms a 30° angle at the center and has a mass of x grams. Which expression is equivalent to the mass of the entire cylindrical cake?

A. $8x$

B. $12x$

C. $16x$

D. $20x$

5

$$x^2 + y^2 - 10x + 2y + 17 = 0$$

A circle with the equation above is graphed on the xy-plane. How many units from the y-axis is the circle at its closest point?

A. 2

B. 3

C. 4

D. 5

6

Square ABCD is located on the xy-plane where A is located at $(3,8)$ and B is located at $(1,-2)$. Which of the following could be the coordinates of point C?

A. $(9,-4)$

B. $(11,-4)$

C. $(13,-4)$

D. $(15,-4)$

7

What is the perimeter of an equilateral triangle with area of $9\sqrt{3}$?

A. 9

B. 18

C. 27

D. 36

8

A cylinder has a volume of 100 cubic centimeters. If the radius of its base increases by a factor 5, what is the new volume in cubic centimeters?

A. 500

B. 1,500

C. 2,500

D. 3,500

9

$$\sin(4x + 15) = \cos(x)$$

In the equation above, x is the measure in degrees of an acute angle. What must be the value of x?

A. 5

B. 10

C. 15

D. 20

10

$$(i^{22})^3$$

Given that i is the imaginary number $i = \sqrt{-1}$, which of the following expressions is equivalent to the one above?

A. -1

B. 0

C. 1

D. i

11

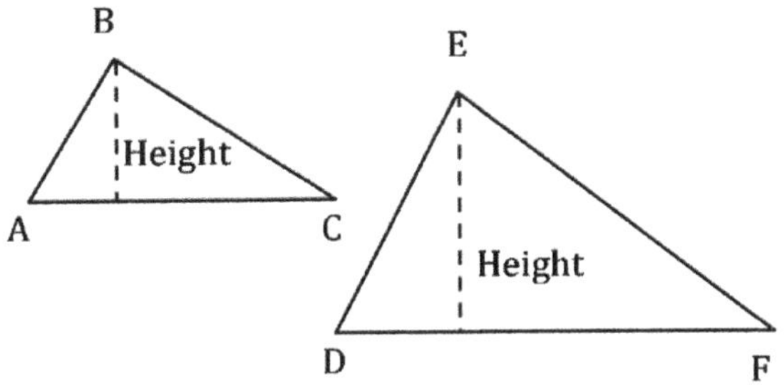

Triangle ABC has a perimeter of 24 feet and triangle DEF has a perimeter of 40 feet. If $\angle A = 64°$, $\angle B = 50°$, $\angle E = 50°$, $\angle F = 66°$ then how many times larger is the height of triangle DEF than the height of triangle ABC?

12

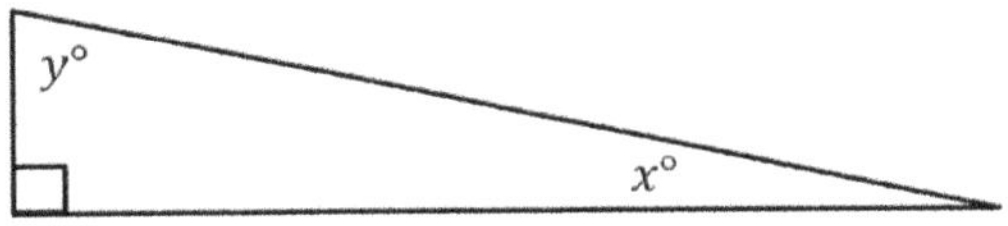

In the triangle above, the cosine of $x°$ is 0.96. What is the tangent of $y°$?

13

Tabitha must modify the dimensions of a rectangular prism such that its volume does not change. If she doubles both the height and width of the prism, by what factor must she multiply the length?

14

The exterior angle of a regular polygon measures 40 degrees. How many sides does the polygon have?

15

$$(x - 3)^2 + (y + 2)^2 = 40$$

A circle on the coordinate plane has the equation shown above. If point P is on the circle at $(1, b)$, then what is a possible positive value of b?

Part 2: Calculator

19 Minutes, 15 Questions

1

Points A and B are endpoints to the diameter of circle P and have the coordinates $(-4,-4)$ and $(6,4)$, respectively. If line m is tangent to circle P at point A, then what is the equation of line m?

A. $y = -\frac{5}{4}x - 9$

B. $y = -\frac{5}{4}x + 9$

C. $y = \frac{5}{4}x + 1$

D. $y = \frac{5}{4}x - 1$

2

Given that $\sin\theta = 0.2$, what could be the value of $\cos\theta$?

A. $\frac{\sqrt{6}}{5}$

B. $\frac{2\sqrt{6}}{5}$

C. $\frac{1}{5}$

D. $\frac{2}{5}$

3

What is the area of the largest circle that can fit inside the lines $y = x$ and $y = x + 8$?

A. 2π

B. 4π

C. 6π

D. 8π

4

The maximum distance between any two vertices of a regular hexagon is 20 meters. How many square meters is the area of this hexagon?

A. $100\sqrt{3}$

B. $125\sqrt{3}$

C. $150\sqrt{3}$

D. $175\sqrt{3}$

5

Triangle ABC is similar to triangle DEF. The area of triangle ABC is 5 ft^2 and the area of triangle DEF is 40 ft^2. If the perimeter of triangle ABC is x then what expression is equivalent to the perimeter of triangle DEF?

A. $2x$

B. $2x\sqrt{2}$

C. $8x$

D. $8x\sqrt{2}$

6

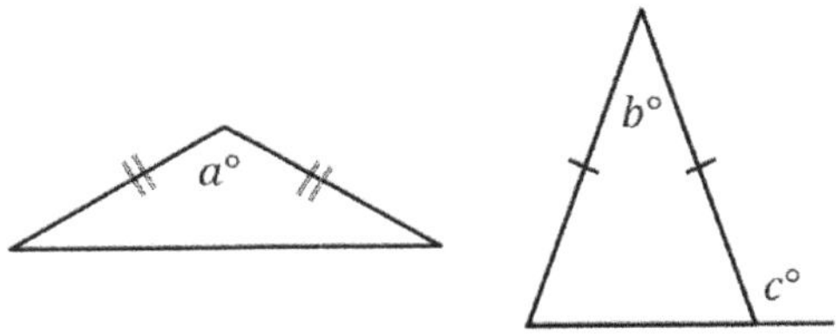

Two isosceles triangles are shown above. If a and b are supplementary and $c = 100°$, then what is the value of a?

A. 130

B. 140

C. 150

D. 160

7

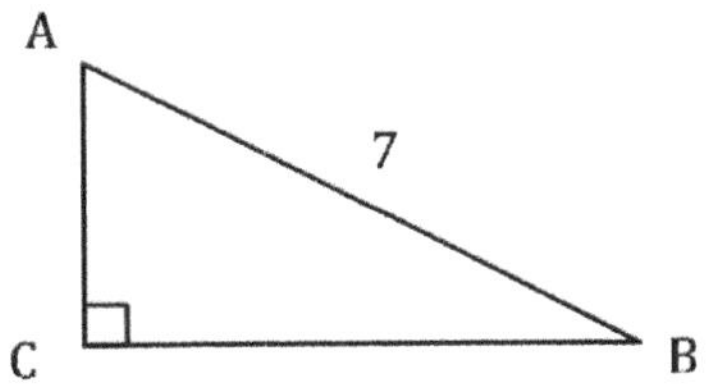

In the right triangle ABC shown above, AB $= 7$ and cos A $= 1/4$. What is the value of sin B ?

A. $1/7$

B. $1/4$

C. $\sqrt{15}/7$

D. $\sqrt{15}/4$

8

In the trapezoid QRST, QR $\parallel$ ST and Q is a right angle. If QT is half of RS, then what is the measure of acute angle R?

A. 30°

B. 45°

C. 60°

D. 75°

9

In which case will the expression $\sin\theta - \cos\theta$ always be postivie?

I. $0 < \theta < \dfrac{\pi}{2}$

II. $\dfrac{\pi}{2} < \theta < \pi$

III. $\pi < \theta < \dfrac{3\pi}{2}$

A. I only

B. II only

C. I and II

D. I, II, and III

10

Circle P contains the central angle MPN, which measures $2\pi/3$ radians. If the minor arc MN has a length of 8, then what is the circumference of circle P?

A. 12π

B. 24π

C. 12

D. 24

11

If $a\pi$ radians is equal to 432°, then what is the value of a?

12

The surface area of a right cylinder is $a^2x\pi$ square feet. If its circular base has a radius of $4a$ feet and the cylinder has a height of $10a$ feet, then what is the value of x?

13

The average measure of an interior angle in a polygon is between 145 and 148 degrees. How many sides does the polygon have?

14

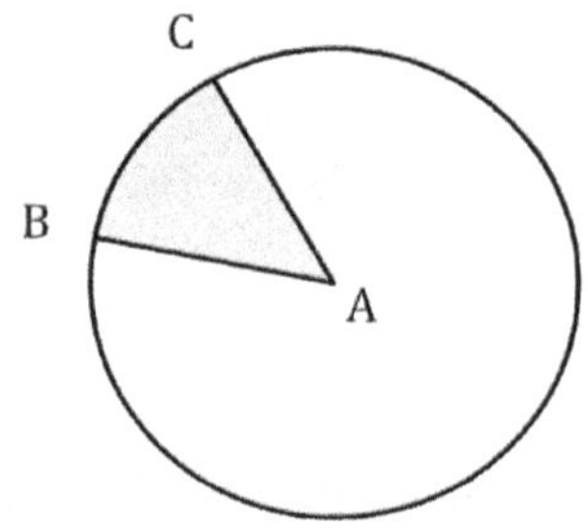

Circle A has a radius of 9 meters and minor arc BC has a length of 5 meters. What is the area in square meters of the shaded sector?

15

$$(7-2i)(4-11i)=a+bi$$

What is the value of a in the equation above, given that $i=\sqrt{-1}$?

Explanations

Remaining Topics Exam: Part 1

1. C

Draw in an altitude that runs through point N, perpendicular to base LM. Doing so creates our two special right triangles. Use the known side ratios of the 45-45-90 and 30-60-90 right triangles to determine the length of NL

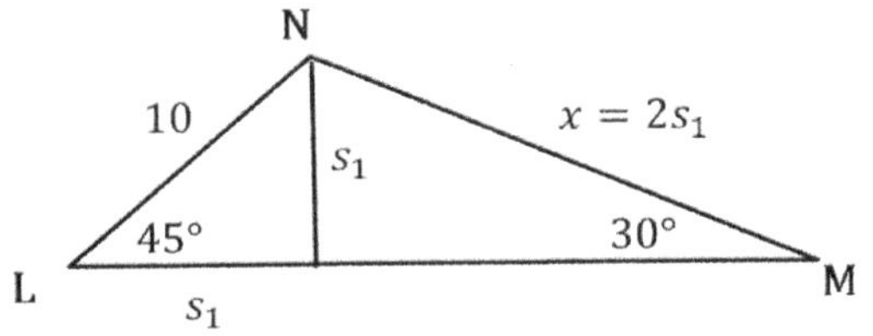

Since LN is the hypotenuse of the isosceles right triangle, it is equal to $s_1\sqrt{2}$. And since s_1 is also the shortest side of the 30-60-90 triangle, MN must be double s_1.

$$10 = s_1\sqrt{2}$$

$$s_1 = \frac{10}{\sqrt{2}} \quad \text{or} \quad 5\sqrt{2}$$

$$x = 10\sqrt{2}$$

2. C

Use the known side ratios of a 30-60-90 right triangle to determine the side length of the square. Since the shortest side (across the 30-degree angle) is 10, the second leg (across the 60-degree angle) is $10\sqrt{3}$. This second leg is also the side of the square, so squaring it gives the squares area.

$$\text{Area} = \left(10\sqrt{3}\right)^2$$

$$\text{Area} = 300$$

3. A

Use the distributive property to expand out the expression, then simplify by combining like-terms.

$$3(1 - 2i) - 8(1 - 2i)$$

$$3 - 6i - 8 + 16i$$

$$-5 + 10i$$

4. B

Identify the fraction of the entire cake that the slice makes up by using the central angle of 30-degrees and the total 360-degrees in a full circle.

$$\frac{30}{360}(\text{Mass of Entire Cake}) = x$$

$$\frac{1}{12}(\text{Mass of Entire Cake}) = x$$

$$\text{Mass of Entire Cake} = 12x$$

5. A

Bring the equation of the circle into its standard form by completing the square. Identify its center and radius to determine how close it gets to the y-axis.

$$x^2 + y^2 - 10x + 2y + 17 = 0$$

$$(x^2 - 10x + 25) + (y^2 + 2y + 1) = -17 + 25 + 1$$

$$(x - 5)^2 + (y + 1)^2 = 9$$

Center: $(5, -1)$ Radius: 3

$$5 - 3 = 2$$

The circle is only 2 units away from the y-axis at its closest point.

6. B

AB and BC are consecutive sides of Square ABCD and, therefore, must intersect at right angles. That means that the slopes from A to B and B to C are opposite reciprocals. Use the slope of BC to find an appropriate coordinate pair for point C.

$$\text{Slope}_{AB} = \frac{8-(-2)}{3-1}$$

$$\text{Slope}_{AB} = \frac{10}{2}$$

$$\text{Slope}_{BC} = -\frac{2}{10}$$

From point B, move down 2 and right 10 to reach point C at $(11, -4)$. Here it is best to not simplify the slope because this exact rise and run not only maintains that the sides are perpendicular, but us also congruent in length.

7. B

Recall that any equilateral triangle can be broken down into two $30-60-90$ triangles by using an altitude, which allows to relate the base and height to each other.

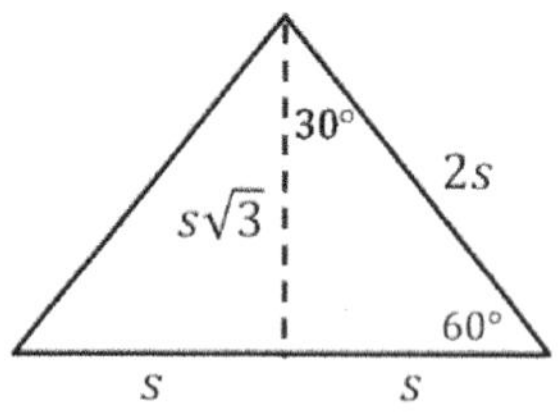

$$\text{Area} = \frac{1}{2}(2s)(s\sqrt{3})$$

$$\text{Area} = s^2\sqrt{3}$$

$$9\sqrt{3} = s^2\sqrt{3}$$

$$9 = s^2$$

$$s = 3$$

$$\text{Perimeter} = 6\text{s} \quad \text{or} \quad 18$$

8. C

Dissect the volume formula of cylinder and keep only the variables that change while maintaining the integrity of the equation. Input the change into the equation to identify the of overall factor of change for the volume. Use that factor to reach the new volume.

$$\text{Volume of Cylinder} = \pi r^2 h$$

$$\text{Volume of Cylinder} \propto r^2$$

$$\text{Volume of Cylinder} \propto (5)^2 \quad \text{or} \quad 25$$

The factor of change is 25. Multiply the original volume by 25 to get 2,500.

9. C

Sine and Cosine are equal when their angles are complementary. Therefore, the sum of $4x + 15$ and x is 90.

$$4x + 15 + x = 90$$

$$5x + 15 = 90$$

$$5x = 75$$

$$x = 15$$

10. A

Since the values of i are cyclical in regards to the pattern we see with its whole number exponents, we should find the remainder of its overall exponent when it is divided by four.

$$(i^{22})^3$$

$$i^{66}$$

Four goes into 66 sixteen times with a remainder of 2. Therefore, $i^{66} = i^2 = -1$.

11. 5/3

The triangles are similar since they have the same angle measurements. Take the ratio of the perimeters to find how many times larger the height of the larger triangle is to the height of the smaller one.

$$\frac{40}{24} = \frac{5}{3}$$

12. 24/7

An effective strategy when confronted with trigonometric values in decimal form is to convert them into fractions. Many times, these fractions will lead to a known Pythagorean Triple.

$$\cos(x) = \frac{\text{adjacent}}{\text{hypotenuse}}$$

$$\cos(x) = 0.96 \quad \text{or} \quad \frac{24}{25}$$

$$\text{Pythagorean Triple: } 7 - 24 - 25$$

$$\tan(y) = \frac{\text{opposite}}{\text{adjacent}}$$

$$\tan(y) = \frac{24}{7}$$

13. 1/4

Use the volume formula of a rectangular prism and set the original volume equal to the new one, since the volume is to remain the same. Apply the factor a to the length, then solve.

$$lwh = l'w'h'$$

$$lwh = (al)(2w)(2h)$$

$$lwh = 4a * lwh$$

$$1 = 4a$$

$$a = 1/4$$

14. 9

The exterior angles of a polygon must always add up to 360-degrees. Determine how many times 40 goes into 360.

$$40n = 360$$

$$n = 9$$

15. 4

Use the given point $(1, b)$ in the circle's equation by making the substitutions $x = 1$ and $y = b$. Solve for b and report the positive value.

$$(x-3)^2 + (y+2)^2 = 40$$

$$(1-3)^2 + (b+2)^2 = 40$$

$$(-2)^2 + (b+2)^2 = 40$$

$$4 + (b+2)^2 = 40$$

$$(b+2)^2 = 36$$

$$b + 2 = \pm 6$$

$$b = -2 \pm 6$$

$$b = -8 \quad \text{or} \quad 4$$

Explanations

Remaining Topics Exam: Part 2

1. A

The slope of the tangent line is perpendicular to the radius at the point of tangency. Find the slope of the radius using the given endpoints, then use the opposite reciprocal and point A to find the equation of the tangent line. Also, remember to narrow your selection down to the answer choices that work for point A located at $(-4,-4)$.

Find the slope of the tangent line:

$$m_{AP} = \frac{-4-4}{-4-6} \quad \text{or} \quad \frac{4}{5}$$

$$m_{\perp} = -\frac{5}{4}$$

Find the equation of the tangent line:

$$y - y_1 = m(x - x_1)$$

$$y + 4 = -\frac{5}{4}(x + 4)$$

$$y = -\frac{5}{4}x - 9$$

2. B

Use a right triangle to determine the value of $\cos(\theta)$. It is useful see the value of sine in fraction form. With the aid of the Pythagorean Theorem, find the second leg of the triangle, or the adjacent side.

$$\sin(\theta) = 1/5$$

$$\text{Opposite} = 1, \text{Hypotenuse} = 5, \text{Adjacent} = 2\sqrt{6}$$

$$\cos(\theta) = \frac{2\sqrt{6}}{5}$$

3. D

To find the area of the largest circle that can fit between the two parallel lines, first find the distance between them. Distance is always measured at the perpendicular, so use a perpendicular line to map out the distance between the points of intersection between the perpendicular and parallel lines.

The best line to use is $y = -x$ because it crosses the line $y = x$ at the origin, $(0,0)$. It then crosses the line $y = x + 8$ at $(-4,4)$. The distance between these points is $4\sqrt{2}$, which would also be the diameter of the circle. Therefore, the radius of the circle is $2\sqrt{2}$.

$$\text{Area} = \pi r^2$$

$$\text{Area} = \pi\left(2\sqrt{2}\right)^2$$

$$\text{Area} = 8\pi$$

4. C

A hexagon can be divided into six equilateral triangles. If the maximum distance between any two vertices is 20, then the side lengths of these equilateral triangles must be 10. Remember that equilateral triangles can be divided into two 30-60-90 triangles. Use the diagrams below to find the area of first each triangle in the hexagon, then the entire hexagon.

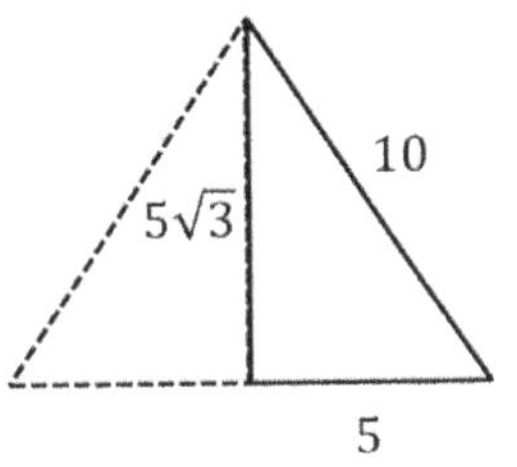

$$\text{Area} = \frac{1}{2}bh$$

$$\text{Area} = \frac{1}{2}(10)(5\sqrt{3})$$

$$\text{Area} = 25\sqrt{3}$$

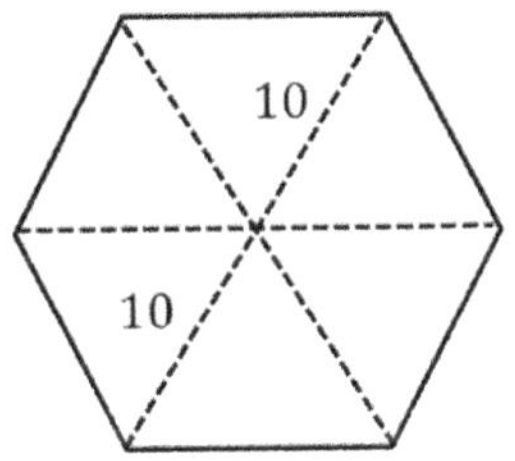

$$\text{Total Area} = 6(25\sqrt{3})$$

$$\text{Total Area} = 150\sqrt{3}$$

5. B

Use a proportion that relates the ratio between the two perimeters and areas. Recall that these ratios help determine the scale factor, which is squared when relating the areas to each other.

$$\frac{\text{Perimeter of ABC}}{\text{Perimeter of DEF}} = \sqrt{\frac{\text{Area of ABC}}{\text{Area of DEF}}}$$

$$\frac{x}{y} = \sqrt{\frac{5}{40}} \quad \text{or} \quad \frac{x}{y} = \sqrt{\frac{1}{8}}$$

$$\frac{x}{y} = \frac{1}{2\sqrt{2}}$$

$$y = 2x\sqrt{2}$$

6. D

The base angles of the second triangle are 80°, which means that b must equal 20. Since a and b must add up to 180°, a equals 160.

7. B

Since A and B are complementary, $\cos(A) = \sin(B)$, so choice B is correct.

8. A

The dimensions of the trapezoid can be used to create a 30-60-90 triangle, which has known dimensions. The acute angle R must 30°.

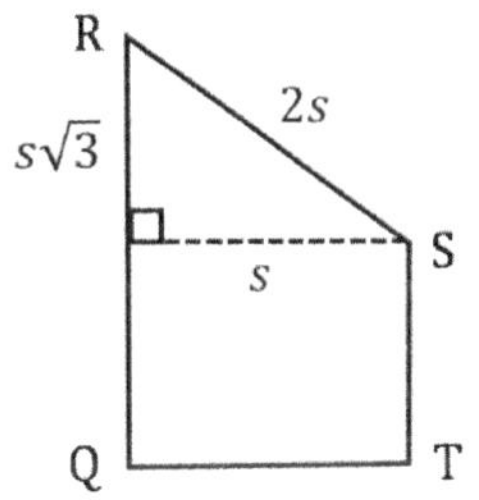

9. B

A positive number minus a negative number is always positive. Therefore, choose the interval that maps out Quadrant II, where $\sin(\theta)$ is positive and $\cos(\theta)$ is negative.

10. D

Use the arc length formula in radians to create an equation, then multiply by both sides by three .

$$\text{arc length} = r\theta$$

$$8 = \frac{2\pi}{3}r$$

$$24 = 2\pi r$$

$$\text{Circumference} = 2\pi r \quad \text{or} \quad 24$$

11. 12/5, 2.4

Use the equivalency statement $180° = \pi\, rad$ to convert degrees into radians.

$$432° \times \frac{\pi\, rad}{180°} = \frac{12}{5}\pi\, rad$$

12. 112

Use the surface area equation of a cylinder as a template to plug in all given information. Simplify by following order of operations and combining like-terms.

$$\text{Surface Area} = 2\pi r^2 + 2\pi rh$$

$$\text{Surface Area} = 2\pi(4a)^2 + 2\pi(4a)(10a)$$

$$\text{Surface Area} = 32a^2\pi + 80a^2\pi$$

$$\text{Surface Area} = 112a^2\pi$$

$$x = 112$$

13. 11

Use the interior angle formula to relate the given constraints. Solve the compound equality to find the number of side lengths that the polygon contains.

$$145n < 180(n-2) < 148n$$

$$145n < 180n - 360$$

$$-35n < -360$$

$$n > 10.3$$

$$180n - 360 < 148n$$

$$32n < 360$$

$$n < 11.25$$

The polygon must have 11 sides.

14. 45/2, 22.5

Break down the sector area formula in radians to see how we can plug in the radius and arc length $(r\theta)$ to find it.

$$\text{Sector Area} = \frac{1}{2}r^2\theta$$

$$\text{Sector Area} = \frac{1}{2}r(r\theta)$$

$$\text{Sector Area} = \frac{1}{2}(9)(5)$$

$$\text{Sector Area} = \frac{45}{2} \quad \text{or} \quad 22.5$$

Another way to approach this problem is to create a proportion between the arc length and sector area relative to the circumference and total area, respectively.

15. 6

Expand the expression and apply the substitution that $i^2 = -1$. Combine like terms to determine the value of a.

$$(7-2i)(4-11i)$$

$$28 - 8i - 77i + 22i^2$$

$$28 - 85i - 22$$

$$6 - 85i = a + bi$$

$$a = 6$$

Many calculators that students use on the test also have an i function. In the home screen of the calculator, type out the expression given, then press enter, and an answer will appear in standard form. In TI-84 calculators, the i button is where the decimal is located. Press 2nd and then the decimal to reach it.

STRATEGY GUIDE

Introduction

In addition to knowing the content, students looking to up their scores in the math section must master the art of test taking. In this section, students will review the strategies needed to correctly answer test-specific questions more efficiently and use patterns in language and numbers to better approach problems. The idea here is simple: do less to score high. While reviewing several unique questions that define the current test, pay close attention to how phrasing and patterns dictate approach. Solving strategy is an implicit testing point because it directly relates to a student's time constraints, susceptibility to arithmetic and process errors, and skill fluency. However, more than anything else, the test reflects a student's ability to bubble in correct answers. For students to do this successfully, they need understand correct problem-solving approaches and utilize time-saving techniques to effectively exploit the vulnerability of multiple-choice testing.

5.1 ANSWER CHOICES

a. Numeric Considerations

Often, the answer choices in the math section look very similar. A key strategy in answering multiple-choice questions is to find the point of difference among all the answer choices that will allow us to choose the right answer using only that one key piece of information. Looking for one aspect of an equation rather than solving the problem in its entirety will save both time and energy, as well as keep us from making "silly" mistakes. Essentially, in multiple-choice testing, we do not need the full answer to get the right answer. And in many cases, doing more work is not actually beneficial as it takes more time and leaves more room for errors.

The idea of using answer choices to dictate approach is particularly helpful when answering questions regarding systems of equations, where the answer choices include solution sets. In the subsequent example do the following: read the question, look at the answer choices, and determine which variable on its own can be used to find the right answer.

Q+A

Identify Trends **No Calculator**

$$3x - 2y = 17$$

$$2x + 3y = 7$$

The equations above are graphed on the xy-plane. Which one of the following (x, y) points shows where the two equations intersect?

A. $(-2, 5)$

B. $(-1, 6)$

C. $(5, -2)$

D. $(5, -1)$

Workspace

Solution Outline

Step 1: Analyze the answer choices. Since the x-coordinate of 5 repeats in answer choices C and D, it is not a point of difference that is enough to answer the question, because if we solve for x, what happens if we get 5? We will have to go back and solve for y. Instead, we can solve for y right from the start. Since no y-coordinate repeats itself, knowing the value of y is all that is necessary to answer the question.

Step 2: Use the elimination method to solve the system for the y-coordinate. Make sure to eliminate the x-variable.

$2(3x - 2y = 17)$ or $6x - 4y = 34$

$-3(2x + 3y = 7)$ or $-6x - 9y = -21$

Adding the equations: $-13y = 13$

Therefore, $y = -1$

Answer: D

Notes

Also note that in the above example, we can directly plug the answer choices into the given equations to select the one that works for both. However, this approach can be too time-consuming to have any extended practicality. Trends in answer choices will not only help in determining what element to solve for in a question, but can help in knowing which answer choices to look at altogether. A great example of this occurs in solving multiple-choice question with linear inequalities. When asked for a possible value of just one variable in a bivariate system of linear inequalities, the correct answer will only ever be the least or greatest value.

Q+A

Eliminate Answer Choices **No Calculator**

$$3x - 2y > 17$$

$$2x + 3y \leq 7$$

Which of the following y-values is in the solution set of the system of linear inequalities above?

A. -2

B. -1

C. 3

D. 5

Workspace

Solution Outline

Step 1: Analyze the answer choices. Since the questions asks for a possible value of a variable in a system of linear inequalities, the correct answer choice must be either the lowest value or the highest one. Therefore, only decide between A and D. Also, since the questions specifically asks for a y value, solve the system for y.

Step 2: Use elimination to solve the question. Since the equations have the same structure, this will be the fastest solving method. To solve for y, eliminate x.

$2(3x - 2y > 17)$ or $6x - 4y > 34$

$-3(2x + 3y \leq 7)$ or $-6x - 9y \geq -21$

Adding the inequalities: $-13y > 13$

Therefore, $y < -1$

Answer: A

Notes

A few notes here from the last example: 1) We can also treat the inequalities as equal signs and solve as equations. Then sketch the graphs on the coordinate plane and shade in appropriate area to identify the full solution set. 2) If choosing to keep inequalities while solving through elimination, only add the greater than sides together and the less than sides together.

Lastly, use answer choices when solving equations with square roots that require squaring both sides to solve for the variable. The reason being that the answer choices already give us a list of values to check. If we choose to answer the question without using the answer choices, not only will we have to solve the equation – often leading to two possible values – but we will also have to check that the values actually work in the original equation. Why not skip out on the solving and just get to the checking?

Q+A

Plug in Answer Choices **No Calculator**

$$\sqrt{2x + 5} = x + 1$$

Which of the following answer choices represents the solution set to the equation above?

A. $\{-2\}$

B. $\{0\}$

C. $\{2\}$

D. $\{-2, 2\}$

Workspace

Solution Outline

Step 1: Plug in answer choices to see which values work satisfy the given equation. Remember, the output to the square root operation is always positive and the correct answers often create perfect squares inside the radical. In addition, only values of zero or more can be square rooted. Choices A and D do not work because -2 makes the right side of the equation negative. Choice B does not work because $\sqrt{5} \neq 1$. Therefore, Choice C must be correct.

Answer: C

Notes

b. Descriptive Statistics

Section four of the test often involves descriptive statistics that tests us on our ability to determine the validity of a claim or form a reasonable conclusion following the completion of a well-designed study. What we need to keep in mind when making inferences based off of limited data is that we should state our conclusions as reasonably as possible. Reasonable answers often engage in neutral language that allow for some possibility of being wrong. After all, we do not know exactly what will happen outside the confines of a preliminary study. So, in these lengthier descriptive statistic questions, correct answers often avoid strong language. For the most part, students can cross out answer choices using strong language, which is also a helpful tactic that can be used in the critical reading section. Use Table 5.11 to review examples of strong and neutral language.

Table 5.11 | Strong and Neutral Language Summary

Strong Language			Neutral Language		
Likely WRONG			Likely CORRECT		
No, never, must, only, causes, certain, only,	absolutely, will not, without a doubt	definitely, exactly...	Might, could, correlates to, possible,	may, is associated with, plausible	seems, appears, likely...

Q+A

Descriptive Statistics

Calculator

A city council set out to determine the favorability of a new proposal to restore and repurpose the buildings in the city's historic district. The council randomly surveyed 250 of the city's residents and found that 72% view the proposal favorably, 16% view the proposal unfavorably, and the remainder of residents felt unsure. The proposal will be on the ballot in the next election where it needs 50% of the electorate's support to pass. Which of the following is the most reasonable statement according to the results of the study?

A. The proposal will likely pass in the next election

B. The proposal will definitely pass in the next election

C. Exactly 72% of all the city's residents support the proposal

D. No interpretation can be made from the survey results

Workspace

Solution Outline

Step 1: Cross out answer choices with overly strong language. Choices B and C contain "definitely" and "exactly", so they are the least likely to be correct. Choice D also makes a big claim that "No interpretation can be made." Only Choice A contains reasonable language (e.g. likely) that fits the results of the survey.

Answer: A

Notes

5.2 HELPFUL SHORTCUTS

a. Working with Expressions

Working with expressions as whole units is often more powerful than working with singular variables. Get comfortable working with expressions by using them to make substitutions and seeing their multiples. Another reason why this is so important is that the test often asks students for the value of an expression, rather than an individual variable. Moreover, the value of an expression can often be found without even knowing the value of the variables that make up the expression.

Q+A

Expression Value **No Calculator**

$$\frac{1}{2}x - \frac{3}{4} = 2$$

Using the equation above, what is the value of $4x + 1$?

A. -2

B. 9

C. 14

D. 23

Workspace

Solution Outline

Step 1: Compare the x terms in the given equation and the desired expression. The factor of change is 8. Therefore, multiply the equation by 8 to generate the $4x$ term.

$$8\left(\frac{1}{2}x - \frac{3}{4} = 2\right)$$

$$4x - 6 = 16$$

Step 2: Add both sides by 7 to generate $4x + 1$.

$$4x + 1 = 23$$

Answer: D

Notes

Identifying repeated expressions is another useful tool because substitutions can be made to condense the workload. In many cases, distinct expressions are placed glaringly in parentheses, guiding students into making a substitution. The benefit to using this strategy is that it cuts down on the amount of algebra involved. Although we can still expand and simplify the equation as is, making a quick substitution can often craftily bring us into a similar solving position without all the erroneous steps. In the subsequent example, identify the repeated expression and think about what substitution should be made to avoid extra work. Then, try to solve the problem.

Q+A

Expression Substitution **No Calculator**

$$(x+7)^2+4(x+7)+4=0$$

Which of the following is a solution to the equation above?

A. -10

B. -9

C. -2

D. 13

Workspace

Solution Outline

Step 1: Make the substitution $u = x + 7$

$u^2 + 4u + 4 = 0$

Step 2: Solve for u by factoring.

$(u+2)^2 = 0$

$u = -2$

Step 3: Solve for x

$x + 7 = -2$

$x = -9$

Answer: B

Notes

The test also often includes questions on linear and quadratic equations or their systems that ask us to find an expression using their solutions. However, it is often not necessary to find these solutions to know the value of the expression in question. For linear systems, be sure to look within and between the equations to see if the expression can be generated without solving the system. If solving the system can be avoided and a direct answer can be found, we will save time and be even more confident in our answer. The less work we do to reach our answer, the less room there is for us to make any arithmetic errors.

Q+A

Generating Expressions from Equations

No Calculator

Lines j and k are graphed on the xy-plane and both go through the point (a, b). The x and y intercepts of line j are -3 and 3, respectively. Line k contains the points (2,5) and $(4, -1)$. What is the value of $b - a$?

A. -4

B. 1

C. 3

D. 9

Workspace

Solution Outline

Step 1: Find the equation of line j in slope intercept form. The slope is 1 ($m = 1$) and the y-intercept is 3 ($b = 3$).

$$y = mx + b$$

$$y = x + 3$$

Step 2: Before moving onto line k, see if the expression $b - a$ can be generated understanding that the cooridtne (a, b) implies that $x = a$ and $y = b$. Make these substitutions into the equation found in Step 1.

$$b = a + 3$$

$$b - a = 3$$

Step 3: Since we generated the expression in question and know its value, there is no need to use line k or solve the sytem.

Answer: C

Notes

b. Number of Solutions

The test often asks students about the number of solutions to systems of linear and quadratic equations. In the case of linear systems, this pivots around the slopes of the lines involved. Lines with different slopes cross once – therefore there is one solution. Lines that have the same slope may be parallel (no solution) or the same line (infinitely many solutions). Revisit section 1.2 Linear Systems to review these ideas in greater detail.

Q+A

Number of Solutions to a Linear System **No Calculator**

$$\frac{3}{2}x + \frac{1}{6}y = \frac{5}{3}$$

$$9x + y = 15$$

How many (x, y) solutions does the system above contain?

A. Zero

B. One

C. Two

D. More than two

Workspace

Solution Outline

Step 1: Multiply the first equation by 6.

$$6\left(\frac{3}{2}x + \frac{1}{6}y = \frac{5}{3}\right)$$

$$9x + y = 10$$

Step 2: Compare this equation to the second. Since the values of A and B (coming from the standard form of linear equations: $\mathrm{A}x + \mathrm{B}y = \mathrm{C}$) is the same in both equations, the lines have the same slope. Since the values of C are different, the lines do not overlap. Instead, they run parallel to each other. Therefore, they do not cross on the xy-plane and the system has no solution.

Answer: A

Notes

With quadratics, the number of solutions comes down to the value of the discriminant: $b^2 - 4ac$. There will be two solutions or real zeros when the discriminant is positive, one solution when it is zero, and no solution when it is negative. Revisit section 2.3 Function Properties to review these cases.

Q+A

Using the Discriminant **No Calculator**

$$y = 5x - 2$$

$$y = x^2 + 10x + 3$$

How many (x, y) solutions does the system of equations above contain?

A. Zero

B. One

C. Two

D. More than two

Workspace

Solution Outline

Step 1: Set the equations equal to each other $(y_1 = y_2)$ and bring terms to one side.

$x^2 + 10x + 3 = 5x - 2$

$x^2 + 5x + 5 = 0$

Step 2: Use the discriminant: $b^2 - 4ac$. Remember, to use the discriminant, the equation of the quadratic must be in the form $ax^2 + bx + c = 0$.

$a = 1, b = 5, c = 5$

$5^2 - 4(1)(5)$

$25 - 20 = 5$

Step 3: Since the discriminant is positive, there are two real solutions.

Answer: C

Notes

c. Root Sum and Product Formulas

Quadratic functions and equations have special formula for their roots. Though the sum and product formulas were mentioned in section 2.3b Quadratic Functions, we have placed them again here because they prove to be an effective shortcut, especially when working with quadratics that cannot be easily graphed on the xy-plane or factored. Remember, other ways we can reference real roots include zeros and x-intercepts, as well as – in some cases – solutions and factors.

Importantly, for **quadratic equations and systems**, the test often asks about the sum or product of the solutions. Recall that these values can be found when the quadratic is in $ax^2 + bx + c = 0$ form.

Sum of solutions: $-b/a$

Product of solutions: c/a

Q+A

Root Product Formula **No Calculator**

$$f(x) = x + 3$$

$$g(x) = 2x^2 - 1$$

The functions $f(x)$ and $g(x)$ below intersect at points (m, n) and (p, q). What is the product of m and p?

A. -3

B. -2

C. 1

D. 5

Workspace

Solution Outline

Step 1: Set the functions equal to each other and set equation equal to zero.

$2x^2 - 1 = x + 3$

$2x^2 - x - 4 = 0$

Step 2: Understand that we are multiplying the x-coordinates of the solutions to a quadratic system, so we can use the Product of solutions formula: c/a.

$c = -4$

$a = 2$

Product $= c/a$ or -2

Answer: B

Notes

5.3 CALCULATOR INFORMATION

a. Approved Calculators

Remember to come to the test prepared with an approved calculator that is fully charged. Bring a replacement calculator or fresh batteries appropriate for your calculator as a precaution. Use Table 5.31 to read through the list of currently approved calculators. Double check on the College Board website for any updates.

Table 5.31 | Approved Calculators

Texas Instrument	TI-73 TI-80 TI-81 TI-82 TI-83 TI-83 Plus TI-83 Plus Silver TI-84 Plus	TI-84 Plus CE TI-84 Plus Silver TI-84 Plus C Silver TI-84 Plus T TI-84 Plus CE-T TI-85 TI-86 TI-89	TI-89 Titanium TI-Nspire TI-Nspire CX TI-Nspire CX II TI-Nspire CX II-T TI-Nspire CM-C TI-Nspire CAS TI-Nspire CX CAS	TI-Nspire CX II CAS TI-Nspire CX II-T CAS TI-Nspire CM-C CAS TI-Nspire CX-C CAS TI-Nspire CX II-C CAS
Casio	FX-6000 series FX-6200 series FX-6300 series FX-6500 series FX-7000 series FX-7300 series FX-7400 series FX-7500 series	FX-7700 series FX-7800 series FX-8000 series FX-8500 series FX-8700 series FX-8800 series Graph25 series FX-9700 series	FX-9750 series FX-9860 series CFX-9800 series CFX-9850 series CFX-9950 series CFX-9970 series FX 1.0 series Algebra FX 2.0 series	FX-CG-10 FX-CG-20 series FX-CG-50 Graph35 series Graph75 series Graph95 series Graph100 series FX-CG500
HP	HP-9G HP-28 series HP-38G	HP-39 series HP-40 series	HP-48 series HP-49 series	HP-50 series HP Prime
Sharp	EL-5200 EL-9200 series	EL-9300 series	EL-9600 series	EL-9900 series
Radio Shack	EC-4033	EC-4034	EC-4037	
Other	Datexx DS-883	Micronta	NumWorks	Smart2

b. Calculator Functions

Previous Answer

Many calculators have an option to use the answer from your last calculation. For TI 83-84 calculators, just pressing an operation key will apply it to the previous answer, indicated by the abbreviation ANS. Additionally for TI 83-84 calculators, we can prompt the answer at any point by pressing [2nd], [ANS]. The [ANS] key is the negative sign key found at the bottom of the key pad. Many calculators will often allow for retrieval of previous answers simply by navigating with the arrow keys.

Re-Enter Function

Many calculators have an option to bring down calculations or answers made previously in the calculator by navigating up to it using the arrow keys and pressing enter. If your calculator does not do this or you accidentally cleared your home screen, there are other options without having to retype the calculation in your calculator. For TI 83-84 calculators, press [2nd], [ENTRY]. The [ENTRY] key is the [ENTER] key found at the bottom of the key pad.

Store Variables

Many calculators have an option to store values for a variable of your choice. For TI-83-84 calculators, use the [STO→] next to the number 1 on the key pad. Type the value, press [STO→], then choose variable of your choice. For example, to store the value of 2.613 as x, type 2.613, press [STO→], then $[x, \mathrm{T}, \theta, n]$ and enter. This option makes calculating the values of expressions and functions much simpler, and allows us to repeat the process with another input value by storing a new number, then using the Re-Entry Function.

Simplified Fraction

Many calculators have an option to create simplified fractions from other fractions or decimals. For TI-83-84 calculators, press [MATH], 1. Frac after the number of your choice and then enter. Since many answer choices are given in simplest fraction form, this makes finding your answer even easier.

Imaginary Numbers

Many calculators have an option to use the imaginary number, $i = \sqrt{-1}$. For TI-83-84 calculators, press [2nd], [i]. The i key is the same as the decimal key at the bottom of the key pad. This means your calculator can simplify imaginary numbers into standard form. For example, type $(5 + i)/(1 - i)$ in your home screen and press enter. The calculator simplifies it to $2 + 3i$.

Graphing

Use a graph of bivariate equations and systems to find unique characteristics and solutions. For example, graphing is a powerful tool we can use to find function values and intersections. Type functions into Y1, Y2, Y3... and graph them. In TI 83-84 calculators, use the [Trace] and [2nd], [Calc] features to locate distinct points, like intersections, zeros, and max/mins. Follow calculator instructions as directed.

Edit Lists

Use lists when given raw data from charts and frequency tables to find their one-variable statistics, like mean, median, and quartiles. In the TI 83-84 calculators, press [STAT], then 1. Edit, and proceed to manually enter data into L1. If numbers are associated with a certain frequency, type the associated frequency into L2. Calculate one-variable statistics by pressing [STAT] and scroll right to CALC to press 1. 1-Var Stats. Make sure to include L1 for the data set and L2 for the frequency list if there is one. The home screen will then display values that describe the data set.

5.4 TIME MANAGEMENT

a. General Strategies

There is no guessing penalty, so make sure to answer every question. Every question is worth the same amount, so there is no need to strain over difficult questions and lose out on time to answer questions you can definitely get right. Sometimes, the grid-in questions are easier than the multiple-choice questions. When it comes down to it, you have a better likelihood of guessing the right answer to a multiple-choice question than a grid-in question. This means that you may choose to skip the difficult multiple-choice questions to tackle the grid-in ones. That way, if you are running low on time and run into a guessing situation, you are then guessing on multiple-choice questions rather than grid-in ones. This is your test, so you can complete the problems within a section in the order you prefer. Try not to let the more difficult questions drain your time. In the end, hard questions matter the least since all the questions are worth the same amount of points and the harder ones often consume more of your time. Harder tests have a more forgiving curve. If you run into a section you do not feel confident about, you are not alone. These tests are meant to challenge you, so put your best foot forward and make sure to answer every question, even if it is just a guess. It can seem intimidating at times to have to solve difficult problems in just over one minute. What we should realize at this point, then, is that all questions CAN be answered in just over a minute. This means that the questions are not as complex and convoluted as they may first appear. Stay calm and stick to what you know.

b. Reminders

Test Day can be a very stressful experience; however, you should go in feeling confident in your ability to perform well. More than feeling confident, you should feel ready. Not only should you come prepared with the right knowledge, but the right tools! Gather everything you will need to enter your testing facility and take the test. Below is a small checklist of items that you should bring with you on Test Day. Check the College Board website to learn more about any other items that may be useful or needed to take the test that you have devoted so much time towards.

What to Bring

- Admission Ticket
- Valid Photo ID
- No. 2 Pencils
- Erasers
- Approved Calculator
- Backup Calculator
- Extra Batteries
- Watch
- Drink and Snack (for breaks)

b.

FULL-LENGTHS

Practice Test 1

Part 1: No Calculator

Part 2: Calculator

Solutions

Practice Test 2

Part 1: No Calculator

Part 2: Calculator

Solutions

PRACTICE TEST 1

DIRECTIONS

For questions 1-15, solve every problem and either circle or record the best answer from the answer choices provided. **For questions 16-20**, solve every problem and record your answer in the appropriate format. On the real test, you must grid in your answers according to the directions. You may use any available space in your workbook for scratch work.

NOTES

1. Calculator usage is NOT allowed in Part 1.
2. Assume real numbers unless otherwise indicated.
3. Figures may not be drawn to scale.
4. All figures are planar, unless otherwise indicated.
5. Domains represent all real numbers of for which f(x) is also a real number.

REFERENCE

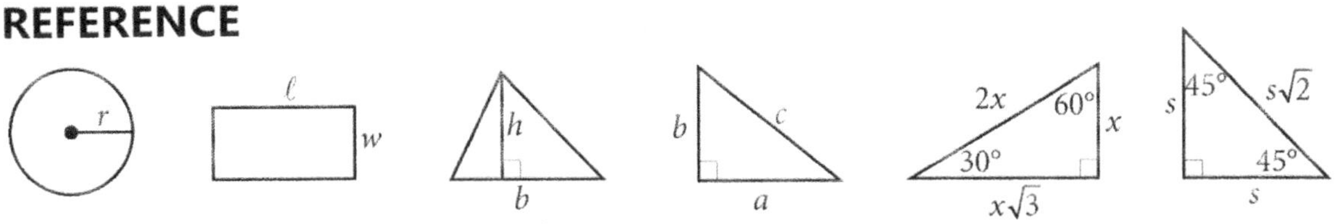

$A = \pi r^2$ $C = 2\pi r$ | $A = \ell w$ | $A = \frac{1}{2}bh$ | $c^2 = a^2 + b^2$ | Special Right Triangles

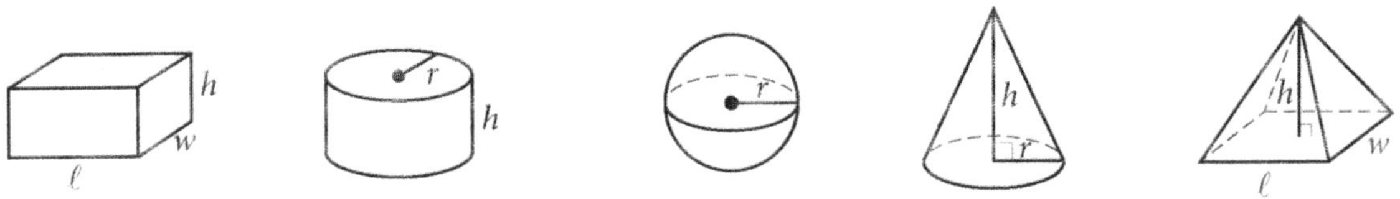

$V = \ell wh$ | $V = \pi r^2 h$ | $V = \frac{4}{3}\pi r^3$ | $V = \frac{1}{3}\pi r^2 h$ | $V = \frac{1}{3}\ell wh$

The arc of a circle contains 360 degrees.

The arc of a circle contains 2π radians.

The interior angles of a triangle add up to 180 degrees.

Part 1: No Calculator

25 Minutes, 20 Questions

1

Darius spent a total of $180 to rent equipment to paddleboard during a weekend vacation at the beach. In addition to his rental equipment, he purchased x passes to board a water taxi to quickly get him to various locations on the beach shore. If each pass on the water taxi costs $12, which expression is equivalent to the amount of dollars Darius spent on his equipment and passes?

A. $12x$

B. $192x$

C. $192 + x$

D. $180 + 12x$

2

$$(2a + 5b)^2$$

Which is equivalent to the expression above, for all real values of a and b?

A. $4a + 25b$

B. $4a^2 + 25b^2$

C. $4a^2 + 10ab + 25b^2$

D. $4a^2 + 20ab + 25b^2$

3

Which of the following equations represents a line that is parallel to $5x - 2y = 10$ and passes through the point (2,3)?

A. $5x - 2y = 3$

B. $5x - 2y = 4$

C. $2x + 5y = 3$

D. $2x + 5y = 19$

4

$$3(x - 2m) = nx + 24$$

In the equation above, m and n are constants. If the equation has infinitely many solutions, then what must be the value of m?

A. -4

B. -2

C. 2

D. 4

5

$$(x-1)(x+1)+4(x+1)$$

Which of the following is equivalent to the expression above?

A. $(x+1)(x+3)$

B. $(x+1)(x+4)$

C. $(x-1)(x-3)$

D. $(x-1)(x-4)$

6

Line l has an x-intercept at 5 and y-intercept at 3. If a and b represent positive integers, which of the following CANNOT be a point on line l?

A. (a, b)

B. $(-a, b)$

C. $(-a, -b)$

D. $(a, -b)$

7

x	1	3	8
$f(x)$	a	b	c

In the table above, $a, b,$ and c represent positive integers, where $b < a < c$. Which of the following could be the equation of the function, $f(x)$?

A. $f(x) = |x-4|$

B. $f(x) = (x-4)^3$

C. $f(x) = x-4$

D. $f(x) = \sqrt{x-4}$

8

$$x^2 - 8x = 2$$

What is the sum of the solutions to the equation above?

A. 2

B. 4

C. 6

D. 8

9

$$a(t) = 315(1.05)^t$$

Since measurements began at the Mauna Loa Observatory in 1958, the average amount of amount of CO_2 in the atmosphere has been increasing at an increasing rate and can be modeled by the Keeling Curve. In the equation above that approximates the Keeling Curve, $a(t)$ estimates the amount of CO_2 in ppm in the atmosphere t decades after 1958. What is the best interpretation of the value 315 in the equation?

A. The estimated amount of atmospheric CO_2 in ppm in 1958

B. The estimated number of months for a 5% increase in CO_2 in the atmosphere

C. The estimated number of decades for a 5% increase in CO_2 in the atmosphere

D. The estimated increase of CO_2 in ppm every decade after 1958

10

$$y = x^2$$

$$y = 5x + 6$$

The equations above are graphed on the xy-plane and intersect at the coordinate point (a, b) What could be a value of b?

A. -1

B. 1

C. 5

D. 6

11

$$\frac{1}{x-2} - \frac{1}{2x-4} = \frac{a}{x-2}$$

The equation above is satisfied for all values of x, such that $x \neq 2$. What is the value of the constant a in the equation?

A. $-1/2$

B. $1/2$

C. 1

D. 2

12

Which of the following is the equation of a circle on the xy-plane that has a radius of 5 and crosses through the origin?

A. $x^2 + y^2 = 5$

B. $x^2 + y^2 = 25$

C. $(x-3)^2 + (y-4)^2 = 5$

D. $(x-3)^2 + (y-4)^2 = 25$

13

What is the sum of all possible values of b for which the polynomials $x^2 + 7x + 10$ and $x^2 + 12x + b$ have a root in common?

A. 20

B. 35

C. 55

D. 90

14

$$f(x) = \sqrt{2x} - 4$$

If the function f given above is graphed on the xy-plane, which of the following coordinate pairs represents its x-intercept?

A. $(0, f(0))$

B. $(0, -4)$

C. $(8, f(0))$

D. $(8, f(8))$

15

$$\frac{1-5i}{1+2i}$$

The expression above is equal to the complex number $a + bi$, where a and b are constants and $i = \sqrt{-1}$. What is the value of a?

A. $-9/5$

B. $-7/5$

C. $7/5$

D. $9/5$

16

$$10x - 3y = 5$$

$$x + y = 7$$

The system above contains the (x, y) solution of $(2, b)$ where b is a positive constant. What is the value of b?

17

$$\sqrt[3]{x^2} = x^a(\sqrt{x})$$

If the equation above is true for all values of x in the domain of whole numbers, then what is the value of a?

18

$$p(x) = a(x + 8)(x + 1)$$

When $p(x)$ is divided by x, the remainder is 2. What is the value of the constant, a?

19

Line m crosses through the origin and contains the point $(b, b - 5)$. If the slope of line m is $2/3$, what is the value b?

20

$$\sqrt{x + 2} = x - 10$$

What is one solution to the equation above?

PART 2: CALCULATOR

DIRECTIONS

For questions 1-30, solve every problem and either circle or record the best answer from the answer choices provided. **For questions 31-38**, solve every problem and record your answer in the appropriate format. On the real test, you must grid in your answers according to the directions. You may use any available space in your workbook for scratch work.

NOTES

1. Calculator usage is allowed in Part 2.
2. Assume real numbers unless otherwise indicated.
3. Figures may not be drawn to scale.
4. All figures are planar, unless otherwise indicated.
5. Domains represent all real numbers of for which f(x) is also a real number.

REFERENCE

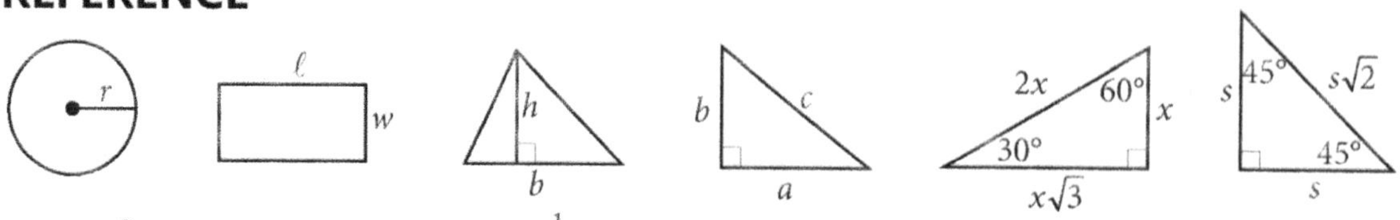

$A = \pi r^2$ $C = 2\pi r$ | $A = \ell w$ | $A = \frac{1}{2}bh$ | $c^2 = a^2 + b^2$ | Special Right Triangles

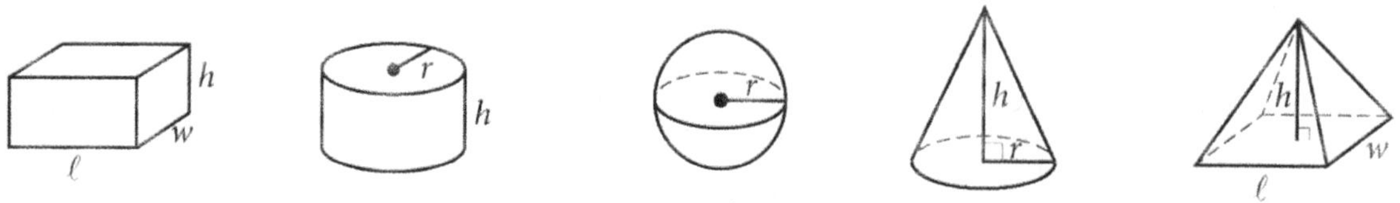

$V = \ell wh$ | $V = \pi r^2 h$ | $V = \frac{4}{3}\pi r^3$ | $V = \frac{1}{3}\pi r^2 h$ | $V = \frac{1}{3}\ell wh$

The arc of a circle contains 360 degrees.

The arc of a circle contains 2π radians.

The interior angles of a triangle add up to 180 degrees.

Part 2: Calculator

55 Minutes, 38 Questions

1

If $2x + 5 = 12$, then which of the following must be true?

A. $3x = 8$

B. $6x = 21$

C. $9x = 34$

D. $12x = 45$

2

If $x^{\frac{3}{2}} = 27$, what is the value of $x^{\frac{1}{2}}$?

A. 3

B. 9

C. 18

D. 36

3

A brick-and-mortar retail store has fixed costs totaling \$3,500 every month. The remaining business costs come from replenishing the store's inventory. If every box of inventory n the store requires costs \$150, which equation models the total yearly cost $c(n)$ in thousands of dollars for this retail store?

A. $c(n) = 3.5 + 0.15n$

B. $c(n) = 42 + 0.15n$

C. $c(n) = 3{,}500 + 150n$

D. $c(n) = 42{,}000 + 150n$

4

There is a total of 32,000 tickets available to attend a music festival. Each participating business is able to obtain a maximum of 460 tickets for its employees. If 25 businesses participate in the program, what is the smallest number of tickets that remain for individual purchase?

A. 5,000

B. 11,500

C. 16,000

D. 20,500

5

A circular farm enclosure for animals is 144π square yards. What is the length in yards around the enclosure?

A. 12

B. 24

C. 12π

D. 24π

6

A bicycle tire manufacturer takes a large random sample of its tires to study their durability. It finds the average bike tire on its production line lasts 5,200 kilometers with a margin of error of 250 kilometers. What is the most reasonable interpretation of these results?

A. The bicycle tires that this manufacturer produces generally last between 250 and 5,2000 kilometers.

B. The bicycle tires that this manufacturer produces always last between 250 and 5,200 kilometers.

C. The bicycle tires that this manufacturer produces generally last between 4,950 and 5,450 kilometers.

D. The bicycle tires that this manufacturer produces always last between 4,950 and 5,450 kilometers.

7

$$y = \frac{1}{2}x - 3$$

The equation above describes line p on the xy-plane. Another line, q, is perpendicular to line p, but has a y-intercept of 7. At what point to do these two lines intersect?

A. $(4, -1)$

B. $(-4, 1)$

C. $(1, -4)$

D. $(-1, 4)$

8

A public charging station for electric vehicles costs $0.15 for every kilowatt hour. How much will it cost a customer using this station to charge up a car with 18 kilowatt hours of energy?

A. $2.70

B. $4.15

C. $17.85

D. $120.00

9

$$f(x) = (x - 6)^3 - 1$$

The cubic function f provided above is shifted up 4 units and left 5 units to form a new function, g. Which of the following defines $g(x)$?

A. $g(x) = (x - 11)^3 + 3$

B. $g(x) = (x + 5)^3 + 4$

C. $g(x) = (x - 1)^3 + 3$

D. $g(x) = (x + 1)^3 + 4$

10

$$3x - y = 7$$

$$x - 3y = 5$$

Using the system of equations shown above, what is the value of $x - y$?

A. 0

B. 1

C. 2

D. 3

11

A university survey shows that 32% of its students are taking a science course. Of those taking a science course, two-thirds plan to go into medicine. If the university has 8,542 students, approximately how many students are taking a science course and plan to go into medicine?

A. 1,822

B. 2,733

C. 5,695

D. 6,532

12

For which of the following functions can $f(c) = -3$, provided that c is a constant?

A. $f(x) = |x - 3|$

B. $f(x) = x^2 - 3$

C. $f(x) = 2^x - 3$

D. $f(x) = \sqrt{x} + 3$

Use the following information to answer questions 13-15

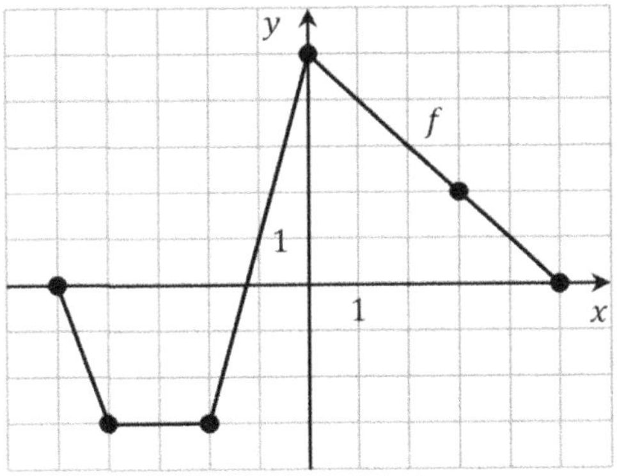

The function f is shown on the xy-plane above. The function g (not shown) is the image of f such that $g(x + 1) = f(x)$.

13

How many times will the line $y = x$ cross the function f if it is drawn on the xy-plane above?

A. 0

B. 1

C. 2

D. 3

14

For how many values of the constant c does $f(c) = -3$

A. 0

B. 1

C. 2

D. more than 2

15

What is the value of $g(4)$?

A. 0

B. 1

C. 2

D. 3

16

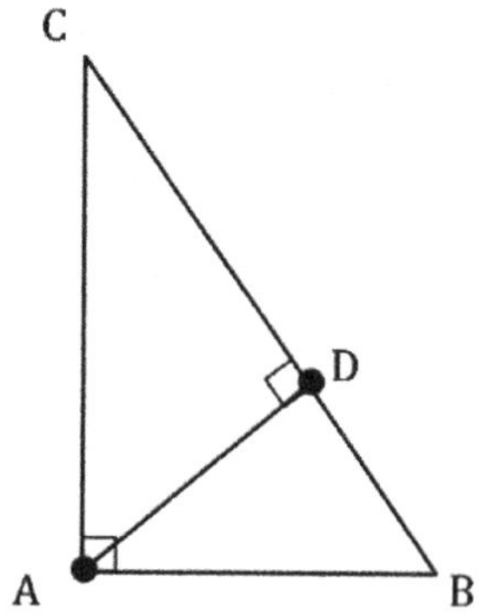

In the diagram above, AB=6 and BC=10. What must by the length of AD?

A. 4.8

B. 5.6

C. 6.4

D. 7.2

17

$$f(x) = 3(x^2 - 5x + 1) + 10(x + k)$$

In the function $f(x)$ above, k is a constant. If $f(x)$ is divisible by $x + 2$, what is the value of k?

A. -4.5

B. -2.5

C. 2.5

D. 4.5

18

$$4(x - 2)^2 = 16a$$

In the equation above, a is a constant. What values of x satisfy the equation above?

A. $2 \pm 2\sqrt{a}$

B. $2 \pm 4\sqrt{a}$

C. $2 \pm 2a$

D. $2 \pm 4a$

Questions 19-21 use the following information

A survey asked respondents to select their favorite of four popular mobile apps: A, B, C, and D. The table below shows the number of respondents, categorized by age range and favorite mobile app.

Number of Users on Popular Mobile Apps

	Age Range				
App	16-22	23-30	31-40	41 & older	Total
A	6,021	4,562	1,178	587	12,348
B	2,305	4,208	896	352	7,761
C	3,033	1,056	652	119	4,860
D	566	2,125	1,023	4,035	7,749
Total	11,925	11,951	3,749	5,093	32,718

19

For which app is the ratio of respondents 16-22 to respondents 41 & older the greatest?

A. App A

B. App B

C. App C

D. App D

20

According to the survey, about how many times as likely is it that a respondent in the 31-40 age group prefers App A than it is a respondent in the 16-22 age group prefers App D?

A. 0.5

B. 2.1

C. 4.8

D. 6.6

21

What is the approximate probability that a respondent from the survey is 31-40 years of age and prefers App C?

A. 0.02

B. 0.09

C. 0.14

D. 0.17

22

Line j has the equation $y = 2x + 5$ and line k has the equation $y = 5x - 3$. When lines j and k are graphed on the xy-plane, their intersection occurs at the point (p, q). What is the value of $q - 2p$?

A. 8/3

B. 5

C. 9

D. 31/3

23

A pet supply company plans to purchase x bags of cat food and y bags of dog food. Each bag of cat food weighs 3.6 pounds and each bag of dog food weighs 5.2 pounds. Which system of inequalities can help the company ensure it purchases at least 15 bags of food at no more than 70 pounds in total?

A. $x + y \geq 15$
$3.6x + 5.2y \leq 70$

B. $x + y > 15$
$3.6x + 5.2y \leq 70$

C. $x + y \geq 15$
$3.6x + 5.2y < 70$

D. $x + y > 15$
$3.6x + 5.2y < 70$

24

Two wheels, A and B, roll a total distance of d miles. Wheel A has an outer radius of R inches and wheel B has an outer radius of r inches. If wheel A is larger than wheel B, which expression can be used to find how many more revolutions wheel B makes than wheel A?

$$1 \text{ mile} = 5280 \text{ feet}$$

A. $\dfrac{5{,}280d}{2\pi}(R - r)$

B. $\dfrac{(5{,}280)(12)d}{2\pi}(R - r)$

C. $\dfrac{5{,}280d}{2\pi}\left(\dfrac{1}{r} - \dfrac{1}{R}\right)$

D. $\dfrac{(5{,}280)(12)d}{2\pi}\left(\dfrac{1}{r} - \dfrac{1}{R}\right)$

25

A researcher collects 2.36 grams of eugenol in his latest experiment and calculates that this was 78% of what was expected. About how many more grams of eugenol would the researcher need to collect to reach 90% of the expected amount of eugenol in this same experiment?

A. 0.4

B. 1.5

C. 2.7

D. 3.0

26

$$3x + 10y = 9$$
$$5x - by = c$$

In the system of equations above, b and c are constants. If the system has infinitely many (x, y) solutions, what is the value of c?

A. 13

B. 14

C. 15

D. 16

27

Which of the following is an equivalent form of the function $f(x) = x^2 - x - 20$ where the x-intercepts can be identified as constants?

A. $f(x) = (x + 5)(x - 4)$

B. $f(x) = (x - 5)(x + 4)$

C. $f(x) = x(x - 1) - 20$

D. $f(x) = x^2 - (x + 20)$

28

A circle on the xy-plane is centered on the origin and contains the point $(4\sqrt{2}, 4\sqrt{2})$. What could be an x-intercept of this circle?

A. 5

B. 6

C. 7

D. 8

29

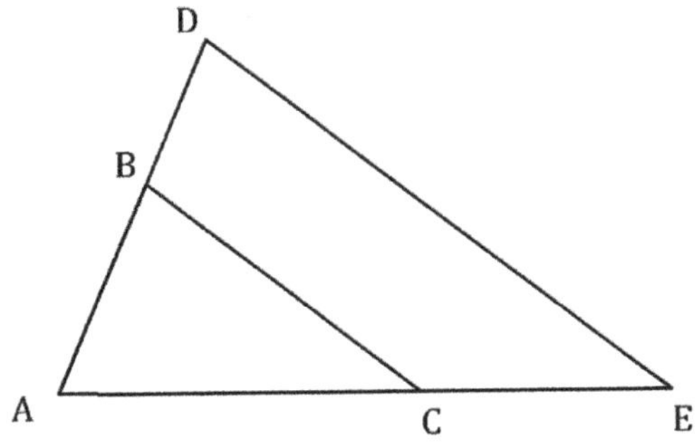

In the diagram, $\overline{BC} \parallel \overline{DE}$, $AC = 12$, and $CE = 8$. If the perimeter of triangle ABC is 30, what is the perimeter of triangle ADE?

A. 40

B. 45

C. 50

D. 55

30

The number N of bacteria species in certain plot of soil has been declining over the course of m months according to the model below.

$$N = 5{,}480(0.97)^{\frac{m}{2}}$$

Which of the following statements is true according to this model

A. The number of bacteria species declines by 3% every month

B. The number of bacteria species declines by 3% every 2 months

C. The number of bacteria species declines by 97% every month

D. The number of bacteria species declines by 97% every 2 months

31

$$f(x) = \frac{1}{(x-3)^2 + 5(x-3) + 6}$$

What value of the constant c is $f(c)$ undefined? More than one answer is possible.

32

Researchers investigating a whale population in the Pacific Ocean initially tagged female and male whales in a ratio of 3:5. Over the course of their study, they tag 18 more female whales so that the two populations are equal. How many whales in total were tagged by the end of the study?

33

$$a + b > 1$$
$$a - b > 5$$

What is the least possible integer value of a that satisfies the system of inequalities above?

34

$$y = (x-5)^2 + 4(x-5) - 10$$
$$y = x - 5$$

What positive value of x satisfies the system of equations above?

35

Michael jogs on a running trial that increases 2 meter in elevation for every 5 meters traveled along the trail. He starts his jog at 60 meters in elevation and travels 180 meters every 40 seconds. If the equation below models Michael's elevation E in meters x seconds after he begins his run, what is the value of the constant a?

$$E = ax + 60$$

36

Alyssa currently has a 92.4% in her precalculus class with only one assessment left: a project worth 18% of her course grade. What is the minimum percent she must receive on the project to keep her grade at or above a 90%? Round answer to the nearest tenth.

Use the following information for questions 37 and 38

Circle O below contains the points A, B, and C. The length of the minor arc from A to B is 13π meters and x is 75. Points B and C can be connected to form a diameter.

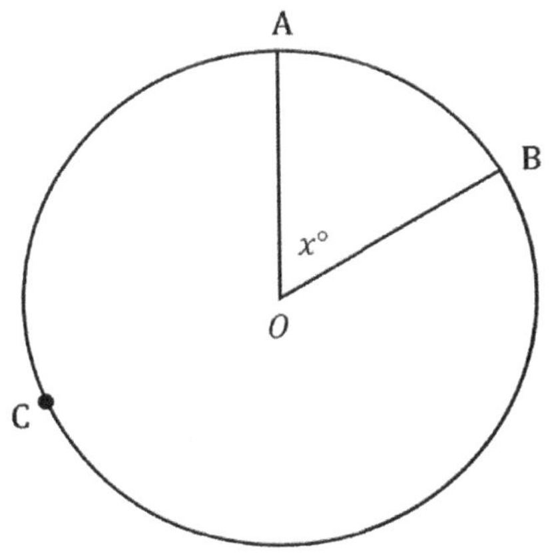

37

If the length of arc ACB is $a\pi$ meters, then what is the value of a?

38

The diagram of circle O is used to model a circular plot of land that will have an irrigation arm extending from the center O to point B. The irrigation arm pivots about center O to steadily deliver water to the entire plot of land. If the irrigation arm travels at a rate of π meters per second, what is the minimum number of full seconds that will be required for it to move past point C, starting from point B?

Explanations

Practice Test 1: No Calculator

1. D

The total cost includes the both the rental equipment and passes. The cost rental equipment is known, while the cost of the passes is dependent on the number purchased.

Cost of Rental Equipment: \$180

Cost of Passes: $12x$

Total Cost: $180 + 12x$

2. D

Expand out the full expression by first rewriting it to show that the base is multiplied to itself. Simplify by combing like-terms.

$$(2a + 5b)^2$$
$$(2a + 5b)(2a + 5b)$$
$$4a^2 + 10ab + 10ab + 25b^2$$
$$4a^2 + 20ab + 25b^2$$

3. B

Since the lines are parallel, we can keep the $Ax + By$ portion of the given equation exactly the same because it is the ratio of A to B that determines the slope. Use the known point to find C.

$$5x - 2y = C$$
$$5(2) - 2(3) = C$$
$$10 - 6 = C$$
$$4 = C$$
$$5x - 2y = 4$$

4. A

Since the equation has infinitely many solutions, both sides must contain the same corresponding terms. Distribute the left-hand side of the equation to create corresponding terms that we can match.

$$3(x - 2m) = nx + 24$$
$$3x - 6m = nx + 24$$
$$-6m = 24$$
$$m = -4$$

5. A

Since all the answer choices are in factored form, we can observe the shared factor of $x + 1$ between the two quantities being added together, as shown below. Alternatively, we can expand fully, then factor the quadratic.

$$(x + 1)(x - 1) + 4(x + 1)$$
$$(x + 1)(x - 1 + 4)$$
$$(x + 1)(x + 3)$$

6. C

The line goes through all quadrants, except for third. This means $(-a, -b)$ cannot be on the line.

7. A

Only Choice A makes b have the smallest output. We can find the values of $a, b,$ and c by plugging in numbers from the chart.

$$a = |1-4| \text{ or } 3$$

$$b = |3-4| \text{ or } 1$$

$$c = |8-4| \text{ or } 4$$

8. D

Bring the equation into $ax^2 + bx + c = 0$ form. The sum of the solutions to a quadratic equation is $-b/a$.

$$x^2 - 8x = 2$$

$$x^2 - 8x - 2 = 0$$

$$-\frac{b}{a} = -\frac{-8}{1} \text{ or } 8$$

9. A

The initial amount of CO_2 in the atmosphere in 1958 can be found by plugging in $t = 0$.

$$a(0) = 315(1.05)^0$$

$$a(0) = 315$$

10. B

Set the equations equal to each other to first find the two possible values for x, then plug these values back into the given equations to find y.

$$x^2 = 5x + 6$$

$$x^2 - 5x - 6 = 0$$

$$(x-6)(x+1) = 0$$

$$x = 6 \text{ or } x = -1$$

$$y = (6)^2 \text{ or } 36$$

$$y = (-1)^2 \text{ or } 1$$

11. B

Multiply through by the common factor $x - 2$ that occurs in all the denominators so that we eliminate it from the equation.

$$\frac{1}{x-2} - \frac{1}{2x-4} = \frac{a}{x-2}$$

$$\left[\frac{1}{(x-2)} - \frac{1}{2(x-2)} = \frac{a}{(x-2)}\right](x-2)$$

$$1 - \frac{1}{2} = a$$

$$a = \frac{1}{2}$$

12. D

Since the right side of the standard form equation for a circle on the xy-plane is the radius squared, only Choices B and D represent circles with a radius of 5. However, since the circle must cross through the origin, it cannot be centered on the origin as in the case of answer choice B. Therefore, Choice D is the correct answer.

13. C

The roots of first quadratic are −5 and −2. Remember that for these values to be roots, they must give an output of zero. This is why another name for roots is zeros.

$$x^2 + 7x + 10$$

$$(x + 5)(x + 2)$$

$$x = -5 \quad \text{or} \quad x = -2$$

Plug these values into the other polynomial and set equal to zero to find values of b. Then, add these values of b to each other to find the sum. The sum is 55.

$$x^2 + 12x + b$$

$$(-5)^2 + 12(-5) + b = 0$$

$$25 - 60 + b = 0$$

$$-35 + b = 0$$

$$b = 35$$

$$(-2)^2 + 12(-2) + b = 0$$

$$4 - 24 + b = 0$$

$$-20 + b = 0$$

$$b = 20$$

14. D

The x-intercept occurs when $y = 0$. Remember, the point (x, y) can also be represented in function notation as $(x, f(x))$.

$$f(x) = \sqrt{2x} - 4$$

$$0 = \sqrt{2x} - 4$$

$$4 = \sqrt{2x}$$

$$16 = 2x$$

$$x = 8$$

$$(8, f(8))$$

15. A

Create an equivalent fraction by using the conjugate of the denominator. Then simplify the expression by combing like-terms and substituting $i^2 = -1$.

$$\frac{1-5i}{1+2i} * \frac{1-2i}{1-2i} = \frac{(1-5i)(1-2i)}{(1+2i)(1-2i)}$$

$$\frac{1 - 5i - 2i + 10i^2}{1 - 4i^2} = \frac{1 - 7i - 10}{1+4} = \frac{-9-7i}{5}$$

$$-\frac{9}{5} - \frac{7}{5}i = a + bi$$

$$a = -\frac{9}{5}$$

16. 5

Since the value of x is already known, we just need to plug it into one of the equations.

$$x + y = 7$$

$$2 + y = 7$$

$$y = 5$$

17. 1/6

Rewrite the radicals in exponential form, the make sure the overall exponents on x are equal on both sides of the equation.

$$\sqrt[3]{x^2} = x^a(\sqrt{x})$$

$$x^{\frac{2}{3}} = x^a x^{\frac{1}{2}}$$

$$x^{\frac{2}{3}} = x^{a+\frac{1}{2}}$$

$$\frac{2}{3} = a + \frac{1}{2}$$

$$a = \frac{1}{6}$$

18. **1/4 or 0.25**

The Remainder Value Theorem states that when you plug in the zero to a linear factor in the form $x \pm c$, then the remainder is equal to the value of the function.

The zero to the factor x is 0. And the remainder, or function value, is 2 when $x = 0$.

$$p(x) = a(x + 8)(x + 1)$$

$$2 = a(0 + 8)(0 + 1)$$

$$2 = 8a$$

$$a = \frac{1}{4} \quad \text{or} \quad 0.25$$

19. **15**

Set the slope between the given point and the origin equal to two-thirds, then solve for b.

$$\frac{b - 5}{b} = \frac{2}{3}$$

$$3(b - 5) = 2b$$

$$3b - 15 = 2b$$

$$b = 15$$

20. **14**

Square both sides to create a quadratic equation that can be solved through factoring. Make sure to check against extraneous solutions.

$$\sqrt{x + 2} = x - 10$$

$$x + 2 = (x - 10)^2$$

$$x + 2 = x^2 - 20x + 100$$

$$0 = x^2 - 21x + 98$$

$$0 = (x - 7)(x - 14)$$

$$x = 7 \quad \text{or} \quad 14$$

However, $x \neq 7$ according to the original equation because the radical would be equal to a negative number, which is never true of the output to an even root operation.

Explanations

Practice Test 1: Calculator

1. B

Generate the desired expression by first isolating the x term, then multiplying to find the value of $6x$.

$$2x + 5 = 12$$

$$2x = 7$$

$$3(2x = 7)$$

$$6x = 21$$

2. A

Cube root – or raise to the power of one-third – to find the value of $x^{\frac{1}{2}}$.

$$x^{\frac{3}{2}} = 27$$

$$x^{\frac{3}{2}} = 3^3$$

$$\left(x^{\frac{3}{2}}\right)^{\frac{1}{3}} = (3^3)^{\frac{1}{3}}$$

$$x^{\frac{1}{2}} = 3$$

3. B

Total cost is the sum of both fixed and variable cost. Since the total cost function must be represented in thousands, divide the given amounts by 1,000.

Fixed Costs: \$3,500 or \$3.5 thousand per month. This is \$42 thousand every year.

Variable Cost: \$150n or \$0.150n thousand

Total Cost: $c(n)$

$$\text{Total Cost} = \text{Fixed Cost} + \text{Variable Cost}$$

$$c(n) = 42 + 0.15n$$

4. D

The smallest number of remaining tickets occurs when the maximum number of tickets have been purchased.

Total Tickets: 32,000

Max Purchased Tickets: $460(25) = 11{,}500$

$$\text{Remaining Tickets} = \text{Total Tickets} - \text{Max Purchased}$$

$$\text{Remaining Tickets} = 32{,}000 - 11{,}500$$

$$\text{Remaining Tickets} = 20{,}500$$

5. D

Find the radius of the circle by using the area equation, then solve use it to solve for the circumference.

$$A = \pi r^2$$

$$144\pi = \pi r^2$$

$$144 = r^2$$

$$r = 12$$

$$C = 2\pi r$$

$$C = 2\pi(12) \quad \text{or} \quad 24\pi$$

6. C

In general, the bicycle tires last 5,200 kilometers, give or take 250 kilometers. However, some tires may not last as long – others might last longer.

$$5{,}200 \pm 250$$

$$4{,}950 \text{ to } 5{,}450$$

7. A

Perpendicular lines have slopes that are opposite reciprocals of each other. Since the slope of line p is $1/2$, the slope of line q is -2.

Line q: $m = -2, b = 7$

Line q equation: $y = -2x + 7$

$$\frac{1}{2}x - 3 = -2x + 7$$

$$x - 6 = -4x + 14$$

$$5x = 20$$

$$x = 4$$

8. A

The total cost is the product of the unit rate and the total number of units.

$$\text{Total Cost} = \text{Rate} \times \text{Quantity}$$

$$\text{Total Cost} = 0.15(18)$$

$$\text{Total Cost} = \$2.70$$

9. C

A shift up of 4 units requires adding 4 to the overall function value; likewise, a shift left of 5 units requires adding 5 directly to x.

$$g(x) = (x + 5 - 6)^3 - 1 + 4$$

$$g(x) = (x - 1)^3 + 3$$

10. D

Remember, the value of an expression can be found without knowing the value of any variable. Always look within and between the equations to determine if there is a shorter route to the correct answer. In this case, add the equations together to reach a multiple of the desired expression.

$$\begin{array}{r} 3x - y = 7 \\ + \quad x - 3y = 5 \\ \hline 4x - 4y = 12 \\ x - y = 3 \end{array}$$

11. B

The number of science students is only a fraction of the total number of students, and the number of pre-medicine science students is only a fraction of that group. This is an "and" type probability question where we should multiply the probabilities together to reach the desired value.

Science Students $= 0.32 \times$ Total Students

Pre-medicine Science Student $= \frac{2}{3} \times$ Science Students

$$\frac{2}{3}(0.32(8{,}542)) \approx 1{,}822$$

12. B

Use lowest output operations: Absolute Values, Even Degree, and Even Root operations have a minimum output of zero, where exponentials only approach zero. Therefore, the only expression that can reach -3 is choice B, because $0 - 3 = -3$.

13. D

The graph crosses the $y = x$ three times when it is sketched on the same coordinate plane.

14. D

Since the function stays constant at -3 from $x = -4$ to $x = -2$, there are infinitely points where $f(c)$ can equal -3.

15. C

Find the value of x that should be used to locate the value of $g(4)$ based on the function f. The input into g is one more than the input into f.

$$g(x+1) = f(x)$$
$$g(4) = f(3) = 2$$

16. A

The right triangle ABC has dimensions $6 - 8 - 10$, where the base $\text{AB} = 6$ and the height $\text{AC} = 8$. However, this same right triangle has another altitude drawn in. The segment AD represents a height to the triangle because it is perpendicular to the side CB. No matter which height and base pair that is used, the area of triangle ABC can only have one value.

$$\frac{1}{2}b_1h_1 = \frac{1}{2}b_2h_2$$
$$b_1h_1 = b_2h_2$$
$$6(8) = 10(\text{AD})$$
$$48 = 10(\text{AD})$$
$$\text{AD} = 4.8$$

17. B

The remainder value theorem states that the remainder a polynomial function divided by the linear factor $x - c$ is equal to the value of the function when $x = c$. Remember, when a polynomial is perfectly divisible by a factor, the reminder is zero. In the context of this problem, $f(-2) = 0$.

$$f(x) = 3(x^2 - 5x + 1) + 10(x + k)$$
$$0 = 3((-2)^2 - 5(-2) + 1) + 10(-2 + k)$$
$$0 = 3(4 + 10 + 1) - 20 + 10k$$
$$0 = 3(15) - 20 + 10k$$
$$0 = 25 + 10k$$
$$10k = -25$$
$$k = -2.5$$

18. A

Isolate x to one side of the equation by using the square root method to solve.

$$4(x-2)^2 = 16a$$
$$(x-2)^2 = 4a$$
$$(x-2) = \pm\sqrt{4a}$$
$$x - 2 = \pm 2\sqrt{a}$$
$$x = 2 \pm 2\sqrt{a}$$

19. C

Divide the number of respondents in the 41 & older age group by the number of respondents in the 16-22 age group for each app, then choose the app that led to the greatest quotient.

For App C

$$\frac{3{,}033}{119} \approx 25.5$$

20. D

Compute each conditional probability, then take the ratio between them to determine how many times as likely it is.

$$P(\text{App A} \mid \text{Age } 31-40) = \frac{1{,}178}{3{,}749}$$

$$P(\text{App D} \mid \text{Age } 16-22) = \frac{566}{11{,}925}$$

$$\frac{P(\text{App A} \mid \text{Age } 31-40)}{P(\text{App D} \mid \text{Age } 16-22)} \approx 6.6$$

21. A

The question references the total being all survey respondents, so use the total of the value represented by the entire chart.

Part = 652

Total = 32,718

$$\text{Probability} = \frac{652}{32{,}718} \approx 0.02$$

22. B

Recall, when asked to find the value of an expression, first look within and between the equations to see if the desired expression can be found. In this case, only the first equation is needed. Since the point (p, q) is a point of intersection, it falls on each line and works for each equation when plugged in. Use point (p, q) as an (x, y) pair.

$$y = 2x + 5$$

$$q = 2p + 5$$

$$q - 2p = 5$$

23. A

Identify the key terms to find the appropriate inequality for each equation.

Key terms: At least (greater than or equal to) and no more than (less than or equal to).

First Inequality: Number of bags

$$x + y \geq 15$$

Second Inequality: Total weight

$$3.6x + 5.2y \leq 70$$

24. D

With every revolution, the tires travel a distance equal to its circumference. Use this to convert the number of miles traveled to the number of revolutions made for each tire. To find how many more revolutions were made by tire B, subtract the number of revolutions made by tire A from the number of revolutions made by tire A. The unit of length must match between the radius and distance.

Tire B Revolutions:

$$\frac{d\text{ mi}}{1} * \frac{5{,}280\text{ ft}}{1\text{ mi}} * \frac{12\text{ in}}{1\text{ ft}} * \frac{1\text{ rev}}{2\pi r\text{ in}} =$$

$$\frac{5{,}280(12)}{2\pi} * \frac{1}{r}$$

Tire A Revolutions:

$$\frac{d\text{ mi}}{1} * \frac{5{,}280\text{ ft}}{1\text{ mi}} * \frac{12\text{ in}}{1\text{ ft}} * \frac{1\text{ rev}}{2\pi R\text{ in}} =$$

$$\frac{5{,}280(12)}{2\pi} * \frac{1}{R}$$

Additional Revolutions = Tire B − Tire A Revs

$$\frac{5{,}280(12)}{2\pi} * \frac{1}{r} - \frac{5{,}280(12)}{2\pi} * \frac{1}{R}$$

$$\text{Additional Revolutions of Tire B} = \frac{5{,}280(12)}{2\pi}\left(\frac{1}{r} - \frac{1}{R}\right)$$

25. A

The difference between 78% and 90% is 12%. Since the 2.36g of eugenol collected represents 78% of the expected amount, divide 2.36 by 78 to get the amount of eugenol in every 1%. Multiply by 12 to find the amount of eugenol needed to raise the amount collected by 12% of the expected amount of eugenol.

$$\text{Each 1\% of eugenol: } \frac{2.36}{78} \approx 0.03 \text{ grams}$$

$$\text{For 12\% of eugenol: } 0.03(12) = 0.36 \approx 0.4 \text{ grams}$$

26. C

Systems with infinitely many solutions contain equations that can be "matched". Seeing as the x-terms in both equations are known, they can be multiplied to reach their least common multiple, leaving all corresponding terms between the two equations equal as well.

First Equation

$$5(3x + 10y = 9)$$

$$15x + 50y = 45$$

Second Equation

$$3(5x - by = c)$$

$$15x - 3by = 3c$$

Therefore

$$3c = 45$$

$$c = 15$$

27. B

Only functions in factored form contain constants that represent x-intercepts. The factored form of the function is $f(x) = (x - 5)(x + 4)$.

28. D

Use the standard form of a circle to relate points on the circle to the radius. Note, since the circle is centered at the origin, the center (h, k) is $(0,0)$. The x-intercepts always occur with a y-coordinate of zero.

$$(x - h)^2 + (y - k)^2 = r^2$$

$$x^2 + y^2 = r^2$$

$$\left(4\sqrt{2}\right)^2 + \left(4\sqrt{2}\right)^2 = r^2$$

$$32 + 32 = r^2$$

$$64 = r^2$$

$$x^2 + y^2 = 64$$

$$(x_{int})^2 + (0)^2 = 64$$

$$(x_{int})^2 = 64$$

$$x_{int} = \pm 8$$

29. C

The two triangles are similar due to AA similarity. Similar figures are related to each using scale factors, ensuring that the lengths of corresponding sides as well as the perimeter are proportional.

$$\frac{\text{Side 2}}{\text{Side 1}} = \frac{\text{Perimeter 2}}{\text{Perimeter 1}}$$

$$\frac{20}{12} = \frac{x}{30}$$

$$x = 50$$

30. B

The base of exponential models can be used to ascertain to percent change by measuring its deviation from 1. Seeing that the base of 0.97 is 0.03 lower than 1, the base implies a 3% decrease. However, the exponent shows that it takes 2 months to reach this decrease because consecutive multiples of 2 are required to reach the next integer.

31. 0, 1

The x-value c required to make $f(x)$ undefined would be the one that brings the denominator of the fraction to zero because dividing by zero is undefined. Seeing that the expression $(x-3)$ is a repeated motif in the denominator, make the substitution: $u = x - 3$.

$$u^2 + 5x + 6 = 0$$

$$(u+3)(u+2) = 0$$

$$u = -2 \quad \text{or} \quad u = -3$$

$$x = 0 \text{ or } 1$$

32. 90

Use the ratio to represent the initial female population as $3x$ and the initial male population as $5x$. Keep in mind that the total number of whales at the end of the study include the additional 18 female whales required to even out the male and female populations.

$$3x + 18 = 5x$$

$$2x = 18$$

$$x = 9$$

Total Whales = Female + Male

Total Whales = $(3x + 18) + 5x$

Total Whales = $8x + 18$

Total Whales = $8(9) + 18$

Total Whales = $72 + 18 = 90$

33. 4

Add equations together to eliminate b and find the solution set of a. Since $a > 3$, the least possible integer of a is 4.

$$\begin{array}{r} a + b > 1 \\ + \quad a - b > 5 \\ \hline 2a > 6 \\ a > 3 \end{array}$$

34. 7

Substitute y into the first equation, then solve by factoring once equation is set equal to 0. Then use the solutions of y to find x.

$$y = y^2 + 4y - 10$$

$$0 = y^2 + 3y - 10$$

$$0 = (y+5)(y-2)$$

$$y = -5 \text{ or } 2$$

$$x = y + 5$$

$$x = 0, 7$$

Since the questions asks for a positive value of x, only 7 is correct.

35. 9/5 or 1.8

Usue a proportion or similar right triangles to determine the appropriate value.

Distance along trail: 5 meters

Distance in elevation: 2 meters

Rate along trail: 180/40 or 4.5

Rate in elevation: a

$$\frac{5}{2} = \frac{4.5}{a}$$

$$a = 9/5$$

36. **79.1**

Ensure her weighted average is at least a 90% by taking into account that her project grade x is worth 18% of her grade while her current 92.4% is worth the remaining 82%.

$$0.18x + 0.82(92.4) \geq 90$$

$$0.18x + 75.768 \geq 90$$

$$0.18x \geq 14.232$$

$$x \geq 79.0\overline{6}$$

Minimum: 79.1

37. **247/5 or 49.4**

The measure of minor arc AB is 75 degrees. The measure of arc ACB must be 285 degrees because it forms the other side of the circle, which contains 360 degrees in total. Since 285 is 3.8 times larger than 75, the length of arc ACB must be 3.8 times larger than the length arc AB.

$$\text{Length of arcACB} = 3.8 \times \text{Length of arc AB}$$

$$\text{Length of arcACB} = 3.8 \times 13\pi$$

$$\text{Length of arcACB} = 49.4\pi$$

38. **32**

Since the points B and C form the endpoint of a diameter, they encompass half of the circle, covering an arc of 180 degrees. Since 180 is 2.4 times larger than 75, the length of from B to C must be 2.4 times larger than 13π.

$$\text{Length of arc BC} = 2.4 \times \text{Length of arc AB}$$

$$\text{Length of arc BC} = 2.4 \times 13\pi$$

$$\text{Length of arc BC} = 31.2\pi$$

Since the irrigation arm travels π meters every second, it will take 31.2 seconds to cover the semicircle. This will require 32 full seconds (at 31 it comes up just short).

PRACTICE TEST 2

DIRECTIONS

For questions 1-15, solve every problem and either circle or record the best answer from the answer choices provided. **For questions 16-20**, solve every problem and record your answer in the appropriate format. On the real test, you must grid in your answers according to the directions. You may use any available space in your workbook for scratch work.

NOTES

1. Calculator usage is NOT allowed in Part 1.
2. Assume real numbers unless otherwise indicated.
3. Figures may not be drawn to scale.
4. All figures are planar, unless otherwise indicated.
5. Domains represent all real numbers of for which f(x) is also a real number.

REFERENCE

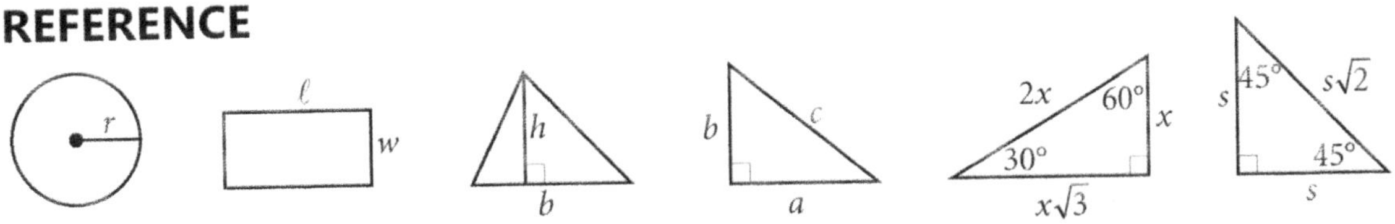

$A = \pi r^2$ $C = 2\pi r$ $A = \ell w$ $A = \frac{1}{2}bh$ $c^2 = a^2 + b^2$ Special Right Triangles

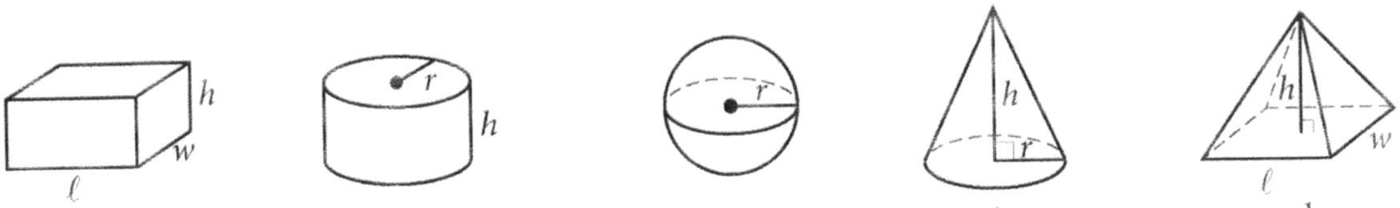

$V = \ell wh$ $V = \pi r^2 h$ $V = \frac{4}{3}\pi r^3$ $V = \frac{1}{3}\pi r^2 h$ $V = \frac{1}{3}\ell wh$

The arc of a circle contains 360 degrees.

The arc of a circle contains 2π radians.

The interior angles of a triangle add up to 180 degrees.

Part 1: No Calculator

25 Minutes, 20 Questions

1

$$3x - 2y = 4$$

$$x + y = -7$$

The equations above are graphed on the xy-plane and intersect at the point (p, q). What is the value of p?

A. -5

B. -2

C. 0

D. 4

2

Which of the following expressions is equivalent to $x^3 + 4x^2 + 3x + 15$?

A. $(x^2 + 3)(x + 4) + 3$

B. $(x^2 + 3)(x + 4) + 3x$

C. $(x^2 - 3)(x + 4) + 27$

D. $(x^2 - 3)(x + 4) + 27x$

3

Line j, when graphed on the xy-plane, passes though the points $(0,3)$ and $(1, -2)$. Line k, when graphed on the xy-plane, contains the origin and the point $(-2, b)$. What value of b will ensure that lines j and k do not intersect?

A. -8

B. -2

C. 4

D. 10

4

$$x - 2\sqrt{x - 1} = 0$$

Which value of x satisfies the equation above?

A. 0

B. 1

C. 2

D. 3

5

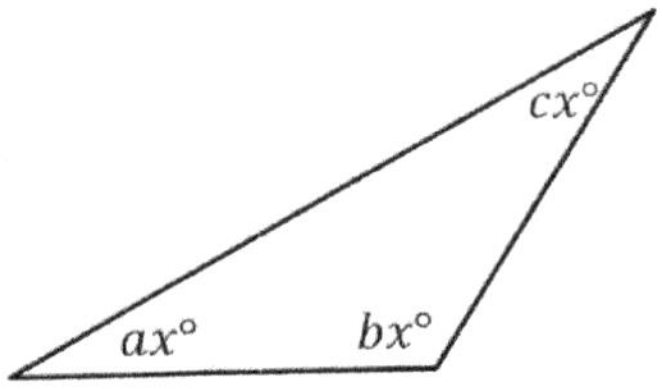

In the triangle above, $a + b + c = 40$. What is the value of x?

A. 4

B. 4.5

C. 6

D. 6.5

6

A children's museum sells two types of tickets. The children's ticket costs \$10 and the adult's ticket costs \$15. In one week, it sells 870 tickets and generates \$10,150 from its tickets sales. If c and d represent the number of children and adult tickets sold, respectively, which system of equations can be used to determine the number of each type of ticket that was sold that week?

A. $c + d = 10{,}150$
$10c + 15d = 870$

B. $c + d = 10{,}150$
$15c + 10d = 870$

C. $c + d = 870$
$10c + 15d = 10{,}150$

D. $c + d = 870$
$15c + 10d = 10{,}150$

7

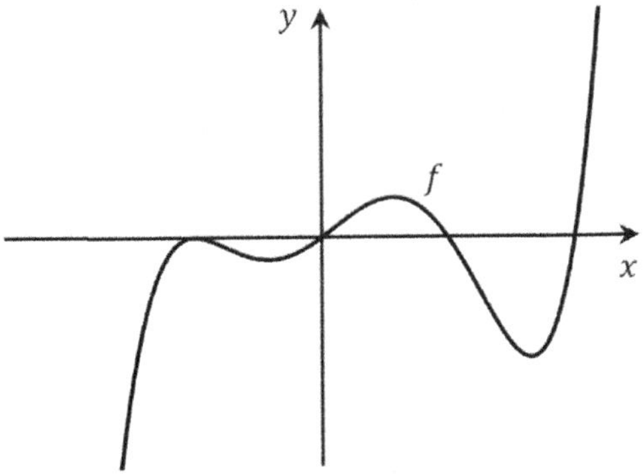

The function f is graphed on the xy-plane above. It has the form $f(x) = x(x - a)(x - b)^2(x - c)$, where a, b, and c are integers. What could be the value of b?

A. -3

B. 0

C. 2

D. 5

8

$$\left(P + \frac{an^2}{V^2}\right)(V - nb) = nRT$$

The Van der Waals equation shown above modifies the Ideal Gas Law to account for the size of each gas molecule and the attractive forces between them. Which of the following equations correctly solves for pressure, P, in the Van der Waals equation?

A. $P = nRT * \dfrac{V^2}{an^2(V - nb)}$

B. $P = \dfrac{nRT}{V - nb} - \dfrac{an^2}{V^2}$

C. $P = nRT - V + nb - \dfrac{an^2}{V^2}$

D. $P = \dfrac{nRTV}{an^2} + nb$

9

$$C(x) = 3.74 + 0.26x$$

The average cost C, in dollars, to rent an electric scooter through a ridesharing app is modeled by the equation above, where x represents every mile ridden on the scooter. What is the best interpretation of the value 0.26 in the equation?

A. Users must pay \$0.26 for every 1 mile ridden

B. Users must pay \$1 for every 0.26 miles ridden

C. Users must pay an upfront fee of \$0.26

D. Users must pay \$3.74 every 0.26 miles ridden

10

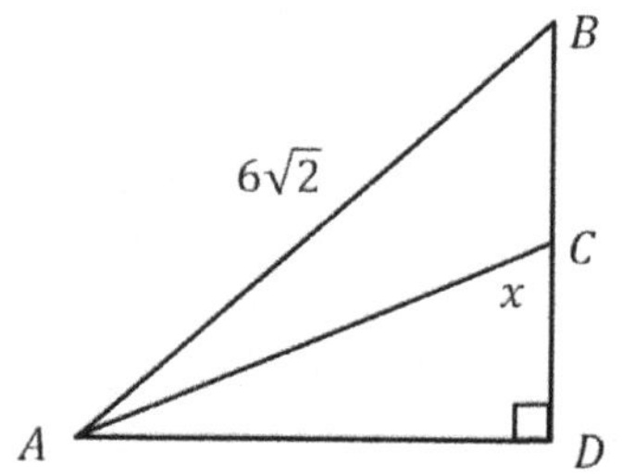

In the triangles above, $\angle BAC + \angle CAD = 45°$ and $x = 60°$. If $\overline{AB} = 6\sqrt{2}$, then what is the length of $\overline{CD}$?

A. 2

B. $2\sqrt{3}$

C. $3\sqrt{2}$

D. 6

11

$$f(x) = x^2 + 1$$
$$g(x) = -x^2 + 4x + 8$$

The functions above are graphed on the xy-plane and intersect at the points (q, r) and (s, t). What is the value of $q + s$?

A. -2

B. -1

C. 1

D. 2

12

$$\frac{p(x)}{ax+5} = 8x^2 - 3x + \frac{1}{ax+5}$$

The polynomial function p has a degree of 3 and leading coefficient of 16. The quotient when $p(x)$ is divided by $ax + 5$ is shown above. What is the value of the constant a?

A. 0

B. 1

C. 2

D. 3

13

$$f(3x + 1) = 2x^3 + 1$$

Using the function above, what is the value of $f(4)$?

A. 3

B. 17

C. 55

D. 129

14

$$5(x-h)+2(x^2+hx+1)=2x^2+13x-18$$

The equation above is true for all values of x. What must be the value of the constant h?

A. -1

B. 4

C. 9

D. 17

15

$$5x-3y=-11$$
$$ax+by=22$$

The system above contains infinitely many solutions. If a and b are constants, what must be the value of b?

A. -10

B. -3

C. 5

D. 6

16

$$(x+2)^2+(y-3)^2=49$$

Circle R takes the equation shown above. When graphed on the xy-plane, how many units up must Circle R be translated such that it is tangent to the x-axis?

17

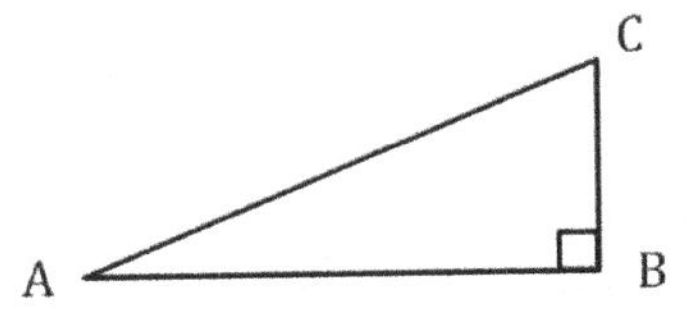

In triangle ABC above, $\angle B = 90°$, $\overline{AB} = 13\sqrt{3}$, and $\overline{AC} = 26$. If $\angle ACB = a\pi$ radians, then what is the value of the constant a?

18

$$f(x)=(x+1)(2x-b)$$

The function f is defined above, where b is a constant. If f has a minimum at $x = 5$, what is the value of b?

19

$$f(x)=\frac{11}{(2x-9)^2+(2x-9)}$$

The domain to the function above is all real numbers except $x \neq a, b$. If b is an integer constant, what is the value of b?

20

$$\frac{3-2i}{1+i}$$

The expression above is equal to the complex number $a + bi$, where a and b are constants and $i = \sqrt{-1}$. What is the value of a?

PART 2: CALCULATOR

DIRECTIONS

For questions 1-30, solve every problem and either circle or record the best answer from the answer choices provided. **For questions 31-38**, solve every problem and record your answer in the appropriate format. On the real test, you must grid in your answers according to the directions. You may use any available space in your workbook for scratch work.

NOTES

1. Calculator usage is allowed in Part 2.
2. Assume real numbers unless otherwise indicated.
3. Figures may not be drawn to scale.
4. All figures are planar, unless otherwise indicated.
5. Domains represent all real numbers of for which f(x) is also a real number.

REFERENCE

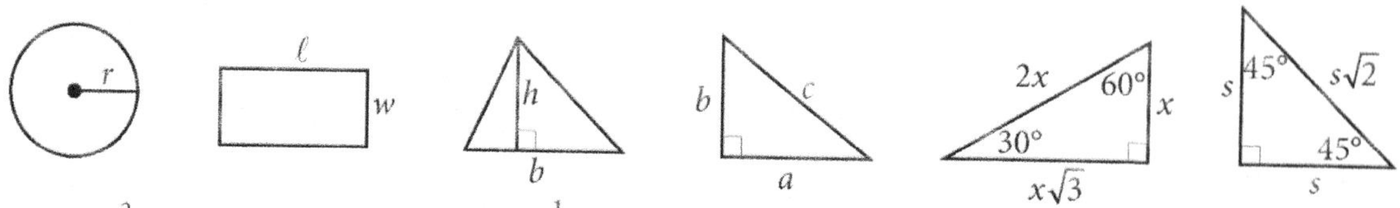

$A = \pi r^2$ $C = 2\pi r$ $A = \ell w$ $A = \frac{1}{2}bh$ $c^2 = a^2 + b^2$ Special Right Triangles

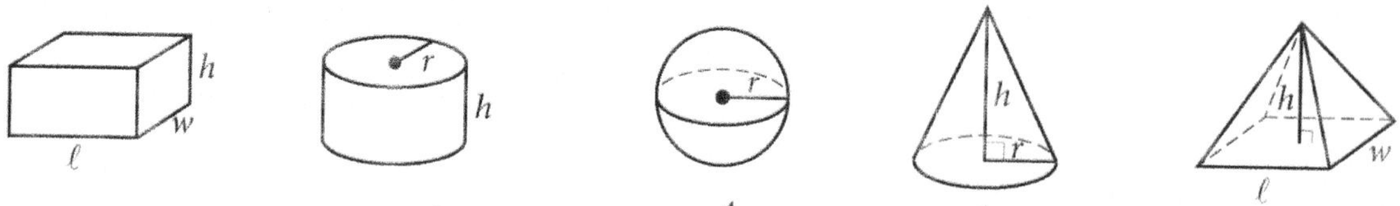

$V = \ell wh$ $V = \pi r^2 h$ $V = \frac{4}{3}\pi r^3$ $V = \frac{1}{3}\pi r^2 h$ $V = \frac{1}{3}\ell wh$

The arc of a circle contains 360 degrees.

The arc of a circle contains 2π radians.

The interior angles of a triangle add up to 180 degrees.

Part 2: Calculator

55 Minutes, 38 Questions

1

If $2m - 1 = 5$, then what is the value of $8m - 5$?

A. 5/4

B. 3

C. 15

D. 19

2

If $a = 3x - 4$ and $b = x + 2$, which of the following expressions is equivalent to $a^2 + ab$?

A. $3x^2 + 2x - 8$

B. $3x^2 - 2x + 8$

C. $12x^2 + 22x - 8$

D. $12x^2 - 22x + 8$

3

A team of scientists wants to decrease the concentration of microplastics (measured in parts per million or ppm) in the Pacific Ocean, as the unwanted plastic negatively affects aquatic ecosystems and human health. If a proposal sought to decrease the current microplastic concentration by a percent every week, then which of the follow best describes the rate of plastic extraction over time? (Assume no new plastic entered the ocean over the course of the extraction period.)

A. Constant

B. Increasing

C. Decreasing

D. Cannot be Determined

4

Which of the following is an equivalent form of the function $f(x) = x^2 + 5x - 24$, from which the x-coordinates of the x-intercepts can be identified if the function were graphed on the xy-plane where $y = f(x)$?

A. $f(x) = x(x + 5) - 24$

B. $f(x) = (x - 3)(x + 8)$

C. $f(x) = (x - 2.5)^2 - 24$

D. $f(x) = (x + 2.5)^2 - 30.25$

5

$$ax^2 + bx + c = (x - 5)(x - 2)$$

In the equation above a, b, and c are constants. If the equation holds true for all values of x, what is the sum of a, b and c?

A. -1

B. 0

C. 4

D. 18

6

Common measures of pressure include atmospheres (atm), kilopascals (kPa) and millimeters of mercury (mmHg). Approximately how any atmospheres of pressure are equivalent to 864mmHg?

$$1\text{atm} = 101.3\text{kPa}$$

$$240\text{kPa} = 1800\text{mmHg}$$

A. 1.14

B. 48.1

C. 87.9

D. 115.2

7

The instructor to dog ratio at a canine training academy is 2 : 3. How many instructors are there at the academy, which has 57 dogs?

A. 19

B. 23

C. 34

D. 38

8

A group of researchers conducts a year-long study to measure the amount of screen time teenagers spend daily on social media apps. The study uses the same methodology for every age group, and finds that teenagers aged 14-15 averaged 4.2 hours of screen time per day on social media apps with a margin of error of ± 0.2, whereas teenagers aged 16-17 averaged 1.3 hours of screen time per day on social media apps with a margin of error of ± 1.2. Which is an appropriate inference from the data collected?

A. The researchers used a smaller sample size for the 14-15 age group, resulting in that group's smaller margin of error.

B. The researchers used a smaller sample size for the 16-17 age group, resulting in that group's larger margin of error.

C. The results demonstrate that all teenagers aged 14-15 use social media apps for longer periods of time than their counterparts in the 16-17 age group.

D. The results demonstrate a causal link between age and social media usage.

9

In trapezoid FGHI, $\overline{FG} \parallel \overline{HI}$. If both $\overline{FG}$ and $\overline{HI}$ triple in length and the height of the trapezoid remains constant, then by what factor will its area increase?

A. 1/9

B. 1/3

C. 3

D. 9

10

$$7x - y = 10$$
$$ax - \frac{1}{3}y = c$$

In the system above, both a and c are constants. If the system has no solution, what must be the value of a?

A. 7/3

B. 7

C. 21

D. Cannot be determined

11

Number of Students at Dover High taking an Advanced Placement math course

	AP Calculus	AP Statistics
Juniors		
Seniors		
Total	87	85

The incomplete table above shows the number of juniors and seniors at Dover High School taking either AP Calculus or AP Statistics. The ratio of juniors taking AP Calculus to juniors taking AP Statistics is 1 : 3, while the ratio of seniors taking AP Calculus to seniors taking AP Statistics is 4 : 1. If there are 87 juniors and seniors taking AP Calculus and 85 juniors and seniors taking AP Statistics, what is the probability that a randomly selected upperclassman is a Junior taking AP Statistics?

A. 0.28

B. 0.40

C. 0.75

D. 0.88

Questions 12-14 require the following:

A student uses the linear relationship between volume and temperature of an ideal gas at constant pressure to compare the experimental value of absolute zero to its accepted value. The student records multiple volume measurements for the same sample of gas at different temperatures and uses a line of best fit to extrapolate absolute zero, which in the case of an ideal gas would coincide with a volume of zero milliliters. The results of the student's experiment are shown in the scatter plot below.

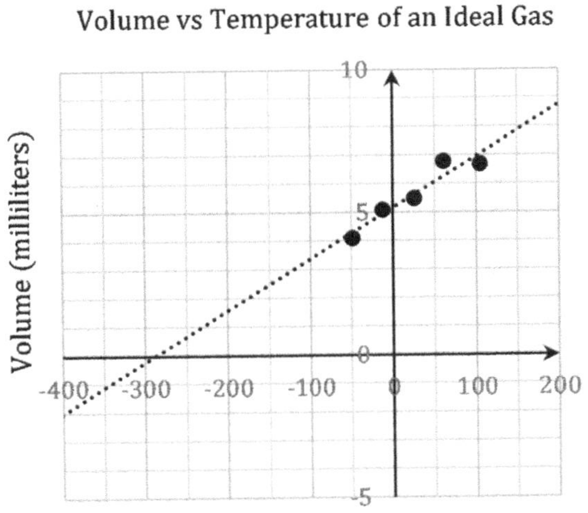

12

$$y = ax + b$$

The equation above represents the line of best fit, where x represents degrees Celsius and y represents milliliters. Which of the following is the best approximation for the constant a?

A. 0.02

B. 0.80

C. 1.00

D. 1.40

13

What is the experimental value of absolute zero when extrapolated using the line of best fit?

A. $-370°C$

B. $-330°C$

C. $-280°C$

D. $-220°C$

14

Assume identical samples of gas with no changes in pressure. According to the line of best fit, about how many times larger is the volume of an ideal gas at $100°C$ than the volume of an ideal gas at $-200°C$?

A. 1.5

B. 4.5

C. 5.5

D. 7

15

Month	Profit (in thousands)
January	26.1
February	6.2
March	18.7
April	22.6
May	26.8
June	29.2
July	18.7
August	22.1
September	49.0
October	25.9
November	31.8
December	23.3

The monthly profits a technology startup company gains in its first year are recorded in the table above. Which two months should be removed to decrease the standard deviation of the data set?

A. January, February

B. November, December

C. March, July

D. February, September

16

$$b^{\frac{5}{3}}b^{\frac{1}{2}}$$

Which is equivalent to the expression above, for all positive values of b?

A. $b\sqrt[6]{b}$

B. $b^2\sqrt[6]{b}$

C. $b\sqrt{b}$

D. $b^2\sqrt{b}$

17

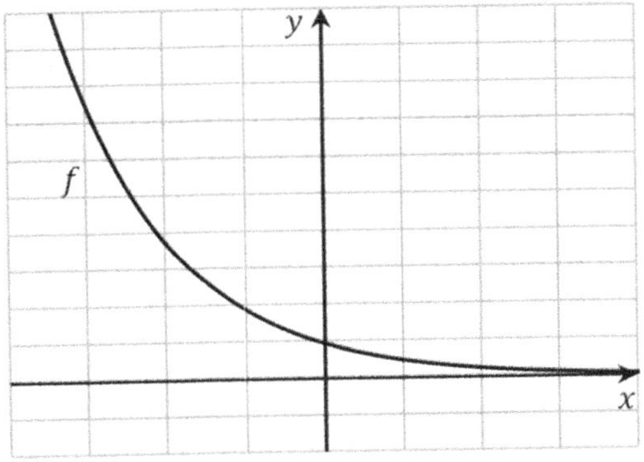

Which correctly describes the slope of the exponential decay model represented by the function f in the coordinate axis above?

A. Decreasing at a decreasing rate
B. Decreasing at an increasing rate
C. Increasing at a decreasing rate
D. Increasing at an increasing rate

18

x	0	1	2	3
y	2	3	5	9

From the data in the chart above, which equation best relates the values of x and y to each other?

A. $y = 2x$
B. $y = 2x + 1$
C. $y = 2^x$
D. $y = 2^x + 1$

19

$$3x^2 + bx = -2?$$

What value of the constant b would make the following equation have two real solutions?

A. 2
B. 3
C. 4
D. 5

20

Which of the following represents the equation of the largest circle that can fit inside the lines $x = 2$ and $x = 4$?

A. $(x-3)^2 + y^2 = 1$
B. $(x-3)^2 + y^2 = 4$
C. $x^2 + (y-3)^2 = 1$
D. $x^2 + (y-3)^2 = 4$

21

The graphs of a linear equation in the form $f(x) = mx + c$ and a continuous, exponential equation in the form $g(x) = ab^x$ on the xy-plane both contain the points (5,7) and (8,1). If the constants $a, b,$ and c are positive and $5 < h < 8$, then which of the following must be true?

A. $f(h) > g(h)$
B. $f(h) < g(h)$
C. $f(h) = g(h)$
D. Cannot be determined

22

Circle A has a radius of 5 meters and arcBC has a length of 8 meters. What value is closest to the number of radians formed by $\angle BAC$?

A. 5/8

B. 8/5

C. 10/8

D. 16/5

23

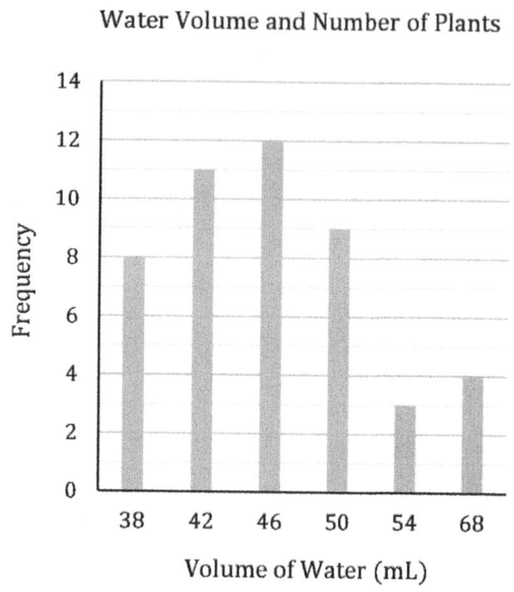

As part of his science project, Orlando collects data on the amount of water he provides to each of his 47 plants. His data is organized in the frequency graph above. Which value is closest to the mean amount of water in milliliters that Orlando gave to the plants in his science project?

A. 46

B. 47

C. 48

D. 49

24

$$C(t) = 4{,}590(1.02)^{\frac{t}{12}}$$

The population of a certain species of butterfly can be modeled by the exponential equation above, where C represents the butterfly population and t represents the number of months since reporting began. Which of the following must be true?

A. The butterfly population grows at a rate of 2% every month

B. The butterfly population grows at a rate of 2% every year

C. After the first month since reporting began, the butterfly population was 4,590

D. After the first year since reporting began, the butterfly population was 4,590

25

On the xy-plane, a rectangle contains vertices located at $(1,-2)$, $(3,3)$, $(6,-4)$, and $(8,a)$. What is the value of a?

A. -11

B. 0

C. 1

D. 2

26

$$y - x > -3$$
$$y + 3x \leq -5$$

In the system of inequalities above, what could be a possible value of x?

A. -1

B. 0

C. 3

D. 8

27

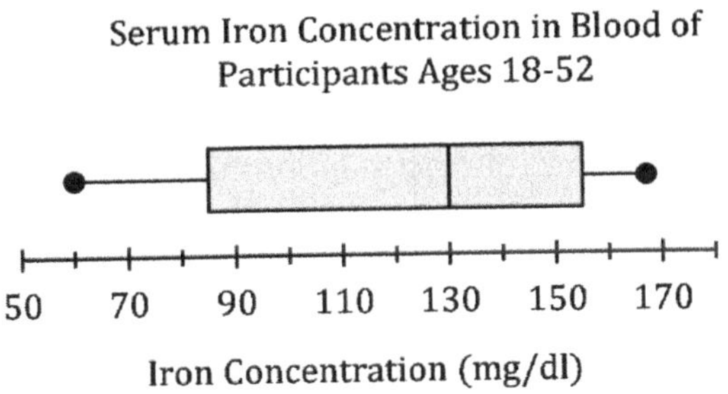

Which of the following is a true statement according to the data on serum iron concentrations presented in the boxplot above?

A. A plurality of the participants had serum iron concentrations between 85 and 130 mcg/dl

B. Almost 170 participants are represented

C. The mean serum iron concentration is greater than the median

D. A quarter of the participants had serum iron concentrations between 155 and 168 mcg/dl

28

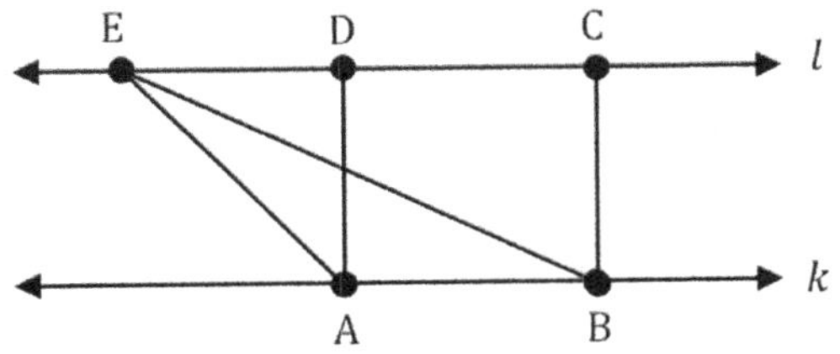

In the figure above, lines l and k are parallel. Square ABCD has a perimeter of 20 feet. What is the area in square feet of triangle ABE?

A. 15/4

B. 15/2

C. 25/4

D. 25/2

29

$$y = \frac{1}{x + 2}$$

When graphed on the xy-plane, the equation above contains the point (a, b). Which cannot be a value of the constant b?

A. -2

B. -1

C. 0

D. 1

30

Value	3	11	23	47
Frequency	6	3	9	18

What is the median value in the frequency table above?

A. 17

B. 23

C. 35

D. 47

31

$$\sin(a°) = \cos(b°)$$

$$a = x + 15$$

$$b = 4x - 5$$

In the system above, a and b are between 0 and 90. What is the value of b?

32

$$y = 2x^2 + 10x + 15$$

What is the minimum value of y in the equation above?

33

A tire of with diameter 52 inches rolls for a total of 3.8 miles. How many full rotations did the wheel complete over this distance?

$$1 \text{ mile} = 5280 \text{ feet}$$

34

$$x^2 - 5x = x - 9$$

How many distinct solutions does the equation above contain?

35

$$2y - x = -10$$

The equation above is graphed on the xy-plane. How many units up must it be translated to contain the point $(5,10)$?

36

$$(x-1)^2 + (y+2)^2 = 16$$

The equation above represents a circle on the xy-plane and contains the tangent line $y = a$. What is the value of a if it represents a positive constant?

Questions 37 and 38 require the following:

A sequence is a list of numbers. The value of the nth term is noted as a_n such that the meaning of $a_5 = 13$ is that the fifth term of a sequence is 13. The recursive formula below will be required to answer questions 37 and 38.

$$a_n = 5(n-1) + a_1$$

37

If $a_2 = 9$, what is a_1?

38

If $a_1 = 3$, what is the difference between the mean and median for the sequence that contains terms a_{10} through a_{15}?

Explanations

Practice Test 2: No Calculator

1. B

Add equations together to eliminate y and find x, which represents the x-coordinate, p.

Use a multiple of the second equation before adding:

$$2(x + y = -7)$$
$$2x + 2y = -14$$

$$\begin{array}{r} 3x - 2y = 4 \\ + \quad 2x + 2y = -14 \\ \hline 5x = -10 \end{array}$$
$$x = -2$$

2. A

Use the answer choices to reach answer. Only choice A is equivalent to the polynomial given.

$$(x^2 + 3)(x + 4) + 3$$
$$x^3 + 4x^2 + 3x + 12 + 3$$
$$x^3 + 4x^2 + 3x + 15$$

3. D

Only parallel lines do not intersect on the xy-plane. Parallel lines share the same slope. Note that line k crosses through the origin, (0,0).

$$m_j = m_k$$
$$\frac{-2-3}{1-0} = \frac{b-0}{-2-0}$$
$$\frac{-5}{1} = \frac{b}{-2}$$
$$b = 10$$

4. C

Use answer choices to quickly reach the correct answer. The correct answer will ensure that the value under the square root is NOT negative. Often, this value is also a perfect square. Only answer choice C satisfies the given equation.

$$x - 2\sqrt{x-1} = 0$$
$$2 - 2\sqrt{2-1} = 0$$
$$2 - 2\sqrt{1} = 0$$
$$2 - 2 = 0$$
$$0 = 0$$

5. B

The interior angles of a triangle add up to 180 degrees.

$$ax + bx + cx = 180$$
$$(a + b + c)x = 180$$
$$40x = 180$$
$$x = 4.5$$

6. C

Use equations to represent both the total number of tickets purchased and the total monetary amount.

Equation for Number of Tickets

Children Tickets: c

Adult Tickets: d

Total: 870

$$c + d = 870$$

Equation for Total Revenue

Children Tickets: $10c$

Adult Tickets: $15d$

Total: 10,150

$$10c + 15d = 10{,}150$$

7. A

Factors and their multiplicities can be used to find the value of and behavior at x-intercepts. Seeing as the factor $(x - b)$ is the only factor that is squared, its corresponding x-intercept contains a bounce. The only x-intercept that the function does not cross the x-axis, but rather bounces off of it, is in the negative region. Therefore, choice A is correct.

8. B

Manipulate the equation to isolate P by first dividing though by the quantity $(V - nb)$.

$$\left(P + \frac{an^2}{V^2}\right)(V - nb) = nRT$$

$$\left(P + \frac{an^2}{V^2}\right) = \frac{nRT}{V - nb}$$

$$P = \frac{nRT}{V - nb} - \frac{an^2}{V^2}$$

9. A

Identify that 0.26 is the slope of the linear function. Use the idea that slope is the change y over the change x to find its meaning in context. Keep in mind to use the units of the y-variable and the units of the x-variable in the analysis.

$$m = \frac{\Delta y}{\Delta x} = \frac{\$0.26}{1 \text{ mile}}$$

It costs \$0.26 for every mile on the scooter.

10. B

Triangle ABD is a 45-45-90 right triangle, meaning $AD = BD = 6$. Triangle ACD is a 30-60-90 right triangle where the leg across from the 60-degree angle is 6. Multiple the usual side lengths of a 30-60-90 triangle by $\sqrt{3}$ to more easily assess the length of CD.

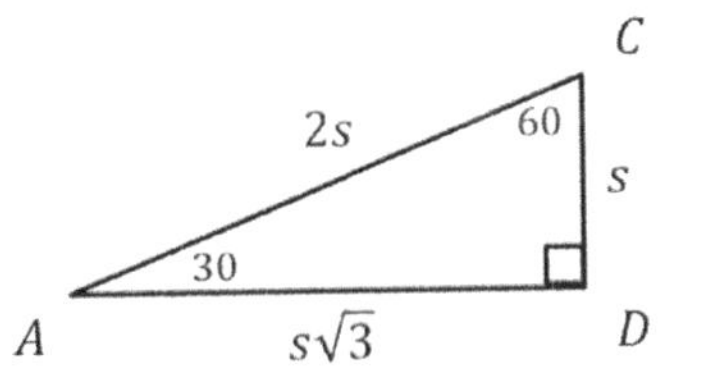

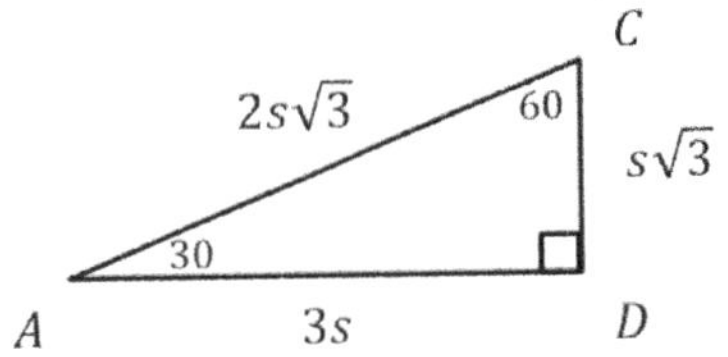

$$3s = 6$$

$$s = 2$$

$$CD = s\sqrt{3} \text{ or } 2\sqrt{3}$$

11. D

Since the x-coordinates of the solutions are being added together, use the Sum of the Roots equation $-b/a$ to quickly get the correct answer. Set the functions equal to each other, then bring all terms to one side so as to get into $ax^2 + bx + c = 0$ form.

$$x^2 + 1 = -x^2 + 4x + 8$$

$$2x^2 - 4x - 7 = 0$$

$$a = 2, \quad b = -4$$

$$-\frac{b}{a} = -\frac{-4}{2} \quad \text{or} \quad 2$$

12. C

The first term of the quotient can be found by dividing the first terms of the numerator and denominator.

$$\frac{16x^3}{ax} = 8x^2$$

$$a = 2$$

13. A

First, find the x-value that makes $3x + 1 = 4$, then use this x-value into the function equation.

$$3x + 1 = 4$$

$$3x = 3$$

$$x = 1$$

$$2x^3 + 1$$

$$2(1)^3 + 1$$

$$3$$

14. B

Expand the left side of the equation, then simplify and match corresponding terms.

$$5(x - h) + 2(x^2 + hx + 1) = 2x^2 + 13x - 18$$

$$5x - 5h + 2x^2 + 2hx + 2 = 2x^2 + 13x - 18$$

$$2x^2 + (2h + 5)x + 2 - 5h = 2x^2 + 13x - 18$$

$$2h + 5 = 13$$

$$2h = 8$$

$$h = 4$$

$$2 - 5h = -18$$

$$-5h = -20$$

$$h = 4$$

Another approach to this type of problem is to plug in our own value for x. For example, if we substitute zero for x, the equation becomes:

$$5(-h) + 2(1) = -18$$

$$-5h = -20$$

$$h = 4$$

15. D

Match the equations through multiplication in a way that makes the constants on the right equal to each other. This requires multiplying the first equation by a factor of -2.

$$-2(5x - 3y = -11)$$

$$-10x + 6y = 22$$

$$ax + by = 22$$

$$b = 6$$

16. 4

Since the circle has a radius of 7 and center with y-coordinate 3, the bottom most point on the circle still lies 4 units below the x-axis. The circle must be shifted up 4 units to be on, or tangent to, the x-axis.

17. 1/3

Since the dimensions of the triangle confirm a 30-60-90 right triangle, $\angle ACB = 60°$. However, the angle measurement must be expressed in radians.

$$60° * \frac{\pi \text{ rad}}{180°} = \frac{1}{3}\pi \text{ radians.}$$

$$a = \frac{1}{3}$$

18. 22

The maximum or minimum of a quadratic always occurs at its vertex. The x-coordinate of the vertex occurs right in the middle of its x-intercepts, which are the zeros of each factor. One way to find the middle of two numbers is to average them out. The zero of the first factor is -1, which is 6 units away from 5. Moving another 6 units away from 5 is 11. This means that the factor $2x - b$ must equal zero when $x = 11$.

$$2x - b = 0$$

$$2(11) - b = 0$$

$$b = 22$$

19. 4

Rational functions often have restricted domains because the function is undefined when its denominator is equal to zero. To find b, set the denominator equal to zero. Use the substitution $u = 2x - 9$ to make the math easier.

$$(2x-9)^2 + (2x-9) = 0$$

$$u^2 + u = 0$$

$$u(u+1) = 0$$

$$u = -1, 0$$

$$2x - 9 = -1$$

$$2x = 8$$

$$x = 4$$

$$2x - 9 = 0$$

$$2x = 9$$

$$x = 4.5$$

Only 4 is correct because it is an integer, whereas 4.5 is not.

20. 1/2

Multiply by the conjugate. Recall that $i^2 = -1$ to simplify the expression completely.

$$\frac{3-2i}{1+i} * \frac{1-i}{1-i} = \frac{3 - 3i - 2i + 2i^2}{1 - i^2}$$

$$\frac{3 - 5i + 2(-1)}{1 - (-1)}$$

$$\frac{(1-5i)}{2} = \frac{1}{2} - \frac{5}{2}i$$

$$a = \frac{1}{2}$$

Explanations

Practice Test 2: Calculator

1. D

Generate a multiple of the given expression, then solve for the desired one.

$$4(2m - 1 = 5)$$
$$8m - 4 = 20$$
$$8m - 5 = 19$$

2. D

Substitute the expressions given for a and b. Note that we can change the desired expression before making these substitutions to minimize the workload.

$$a^2 + ab$$
$$a(a + b)$$
$$(3x - 4)(3x - 4 + x + 2)$$
$$(3x - 4)(4x - 2)$$
$$12x^2 - 6x - 16x + 8$$
$$12x^2 - 22x + 8$$

3. C

Since the proposal aims to decrease the amount of plastic in the ocean by a constant percentage every week, it models an exponential decay function. Exponential decay functions decrease at a decreasing rate, so choice C is correct.

4. B

Only functions in factored form display x-intercepts as constants in their equations. Since choice B is the only function in factored form, it must be correct.

5. C

Expand out the right-hand side of the equation to find the values of the constants.

$$ax^2 + bx + c = (x - 5)(x - 2)$$
$$ax^2 + bx + c = x^2 - 7x + 10$$
$$a + b + c = 1 - 7 + 10$$
$$a + b + c = 4$$

6. A

Use dimensional analysis to convert the units from millimeters of mercury to atmospheres.

$$864 \text{ mmHg} * \frac{240 \text{ kPa}}{1800 \text{ mmHg}} * \frac{1 \text{ atm}}{101.3 \text{ kPa}} \approx 1.14 \text{ atm}$$

7. D

Use a multiple of the given ratio to find total number of instructors.

$$2x \text{ Instructors} : 3x \text{ Dogs}$$
$$3x = 57$$
$$x = 19$$
$$2x = 38 \text{ Instructors}$$

8. B

Studies conducted in the same manner will have smaller margins of error with larger sample sizes and larger margins of error with smaller sample sizes. Since the 16-17 age group had a larger margin of error, it likely had a smaller sample size than the 14-15 age group.

9. C

Understand that the segments that triple in length correspond to the bases 1 and 2 of the trapezoid. Use the area formula of a trapezoid to apply the changes that occur to each base, then see the overall factor of change. Remove any constants in the formula first.

$$A = \frac{1}{2}(b_1 + b_2)h$$

$$A \propto (b_1 + b_2)$$

$$A' \propto (3b_1 + 3b_2)$$

$$A' \propto 3(b_1 + b_2)$$

Since we have a coefficient of 3 left over in the new area formula, the area changes by a factor of 3.

10. A

Systems of linear equations have no solution when the lines they describe are parallel. Since parallel lines have the same slope, which in standard for is dictated by the coefficients A and B from $Ax + By = C$, the ratio of A to B in both equations are proportional

$$\frac{A}{B} = \frac{A}{B}$$

$$\frac{7}{-1} = \frac{a}{-\frac{1}{3}}$$

$$a = 7/3$$

11. B

Fill out the chart to determine the system of equations can be used to find the needed values.

Number of Students at Dover High taking an Advanced Placement math course

	AP Calculus	AP Statistics
Juniors	$1x$	$3x$
Seniors	$4y$	$1y$
Total	87	85

$$x + 4y = 87$$

$$-\quad 4(3x + y = 85)$$

$$-11x = -253$$

$$x = 23$$

$$3x = 69$$

$$\text{Probability} = \frac{69}{172} \quad \text{or} \quad 0.40$$

12. A

Note that the value of a corresponds to the slope of the graph and that the x-axis has lines representing every 50 degrees Celsius, the slope is approximately "up 1, over 50", or 0.02. Therefore, choice A is correct.

13. C

As the information precluding the questions states, absolute zero corresponds to a volume of zero. Therefore, it is graphically interpreted as the x-intercept, which falls just before −300°C. Choice C is correct.

14. B

Find 100 on the x-axis and move up to the line of best fit to see that it reaches a value of 7. Find -200 on the x-axis and move up to the line of best fit to see that it reaches a value of 1.5. Dividing 7 by 1.5 is 4.5

15. D

To decrease the standard deviation, remove the most extreme data points. Since the months of February and September deviate the most from the rest of the data, removing them from the list will decrease the standard deviation.

16. B

Multiplying common bases requires adding their exponents. To add fractions, find the least common denominator and add the numerators keeping the LCD. Convert to radical form. Recall the denominator swings out as the index to the radical.

$$b^{\frac{5}{3}} * b^{\frac{1}{2}}$$

$$b^{\frac{5}{3}+\frac{1}{2}}$$

$$b^{\frac{13}{6}}$$

$$b^{2+\frac{1}{6}}$$

$$b^2 b^{\frac{1}{6}}$$

$$b^2 \sqrt[6]{b}$$

17. A

Moving to the right, the function's slope is becoming less steep, while still decreasing. Therefore, it is decreasing at a decreasing rate.

18. D

Plug in values from the chart to ensure they work for the correct answer choice. Since it is clear the gaps between the y-values are increasing will the gaps between x-values are not changing, the rule cannot be linear. Choices A and B are linear and incorrect. Choice C and D are exponential; however, only Choice D contains a rule that works for the data.

19. D

Quadratics have two real solutions when its discriminant, $b^2 - 4ac$, is positive. Choice D is correct because it has a value of b that ensures the discriminant is greater than zero.

$$3x^2 + bx = -2$$

$$3x^2 + bx + 2 = 0$$

$$a = 3 \mid b = b \mid c = 2$$

$$b^2 - 4ac > 0$$

$$b^2 - 4(3)(2) > 0$$

$$b^2 > 24$$

20. A

To fit inside the lines $x = 2$ and $x = 4$, the circle must be centered right in between at $x = 3$, just as in choices A and B. Since 3 is 1 away from 2 and 4, the radius is one and $r^2 = 1$, as in choice A.

21. A

The function $g(x)$ represents an exponential decay model that would dip below the linear function $f(x)$ within the interval between the points of intersection. Choice A is correct.

22. B

Use the arc length formula for radians to determine the angle.

$$\text{Arc Length} = r\theta$$
$$8 = 5\theta$$
$$\theta = \frac{8}{5} \text{ radians}$$

23. B

Recall that frequency histograms represent an ordered list of values. Use multiplication to expedite the addition process. Or input values and frequencies into L1 and L2, respectively, of your calculator and go to 1-Variable Stats.

$$\frac{38(8) + 42(11) + 46(12) + 50(9) + 54(3) + 68(4)}{47}$$
$$\approx 46.85 \text{ or } 47$$

24. B

Choices C and D are incorrect because 4,590 represents the initial number of butterflies in the population when reporting began. Choice A is incorrect because it takes multiples of 12 months for the exponent to reach the next integer value. Since 12 months is equivalent to one year, choice B is correct.

25. C

The slope between $(1, -2)$ and $(3,3)$ must match the slope between $(6, -4)$ and $(8, a)$. The slope between he first pair is up 5 over 2. To get the y-coordinate, a, move up 5 units from -4 to reach 1.

26. A

Remember, the correct answer to a multiple-choice question about a possible value of a variable in a system of linear inequalities will always be the smallest or largest value. Therefore, only choices A and D are possible. Use a rough sketch of the system to quickly justify the answer. To do this, bring the inequalities into slope-intercept form, then find the overlapping shaded region that represents the solution set. Since this area begins in the negative x region and continues left, choice A must be correct.

27. D

Each region of the box plot represents 25% of the data. Those regions are 60 to 85, 85 to 130, 130 to 155, and 155 to 168. Therefore, choice D is correct. Note the length of these regions does NOT mean there are more data. The mean for this data would be less than the median because the first 50% of the data strings out to a wider range of lower values, creating a distribution that is skewed slightly left.

28. D

The base and height of the square are both 5 and represent the same base and height dimensions for triangle ABE.

$$\text{Perimeter of Square ABCD} = 20$$
$$4s = 20$$
$$s = 5$$

$$\text{Area of Triangle ABE} = \frac{1}{2}bh$$
$$\text{Area of Triangle ABE} = \frac{1}{2}(5)(5)$$
$$\text{Area of Triangle ABE} = \frac{25}{2}$$

29. C

Since the question asks for a constant that represents the y-intercept, the correct answer is therefore a restriction to the range. Fractions can only reach zero if the numerator is zero. Since the numerator is always 1, there is no way for the fraction to be zero. Choice C is correct. Note, choice A is incorrect because it represents a restriction to the domain, or a value the constant a cannot be.

30. C

The table contains 36 data points, found by adding up the frequency column. The 18th and 19th data points are the middlemost values. The 18th data point has a value of 23 and the 19th data point has a value of 47. The average of 23 and 47 is 35.

We can also type the values into List 1 and frequency into List 2 in our calculators, then find the median with the 1-Variable Stats function.

31. 59

Sine and Cosine are equal when they're of complementary angles, which means that for $\sin(a°)$ to equal $\cos(b°)$, $a° + b° = 90°$.

$$a + b = 90$$
$$x + 15 + 4x - 5 = 90$$
$$5x + 10 = 90$$
$$5x = 80$$
$$x = 16$$

$$b = 4x - 5$$
$$b = 4(16) - 5$$
$$b = 59$$

32. 5/2 or 2.5

All quadratic functions contain either a max or min, which occurs at the vertex (h, k). Since the quadratic contains a positive x^2 term, it contains a minimum value at h. To find k, plug h back into the equation.

$$h = -\frac{b}{2a}$$
$$h = -\frac{10}{2(2)}$$
$$h = -\frac{5}{2}$$

$$y = 2x^2 + 10x + 15$$
$$k = 2h^2 + 10h + 15$$
$$k = 2\left(-\frac{5}{2}\right)^2 + 10\left(-\frac{5}{2}\right) + 15$$
$$k = \frac{5}{2}$$

33. 1473

The distance travelled is equal to the number of rotations times the circumference.

Distance Travelled:

$$3.8 \text{ miles} * \frac{5{,}280 \text{ ft}}{1 \text{ mi}} * \frac{12 \text{ in}}{1 \text{ ft}} = 240{,}768 \text{ in}$$

$$\text{Number of Rotations} = n$$

$$\text{Circumference} = \pi d \text{ or } 52\pi \text{ inches}$$

$$\text{Distance} = \text{Number of Rotations} \times \text{Circumference}$$

$$240{,}768 = 52\pi n$$

$$n \approx 1473.8$$

The wheel only made 1473 complete rotations to when it reached the full distance of 3.8 miles.

34. 1

Solve the equation by bring it equal to zero and factoring the quadratic.

$$x^2 - 5x = x - 9$$
$$x^2 - 6x + 9 = 0$$
$$(x - 3)(x - 3) = 0$$
$$x = 3$$

The equation only has one distinct solution.

35. **25/2 or 12.5**

The line must be shifted up to contain the new point $(5,10)$. The line currently contains the point $(5, a)$. Knowing the value of a will allow to see the difference between the two y-coordinates so that we can shift it up by the right amount.

$$2y - x = -10$$
$$2a - 5 = -10$$
$$2a = -5$$
$$a = -2.5$$

The gap between -2.5 and 10 is 12.5, or $25/2$.

36. 2

Since the equation of the circle is in standard form, the center (h, k) can be seen as $(1, -2)$ and the radius is 4. Adding 4 to the -2 brings us to the tangent line touching the top edge of the circle. Therefore, $a = 2$.

37. 4

Knowing that $a_2 = 9$, we have two values to plug into the equation: $n = 2$ and $a_n = 9$. With that, we can solve for the first term, a_1.

$$a_n = 5(n - 1) + a_1$$
$$9 = 5(2 - 1) + a_1$$
$$9 = 5 + a_1$$
$$a_1 = 4$$

38. 0

Sequences that have a linear pattern have the same mean and median. Therefore, there is no difference between the two values, so the answer is zero.

Acknowledgements

We want to take the time now to thank everyone who contributed to the success of this book. Our team had the support of family and friends who encouraged us to deliver effective instruction to students and provide additional resources to educators. A big thank you goes out to them and you. We appreciate your trust in us and will continue to encourage you on your academic and personal journeys. Please visit us at SchwannsReview.com where you can find additional resources, important updates, and our contact information.

Made in the USA
Monee, IL
24 January 2022

89790666R00203